PARAMETER ESTIMATION FOR SCIENTISTS AND ENGINEERS

BICENTENNIAL 1807 WILEY 2007 BICENTENNIAL

The Wiley Bicentennial–Knowledge for Generations

Each generation has its unique needs and aspirations. When Charles Wiley first opened his small printing shop in lower Manhattan in 1807, it was a generation of boundless potential searching for an identity. And we were there, helping to define a new American literary tradition. Over half a century later, in the midst of the Second Industrial Revolution, it was a generation focused on building the future. Once again, we were there, supplying the critical scientific, technical, and engineering knowledge that helped frame the world. Throughout the 20th Century, and into the new millennium, nations began to reach out beyond their own borders and a new international community was born. Wiley was there, expanding its operations around the world to enable a global exchange of ideas, opinions, and know-how.

For 200 years, Wiley has been an integral part of each generation's journey, enabling the flow of information and understanding necessary to meet their needs and fulfill their aspirations. Today, bold new technologies are changing the way we live and learn. Wiley will be there, providing you the must-have knowledge you need to imagine new worlds, new possibilities, and new opportunities.

Generations come and go, but you can always count on Wiley to provide you the knowledge you need, when and where you need it!

William J. Pesce
President and Chief Executive Officer

Peter Booth Wiley
Chairman of the Board

PARAMETER ESTIMATION FOR SCIENTISTS AND ENGINEERS

Adriaan van den Bos

BICENTENNIAL 1807 WILEY 2007 BICENTENNIAL

WILEY-INTERSCIENCE
A John Wiley & Sons, Inc., Publication

Published by John Wiley & Sons, Inc., Hoboken, New Jersey.
Published simultaneously in Canada.

Wiley Bicentennial Logo Richard J. Pacifico

Library of Congress Cataloging-in-Publication Data:

van den Bos, Adriaan, 1936–.
Parameter estimation for scientists and engineers / A. van den Bos.
p. cm.
ISBN 978-0-470-14781-8
1. Engineering—Statistical methods. 2. Parameter estimation I. Title.

TA340 V34 2007
620.001' 519544—dc22 2007019912

10 9 8 7 6 5 4 3 2 1

To Margaretha Flora
Pellikaan

CONTENTS

PREFACE

The subject of this book is estimating parameters of expectation models of statistical observations. The book describes what I consider the most important aspects of the subject for applied scientists and engineers. From experience, I know that this group of users is often not aware of estimators other than least squares. Therefore, one of my purposes is to show that statistical parameter estimation has much more to offer than least squares estimation alone. To resort to least squares estimation almost automatically is, in fact, a purely expectation model oriented approach since the statistical properties of the observations are disregarded. In the approach of this book, knowledge of the distribution of the observations is involved in the choice of estimator. I hope to show that thus the available a priori knowledge may be used more fully to improve the precision of the estimator. A further advantage of the chosen approach is that it unifies the underlying theory and reduces it to a relatively small collection of coherent, generally applicable principles and notions. Moreover, this offers the opportunity to teach the subject in a systematic way.

The book is intended for a broad category of users: applied scientists, engineers, and undergraduate and graduate students. To enhance its suitability as course material and for exercise in general, I have included Problems in Chapters 3–6. Throughout, I have assumed that users have an elementary knowledge of statistics. They should be familiar with notions such as univariate and multivariate distribution, expectation, covariance, and hypothesis testing. In this respect, references such as [25, 20, 24] might be helpful.

If the book is used as course material, there are also options other than using the full text. The first is to skip all (sub)sections dealing with complex parameters or complex observations. A further option is to skip all (sub)sections concerned with exponential families of distributions. A disadvantage of the latter option is that it reduces to some extent the pursued coherence of the material taught. In view of these options, I have tried to make the (sub)sections dealing with complex observations or parameters or with exponential families easily identifiable as such. The same applies to the problems corresponding to these subjects.

The contents of the book may be summarized as follows. In Chapter 1, I present a detailed overview of the book in words, without the use of equations or mathematical formalism. I recommend the reading of this account since it is a first introduction to important terminology and definitions used in the subsequent chapters. The chapter also sketches the relations between the various parts of the book. In Chapter 2, I explain what statistical parametric models of observations are and why we use them. Such models require a description of the distribution of the observations around their expectations. Therefore, in Chapter 3, I present a number of distributions along with some of their characteristics, their Fisher score in particular. For the greater part, Chapter 4 is devoted to Fisher information and the Cramér–Rao lower bound. These are key notions in this book. They will be used to judge the quality of parameter estimators and of experimental designs. In Chapter 5, I discuss the use of the maximum likelihood method and the least squares method for estimating parameters of expectation models. I chose the former method because of its excellent properties and the latter because it is widely used. Chapter 5 also contains a short discussion of model hypothesis testing. In Chapter 6, I explain and compare a number of numerical methods suitable for the parameter estimation problems dealt with in this book. Also, in this chapter, I give my views on how to use these methods in practice. In Chapter 7, I present solutions to the problems of the Chapters 3–6. The book is concluded by four appendices. They are intended to provide the user with a number of useful mathematical and statistical tools and to introduce essential terminology and notation. Finally, I chose to collect and discuss all references of Chapters 2–6 in a Comments and References section at the end of each of these chapters to further the continuity of the text.

ADRIAAN VAN DEN BOS

Department of Applied Physics,
Delft University of Technology

CHAPTER 1

INTRODUCTION

The purpose of this book is to introduce methods for the solution of a general type of parameter estimation problem often met in applied science and engineering. In these disciplines, observations are usually not the quantities to be measured themselves but are related to these quantities. Throughout the book, it will be assumed that this relation is a known function and that the quantities to be measured are parameters of this function. Thus, parameter estimation is the computation of numerical values for the parameters from the available observations. For example, if observations have been made of the radioactive decay of a compound, the function is a multiexponential decay function of time. The parameters are the amplitudes and the decay constants of the exponential components. The parameter estimation problem is here: computing the amplitudes and the decay constants from the observations.

Applied scientists and engineers agree that most observations contain errors. Clearly, the description of observations as function values for known values of the independent variable, called measurement points, is incomplete. If a particular experiment is repeated under the same conditions, the resulting sets of observations will, as a rule, differ from experiment to experiment. These unpredictable fluctuations of the observations are sometimes called nonsystematic errors. An effective way of describing the fluctuations is by means of statistics. This implies that observations are modeled as stochastic variables. The function values are the expectations of these stochastic observations. The fluctuation is the deviation of an observation from its expectation. The function of the observations used to compute the parameters is called estimator. Since the observations are stochastic variables, so is the estimator. Therefore, the estimator has an expectation and a standard deviation.

One particular outcome of the estimator is called estimate. An estimate is a number. The standard deviation of the estimator defines and quantifies its precision. It is a measure of the magnitude of the nonsystematic error of the estimator caused by the fluctuations of the observations. The estimator is said to be more precise as its standard deviation is smaller. The deviation of the expectation of the estimator from the hypothetical true value of the parameter is called the bias of the estimator. The bias of the estimator defines and quantifies its accuracy. It is a measure of the systematic error of the estimator as a result of the fluctuations of the observations. The estimator is said to be more accurate as its bias is absolutely smaller. If the fluctuations of the observations would be absent, the estimator would produce exact values for the parameters.

A completely different source of error are modeling errors. If the expression for the expectations of the observations and the function used by the experimenter are different parametric families of functions, this will lead to systematic errors in the estimated parameters. For example, if the expectations of observations made are the sum of a multiexponential function and a constant representing background radiation while the function used by the experimenter is only the multiexponential function, the parameters of a wrong function will be estimated. Modeling errors are not to be considered systematic errors in the observations. They are caused by the experimenter. On the other hand, if a background is routinely included in the model although it is known that there is no background present, the model is correct but the background parameter is equal to zero and superfluous. This is an example of overparameterization. Overparameterization increases the standard deviation of the estimates of the remaining parameters, which, in our example, are the amplitudes and the decay constants.

The terminology thus introduced enables us to give the following overview of the book.

In Chapter 2, we discuss parametric models of observations. After a short introduction in Section 2.1, this discussion starts in Section 2.2 with deterministic (that is, nonstatistical) parametric models. This is followed, in Section 2.3, by an analysis of traditional parameter estimation methods which are based on the assumption that such deterministic observations really exist. This analysis reveals the need for the statistical parametric models introduced in Section 2.4. In the same section, the term expectation model is introduced for the parametric function that describes the expectations of the observations.

Describing observations as stochastic variables implies that they are defined by probability (density) functions. These are the subject of Chapter 3. After a short introduction in Section 3.1, we define in Section 3.2 some preliminary notions such as the covariance matrix of a set of stochastic variables and the Fisher score vector. Then, in the Sections 3.3–3.5, three important joint distributions of sets of observations are defined and discussed: the multivariate normal, the Poisson, and the multinomial distribution. These belong to an important general class called exponential families of distributions. This class is introduced in Section 3.6. Exponential families will be used throughout the book and will lead to considerable generalizations and simplifications. Statistical properties of the Fisher score vector are discussed in Section 3.7. In Section 3.8, complex stochastic variables are introduced. These are important in a number of disciplines in applied science dealing with complex parameters or complex observations. As an important example, the joint normal distribution of real and complex stochastic variables is discussed in Section 3.9.

Accuracy and precision of parameter estimators are the subjects of Chapter 4. It is essential to know to what extent observations reduce the uncertainty about the hypothetical true value of a parameter. If we make suitable assumptions about the distribution of the observations, this reduction of uncertainty can be quantified using the concept Fisher information. After a short introduction in Section 4.1 and a review of relevant estimation terminology and properties of covariance matrices in Sections 4.2 and 4.3, we introduce this concept in Section 4.4 in the form of the Fisher information matrix. The Fisher information matrix depends on the distribution of the observations. As examples, the expressions for the Fisher information matrix are derived for observations that are normally, Poisson, or multinomially distributed or have a distribution that is an exponential family. The inflow of Fisher information—that is, the contribution of each additional observation to the Fisher information matrix— is also addressed.

The inverse of the Fisher information matrix is called Cramér–Rao lower bound matrix. Under general conditions, unbiased but otherwise unspecified estimators cannot have a variance smaller than the corresponding diagonal element of the Cramér–Rao lower bound matrix. Like the Fisher information matrix, the Cramér–Rao lower bound matrix is a key notion in measurement since it specifies a bound to the precision of unbiased parameter measurement. This is the reason why, in Sections 4.5 and 4.6, we present a detailed account of the Cramér–Rao lower bound matrix and its properties. Also, in Section 4.7, we derive expressions for the Fisher information matrix and the Cramér–Rao lower bound matrix for complex parameters. They simplify applications of these concepts in measurement of complex quantities. In Section 4.8, an expression for the Cramér–Rao lower bound matrix for exponential family distributed observations is derived. Generally, the Cramér–Rao lower bound matrix exists only if the corresponding Fisher information matrix is nonsingular. Then, the parameters are called identifiable. Identifiability is discussed in Section 4.9.

The various expressions for the Cramér–Rao lower bound matrix emerging in this chapter show that this bound is typically a function of variables that may, within certain bounds, be freely chosen by the experimenter. An example are the measurement points. This offers the opportunity to select these free variables so that the bound is influenced in a desired way. This technique, called experimental design, is introduced and explained in Section 4.10. An example is the minimization of the Cramér–Rao lower bound on the variance with which a particular parameter may be estimated. Even if such an optimal design itself is not used, it may act as a reference to which the nonoptimal experimental design used or preferred by the experimenter may be compared.

The Cramér–Rao lower bound presents a limit to the precision of unbiased estimators. It does not indicate how to find estimators that are precise, in the sense of having a precision more or less comparable to the Cramér–Rao lower bound. Also, it does not inform about inaccuracy, that is, bias. These questions are addressed in Chapter 5, devoted to precise and accurate estimation.

After a short introduction in Section 5.1, maximum likelihood estimators are introduced in Section 5.2. Under general conditions, these have attractive properties described in Section 5.3. For us, the most important of these is that, typically, the variance of the maximum likelihood estimator attains the Cramér–Rao lower bound if the number of observations used is large enough. Therefore, under this condition, the maximum likelihood estimator may be rightly called most precise. The maximum likelihood estimators of the parameters of the expectation model for normally, Poisson, and multinomially distributed observations are derived and discussed in Sections 5.4–5.6, and those for exponential family distributed observations are covered in Section 5.7. Earlier in this chapter, we presented an example of a multiexponential decay model with and without an additional parameter representing the background or, equivalently, with a background parameter different from or equal to zero.

In Section 5.8, a statistical test is presented enabling the experimenter to conclude from the available observations if there is reason to reject constraints on the parameters such as an equality to zero. This test, the likelihood ratio test, is subsequently specialized to testing if the expectation model used must be rejected. For exponential families of distributions, a simple general expression is derived for the likelihood ratio used in the latter test.

For normally distributed observations, the maximum likelihood estimator is equivalent to the weighted least squares estimator with the elements of the inverse of the covariance matrix of the observations as weights. However, in practice, the least squares estimator is widely applied to observations of any distribution. This is an additional reason why, in Sections 5.9–5.19, extensive attention is paid to it. After an introduction to least squares estimation in Section 5.9, we discuss in Section 5.10 nonlinear least squares estimation. This is least squares estimation applied to expectation models that are nonlinear in one or more of their parameters. Typically, nonlinear least squares estimators are not closed-form and require iterative numerical treatment. Sections 5.11–5.19 deal with various aspects of linear least squares estimation. This is least squares estimation applied to expectation models that are linear in all parameters. An essential difference with nonlinear least squares is that the estimator is now a closed-form expression. In Section 5.12, we first present the general solution for weighted linear least squares estimation with arbitrary weighting matrix. The most important properties of this estimator are that it is unbiased and linear in the observations. Then, the weighting matrix is presented that for any distribution of the observations yields the most precise weighted linear least squares estimator. It is called the best linear unbiased estimator. Chapter 5 is concluded by introducing recursive linear least squares estimators. These update the value of the estimates of the parameters with each additional observation. Two different versions of this estimator are presented in Section 5.18 and Section 5.19, respectively. The first is an ordinary least squares estimator, that is, its weighting matrix is the identity matrix. The second is suitable for tracking time varying parameters since the weighting matrix is chosen so that observations influence the estimates more as they are more recent.

In the final chapter, Chapter 6, we explain principles and use of iterative numerical function optimization methods relevant to the estimators or experimental designs described in this book. The estimators require either the log-likelihood function to be maximized or the least squares criterion to be minimized. The experimental designs require the Cramér–Rao lower bound matrix to be optimized in the sense of a chosen optimality criterion. These optimization problems have in common that their solution is typically not closed-form and has to be computed iteratively.

After a short introduction in Section 6.1, key notions in numerical function optimization are introduced in Section 6.2. These are: the objective function, which is the function to be optimized; the gradient vector, which is the vector of first-order partial derivatives of the objective function; and the Hessian matrix, which is the matrix of second-order derivatives with respect to the independent variables. In the optimization problems in this book, the objective functions are the log-likelihood function or the least squares criterion as a function of the parameters, or the optimality criterion for the experimental design as a function of free experimental variables. Furthermore, the concept ascent (descent) direction of a vector is defined. Finally, the concepts exact observations, reference log-likelihood function, and reference least squares criterion are introduced in Section 6.2. These facilitate software testing.

Section 6.3 is devoted to the steepest descent (ascent) method. This is a general function minimization (maximization) method. It is not specialized to optimizing least squares criteria or log-likelihood functions. The method converges under general conditions, but its rate of convergence may be insufficient. This is improved by the Newton method discussed in Section 6.4. This is also a general function optimization method, but the conditions under which it converges are less general than those for the steepest descent method.

In Section 6.5, the Fisher scoring method is introduced. Its iteration step is an approximation to the Newton step for maximizing log-likelihood functions of parameters of expectation models. Therefore, the Fisher scoring method is a specialized method.

In Section 6.6, an expression is derived for the Newton step for maximizing the log-likelihood function for normally distributed observations. Because of the particular form of the log-likelihood function concerned, this step is also the Newton step for minimizing the nonlinear least squares criterion for observations of any distribution.

From the Newton step for normal observations, a much simpler approximate step may be derived. The method using this step is called Gauss–Newton method and is the subject of Section 6.7. The Newton steps for maximizing the Poisson and the multinomial log-likelihood functions are discussed in Section 6.8 and Section 6.9, respectively. In Section 6.10, an expression is derived for the Newton step for maximizing the log-likelihood function if the distribution of the observations is an exponential family.

From the Newton step for maximizing the log-likelihood function for exponential families of distributions, a much simpler approximate step may be derived that is used by the generalized Gauss–Newton method. This method is the subject of Section 6.11. In Section 6.12, the iteratively reweighted least squares method is described and it is shown that it is identical to the generalized Gauss–Newton method and the Fisher scoring method if the distribution of the observations is a linear exponential family.

Like the Newton method, the Gauss–Newton method solves a system of linear equations for the step in each iteration. The Levenberg–Marquardt method, discussed in Section 6.13, is a version of the Gauss–Newton method that can cope with (near-)singularity of these equations that could occur during the iteration process.

Section 6.14 is a summary of the numerical optimization methods discussed.

Finally, Section 6.15 is devoted to parameter estimation methodology. A number of intermediate steps are recommended in the process starting with choosing a statistical model of the observations and ending with actually estimating the parameters from experimental observations.

CHAPTER 2

PARAMETRIC MODELS OF OBSERVATIONS

2.1 INTRODUCTION

In science and engineering, observations are usually not the quantities to be measured themselves but are related to these quantities. Throughout the book, it will be assumed that a mathematical model of this relation is available. More precisely, we will assume that this model is a known parametric function and that the quantities to be measured are parameters of this function. Under these assumptions, measurement is estimating parameters of mathematical models of observations.

In Section 2.2, examples of purposes of estimating parameters of mathematical models are given and the notions of target parameters and nuisance parameters are introduced. The models presented in the examples are deterministic and exact. They describe observations as exact function values. The traditional parameter estimation methods introduced in Section 2.3 are based on the existence of such errorless observations. The most important difficulty with these methods is that errors in the measured parameters cannot be explained since errors in the observations are absent. This difficulty vanishes if statistical parameter estimation methods are used. These methods, introduced in Section 2.4, use an exact deterministic model for the expectations of the observations instead of for the observations themselves. The observations are modeled as sample values of stochastic variables that fluctuate around the expectations. Section 2.4 is concluded by a description of advantages of this statistical approach, especially the possibility to compute the errors in the measured parameters as a result of the fluctuations of the observations.

2.2 PURPOSES OF MODEL PARAMETER ESTIMATION

To show that purposes of model parameter estimation may essentially differ, we first present three examples.

■ EXAMPLE 2.1

Radioactive decay model

Suppose that observations are made of radiation produced by a radioactive specimen consisting of two components with different intensities and decay constants. Then, a deterministic description of the observations could be

$$y_n = \alpha_1 \exp(-\beta_1 x_n) + \alpha_2 \exp(-\beta_2 x_n), \tag{2.1}$$

with $n = 1, ..., N$, where y_n is the observation at the nth time instant x_n, N is the number of observations, and the parameters α_k and β_k, $k = 1, 2$, represent the unknown intensities and the decay constants of the components. ■

In Example 2.1, the biexponential description follows from physical laws and the parameters are quantities with a well-defined physical meaning. The purpose is the measurement of these physical quantities.

The next example shows that, even if the applications are physical, the parameters need not have a clear physical meaning at all.

■ EXAMPLE 2.2

Thermocouple calibration model

Thermocouples are electrical devices for the measurement of temperature. They produce an electromotive force (emf) which depends on the temperature. For their calibration, the emf is measured as a function of temperature. The usual description of emf observations is

$$y_n = \alpha_0 + \alpha_1 x_n + \alpha_2 x_n^2 + \cdots + \alpha_K x_n^K, \tag{2.2}$$

with $n = 1, \ldots, N$, where y_n and x_n are the measured emf and the temperature at the nth calibration point, respectively, and the coefficients α_k, $k = 0, \ldots, K$ are the parameters to be computed from the y_n. ■

In Example 2.2, the polynomial description has been chosen because it can be made to fit very well to observations and to accurately describe the values of the emf between the measurement points and that is what it is intended for. However, the polynomial has not been derived from physical laws and, as a result, the parameters have no physical meaning. Therefore, (2.1) represents a *physical model* and (2.2) a *curve-fitting model*. This classification is not strict since combinations of both types of models are also used.

■ EXAMPLE 2.3

Radioactive decay model with background

Suppose that radioactive decay observations are made described by

$$y_n = \alpha_1 \exp(-\beta_1 x_n) + \alpha_2 \exp(-\beta_2 x_n) + \gamma_1 + \gamma_2 x_n \tag{2.3}$$

with $n = 1, \ldots, N$. Then, the difference with the description of the observations in Example 2.1 is the presence of the contribution $\gamma_1 + \gamma_2 x_n$ representing a *background function*. ■

In the description (2.3), both first terms represent a physical model. Both last terms, on the other hand, incorporate otherwise unspecified contributions that are supposed to be described accurately by a linear function. Therefore, both last terms represent a curve-fitting model.

The description (2.3) also shows that there may be different types of unknown parameters. The parameters α_k and β_k are the quantities the experimenter wants to know. They are called *target parameters*. The parameters γ_k, on the other hand are measured only since the observations are not completely described without the background function. Therefore, parameters like the γ_k are called *nuisance parameters*.

Examples 2.1–2.3 show that parametric physical models are a description of the process that generates the observations. These models are intended for the measurement of physical quantities represented by the parameters. On the other hand, parametric curve-fitting models produce a compact parametric description of the observations but have, as a rule, little or no connection with the physical process that generates the observations. The model may also be partly a physical model, partly a curve-fitting model. This illustrates the general rule that in science and engineering the parametric model chosen for the observations should reflect the *purpose* of the measurement.

2.3 TRADITIONAL DETERMINISTIC PARAMETER ESTIMATION METHODS

Parameter estimation has a long, deterministic tradition. Most traditional methods, sometimes still in use, are deterministic in the sense that no statistical assumptions are made with respect to the observations. We give two examples of traditional methods.

■ **EXAMPLE 2.4**

Prony's method

The purpose of Prony's method is to compute the amplitudes $\alpha_k > 0$ and the decay constants $\beta_k > 0$ of the function

$$y(x) = \sum_{k=1}^{K} \alpha_k \exp(-\beta_k x) \tag{2.4}$$

from equidistant observations $y_1, \ldots, y_N$, where y_n is defined as $y(n\Delta)$ with Δ a constant sampling interval. Basic to Prony's method is that the exponential sequences

$$s_n = \exp(-\beta_k n\Delta), \qquad n = 1, \ldots, N \tag{2.5}$$

satisfy the linear difference equation

$$s_n + \lambda_1 s_{n-1} + \cdots + \lambda_K s_{n-K} = 0 \tag{2.6}$$

if the coefficients $\lambda_1, \ldots, \lambda_K$ are such that $\exp(-\beta_k\Delta)$, $k = 1, \ldots, K$ are the roots of

$$u^K + \lambda_1 u^{K-1} + \cdots + \lambda_K = 0. \tag{2.7}$$

The reader may verify this by substituting $\exp(-\beta_k n\Delta)$ for s_n in (2.6). Since y_n is a linear combination of the sequences $s_n = \exp(-\beta_k n\Delta)$ for $k = 1, \ldots, K$, it also satisfies (2.6). This implies that

$$\begin{array}{llllll} y_N & +\lambda_1 y_{N-1} & +\cdots & +\lambda_K y_{N-K} & = 0 \\ y_{N-1} & +\lambda_1 y_{N-2} & +\cdots & +\lambda_K y_{N-K-1} & = 0 \\ \vdots & \vdots & \vdots & \vdots & \vdots \\ y_{K+1} & +\lambda_1 y_K & +\cdots & +\lambda_K y_1 & = 0. \end{array} \tag{2.8}$$

These are $N - K$ linear equations in the K unknown coefficients $\lambda_1, \ldots, \lambda_K$. Therefore, for solving (2.8) for the λ_k, the number of observations N should be at least $2K$. However, in many applications N will be larger than $2K$. Then, the system of equations (2.8) is overdetermined and is solved in the least squares sense. This means that the solution $\hat{\ell} = (\hat{\ell}_1 \cdots \hat{\ell}_K)^T$ for the vector of coefficients $\lambda = (\lambda_1 \ldots \lambda_K)^T$ is taken as that value of $\ell = (\ell_1 \ldots \ell_K)^T$ that minimizes the least squares criterion

$$J(\ell) = \sum_{n=K+1}^{N} (y_n + \ell_1 y_{n-1} + \cdots + \ell_K y_{n-K})^2 \tag{2.9}$$

where the superscript T denotes transposition. This least squares solution is found by differentiating $J(\ell)$ with respect to each of the ℓ_k and equating the result to zero. This produces K linear equations in the K unknown $\hat{\ell}_k$:

$$\hat{\ell}_1 \overline{y_{n-k}\, y_{n-1}} + \cdots + \hat{\ell}_K \overline{y_{n-k}\, y_{n-K}} = -\overline{y_{n-k}\, y_n} \tag{2.10}$$

with $k = 1, \ldots, K$, where

$$\overline{y_{n-p} y_{n-q}} = \frac{1}{N-K} \sum_{n=K+1}^{N} y_{n-p} y_{n-q} \tag{2.11}$$

with $p = 1, \ldots, K$ and $q = 0, \ldots, K$. Below, expressions like (2.11) will be called *mean lagged products*. Next, the solutions $\hat{\ell}_k$ are substituted for the λ_k in (2.7), and the roots $\hat{u}_k$ of the polynomial thus obtained are computed. These roots must be equal to $\exp(-\beta_k \Delta)$. Therefore, the solution $\hat{b}_k$ for β_k is taken as $-(1/\Delta) \ln \hat{u}_k$. This completes the computation of the decay constants. Next, the amplitudes are computed. This is done by least squares fitting

$$\sum_{k=1}^{K} a_k \exp(-\hat{b}_k n\Delta) \tag{2.12}$$

with respect to the elements of $a = (a_1 \ldots a_K)^T$ to the y_n, $n = 1, \ldots, N$. This least squares procedure produces closed-form solutions $\hat{a}_k$ for the α_k and completes the computation of solutions for the amplitudes α_k and the decay constants β_k. ■

In fact, the Prony method such as sketched in Example 2.4 is self-contradictory. On the one hand, the observations are supposed to be exactly describable by a model of the multiexponential parametric family. If that is true, the coefficients λ_k could be solved directly from an arbitrary system of only K equations taken from the system of $N - K$ equations (2.8). On the other hand, a least squares procedure is used operating on all $N - K$ equations. This suggests that there are errors in the observations. In the absence of a model of these errors, it remains unclear to what extent the least squares procedure is helpful. We will return to this question in Section 2.4.

■ **EXAMPLE 2.5**

The method of exponential peeling

The purpose of this method is again to compute the amplitudes and decay constants of the decay model (2.4) from a number of N observations $y(x_1), \ldots, y(x_N)$. Different from the Prony method, the method of exponential peeling does not require the measurement points x_n to be equidistant. Without loss of generality, suppose that $\beta_k < \beta_{k+1}$. Then, β_1 is the smallest decay constant and the tail of the multiexponential function is composed of $\alpha_1 \exp(-\beta_1 x)$ alone. Therefore, if the logarithm of the observations is plotted instead of the observations themselves, the tail is described by

$$\ln y_n = \ln \alpha_1 - \beta_1 x_n \tag{2.13}$$

with $y_n = y(x_n)$. This is recognized as a straight line with intercept $\ln \alpha_1$ and slope $-\beta_1$. In the method of exponential peeling, these quantities are determined graphically. Once the solutions $\breve{a}_1$and $\breve{b}_1$ for α_1 and β_1 have been computed, the contribution $\breve{a}_1 \exp(-\breve{b}_1 x_n)$ is subtracted from the y_n and the procedure is repeated for α_2 and β_2, etc. ■

A difficulty with the method of exponential peeling is how to decide which points in each step constitute the tail. Of course, this choice, which is subjective, influences the result.

The Examples 2.4 and 2.5 illustrate the following general characteristics of traditional parameter estimation methods:

- Errors are absent in the mathematical model of the observations. Undoubtedly, this has to do with the unfamiliarity with mathematico-statistical models of errors and their treatment at the time of invention of these methods.
- The emphasis is usually on computational and conceptual simplicity. Traditional methods are a collection of, often ingenious, tricks intended especially to avoid numerical computation in a time that computers did not yet exist. Of course, computational and conceptual simplicity are valuable assets. However, the advent of the computer and progress in software have made it possible to actually apply computationally demanding and conceptually relatively complicated methods to improve accuracy and precision.
- Traditional methods are model-specific. They apply to a small class of parametric functions only.

There are a number of serious difficulties with traditional parameter estimation methods. The most important of these are the following:

- Since errors in the observations are absent, errors in the estimated parameters cannot be explained.
- These methods are often subjective, those using graphical techniques in particular. By subjectivity we mean the phenomenon that different experimenters applying the same method to the same observations get different results.
- The experimenter cannot exploit the available a priori knowledge about the errors in the observations to improve the precision or accuracy of the estimates of the parameters.

- Since the methods are function-model specific, for every function model a special method has to be developed.

Clearly, there is a need for alternative methods. We will show below that these exist and that they offer substantial advantages over traditional methods. First, we will demonstrate that systematic and nonsystematic errors in estimated parameters can often be computed if errors are included in the model of the observations. We will also show that a priori knowledge about the errors in the observations may be used to improve or even optimize the precision and accuracy of the parameter estimates. This implies that thus the estimation method is inspired by the errors in the observations and not by the parametric function model underlying the observations as in traditional methods. Therefore, these estimation methods are not model-specific. Moreover, their definition will be seen to leave little room for subjective use. Describing observations including errors, as required by alternative methods, is the subject of the next section.

2.4 STATISTICAL PARAMETRIC MODELS OF OBSERVATIONS

2.4.1 The expectation model

In science and engineering, reducing errors in the observations is always worthwhile if the precision and accuracy of the measured parameters are of concern. Both qualities are nearly always of concern but are seldom purposes in themselves. More often, precision and accuracy are required to enable the experimenter to draw conclusions about the parameters, about their magnitude in particular. The modern parametric-model based statistical methods that will be described below are not a substitute for careful instrumentation and measurement intended to reduce errors in the observations. They are complementary to, and a further improvement of, good measurement practice. This means that classical measures to decrease fluctuations in observations, such as cooling and shielding in instrumentation, remain essential. However, even if these measures have been taken, unavoidable and unpredictable fluctuations remain. Observations will typically differ as a result of these fluctuations if we repeat an experiment. We may, therefore, wonder what an experimenter means if he says that his observations are, for example, "sinusoidal" since as a result of the fluctuations they cannot be purely sinusoidal. We think that the statement of the experimenter can be best interpreted using statistics where we use the word "best" since we think that there is no credible alternative. The use of statistics implies that the observations are modeled as stochastic variables. Then, the interpretation of the statement of the experimenter is that the *expectations* of his observations, the theoretical mean values in every measurement point, are described by a sinusoidal function. Generally, in this book, the expectation of the nth observation w_n is supposed to be equal to the value of the known parametric function $g(x;\theta)$ of x at the known nth *measurement point* x_n:

$$\boxed{Ew_n = g(x_n;\theta)} \tag{2.14}$$

where E denotes the mathematical expectation and θ is the vector of exact parameters $(\theta_1 \dots \theta_K)^T$.

The function $g(x;\theta)$ is the *expectation model* of the observations. Throughout the book, the following notation will be used:

$$\boxed{g(\theta) = [g_1(\theta) \dots g_N(\theta)]^T}, \tag{2.15}$$

where $g_n(\theta) = g(x_n; \theta)$ is the expectation model at x_n. The vector of observations w will be defined as

$$\boxed{w = (w_1 \ldots w_N)^T}. \tag{2.16}$$

■ EXAMPLE 2.6

Poisson distributed observations with an exponential expectation model

Suppose that the expectation of the observations $w = (w_1 \ldots w_5)^T$ is described by

$$Ew_n = g_n(\theta) = 100\,\alpha \exp(-\beta x_n). \tag{2.17}$$

Furthermore, suppose that the measurement points are $x_n = 0.25 + 0.75(n-1)$ with $n = 1, \ldots, 5$ and that $\theta = (\alpha\ \beta)^T = (1\ 1)^T$ is the vector of unknown parameters. This expectation model and its values at the measurement points are shown in Fig. 2.1. Next, suppose that the probability of occurrence of a particular value of an observation is described by the Poisson probability function. This probability function describes the probability of occurrence of outcomes of counting processes such as radioactive particle count and is described in Chapter 3. Figures 2.1. (a)–(d) show four independent realizations of simulated observations thus distributed. For each of the realizations, the expectation model and its values at the measurement points are, of course, the same but the observations fluctuate around these values from realization to realization. ■

The deviations of the observations $w = (w_1 \ldots w_N)^T$ from the values of the expectation model at the measurement points

$$\boxed{d_n(\theta) = w_n - Ew_n = w_n - g_n(\theta)} \tag{2.18}$$

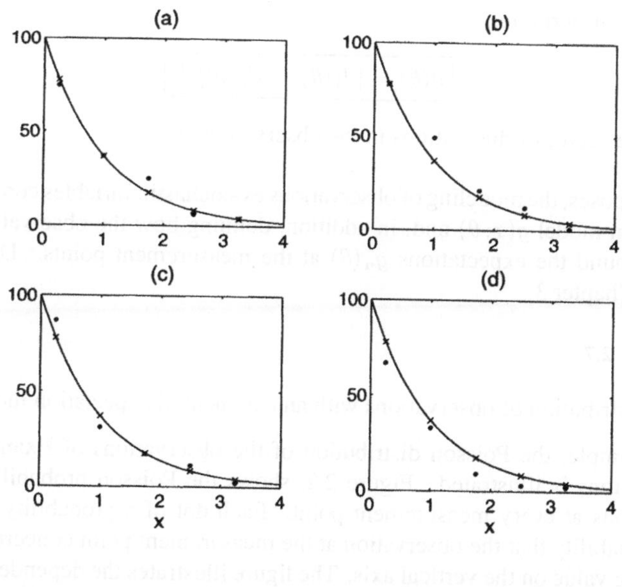

Figure 2.1. Monoexponential expectation model (solid line), the values of the model at the measurement points (crosses), and four sets of Poisson distributed observations (dots). (Example 2.6)

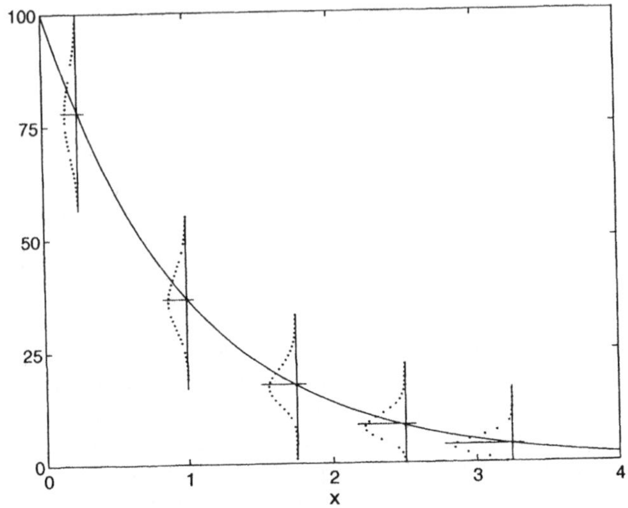

Figure 2.2. Monoexponential expectation model (solid line) and, vertically, the probability functions of the Poisson distributed observations at the measurement points. (Example 2.7)

are often called *nonsystematic errors* reflecting that their expectations are equal to zero:

$$\boxed{Ed_n(\theta) = 0}. \tag{2.19}$$

We will call the $d_n(\theta)$ *fluctuations* since they represent the zero-mean statistical part of the observations. Furthermore,

$$\boxed{d(\theta) = [d_1(\theta) \ldots d_N(\theta)]^T} \tag{2.20}$$

will denote the vector of fluctuations in the observations w.

For our purposes, the modeling of observations as stochastic variables consists of defining the expectation model $g(x;\theta)$ and, in addition, defining how the observations are jointly distributed around the expectations $g_n(\theta)$ at the measurement points. Distributions are discussed in Chapter 3.

■ **EXAMPLE 2.7**

Poisson distribution of observations with an exponential expectation model

In this example, the Poisson distribution of the observations of Example 2.6 around their expectations is illustrated. Figure 2.2. shows the Poisson probability functions of the observations at every measurement point. Each dot of a probability function represents the probability that the observation at the measurement point concerned assumes the corresponding value on the vertical axis. The figure illustrates the dependence of the probability functions on the expectation of the observation. The standard deviation increases as the expectation increases. Furthermore, the probability function becomes increasingly asymmetric as the expectation value decreases, but all probability functions shown have in common that the expectation of the observations is equal to the value of the expectation model at the measurement point concerned. ■

Example 2.7 demonstrates that the statistical properties of the observations made on one and the same expectation model may vary from point to point. In Chapter 5, we will show how this can be dealt with in the estimation process.

■ **EXAMPLE 2.8**

Sinusoidal expectation model in the presence of drift

Suppose that the observations $w = (w_1 \ldots w_N)^T$ have an expectation model that is the sum of (a) a sinusoidal signal with unknown phase angle and angular frequency and (b) a drift component. Then, if it is assumed that the drift is adequately modeled by a quadratic polynomial, the expectation model at sampling instant x_n is described by

$$Ew_n = g_n(\theta) = \alpha \cos(\gamma x_n) + \beta \sin(\gamma x_n) + \delta_0 + \delta_1 x_n + \delta_2 x_n^2 \qquad (2.21)$$

with $\theta = (\alpha\, \beta\, \gamma\, \delta_0\, \delta_1\, \delta_2)^T$. In addition, assumptions have to be made about the distribution of the observations around the expectations. ■

Suppose that, in Example 2.8, the true expectation is described by (2.21) with δ_0, δ_1, and δ_2 not all equal to zero. Furthermore, suppose that the model chosen by the experimenter is sinusoidal without drift. Then, the function describing the true expectations does not belong to the parametric family of functions used by the experimenter. The experimenter will, therefore, estimate the parameters of a wrong model and the estimates of α, β, and γ will systematically deviate from their true values. Such a systematic deviation is not caused by errors in the observations but is introduced by the experimenter. On the other hand, if the model (2.21) is routinely used although it is known that there is no drift, the function describing the true expectations still belongs to the parametric family of functions used. However, the known and zero-valued parameters δ_0, δ_1, and δ_2 are now estimated along with α, β, and γ. Estimating more parameters from a set of observations than needed is highly undesirable as will be explained in Subsection 4.6.5.

Generally, carefully modeling the expectations of the observations is essential. The model chosen should be complete in the sense that it must be capable of representing the true expectations of the observations but with the smallest number of parameters.

A further important statistical property of the observations modeled as stochastic variables is their *covariance*. If the w_n are real, the covariance of the observations w_m and w_n is defined as

$$\text{cov}(w_m, w_n) = E[(w_m - Ew_m)(w_n - Ew_n)]. \qquad (2.22)$$

Therefore, if $m = n$, $\text{cov}(w_m, w_n)$ is equal to the *variance* σ_m^2 where σ_m is the *standard deviation* of w_m. If, for $m \neq n$, $\text{cov}(w_m, w_n) = 0$, the observations w_m and w_n are said to be *uncorrelated*.

The $N \times N$ *covariance matrix* of the observations is defined as

$$\boxed{\text{cov}(w, w) = [\text{cov}(w_m, w_n)]}, \qquad (2.23)$$

where, as before, $w = (w_1 \ldots w_N)^T$. Combining (2.18), (2.19), (2.22), and (2.23) shows that $\text{cov}(w_m, w_n) = E[d_m(\theta) d_n(\theta)] = \text{cov}\,(d_m(\theta), d_n(\theta))$. Therefore,

$$\boxed{\text{cov}(w, w) = \text{cov}\,(d(\theta), d(\theta))}. \qquad (2.24)$$

Expression (2.24) also implies that

$$\operatorname{var} w_n = \operatorname{var} d_n(\theta)\,. \tag{2.25}$$

2.4.2 Advantages of statistical parametric models of observations

The most important consequence of modeling observations as stochastic variables is that parameter values computed from these observations also become stochastic variables. These values will be different for different realizations of the observations. Thus, the parameter measurement problem has become a *statistical parameter estimation problem*. As we will see, this offers the scientist and engineer the opportunity to exploit the vast collection of statistical theories and methods available in mathematical statistics, econometrics, and related fields.

Before we will address the relevant applications of statistical parameter estimation to our problems, we will sketch some of the advantages of this approach. These are the following:

- **The possibility to compute accuracy and precision**

  In the first place, the statistical formulation often enables us to quantify the accuracy and the precision of a chosen estimation method by computing its bias (systematic error) and standard deviation (nonsystematic error). Bias and standard deviation will be discussed in more detail in Section 4.2.

■ EXAMPLE 2.9

Systematic error in Prony's method

Suppose that the model of the observations is

$$\begin{aligned} w_n &= \alpha \exp(-\beta n \Delta) + v_n \\ &= y_n + v_n\,, \end{aligned} \tag{2.26}$$

where $n = 1, \ldots, N$. The v_n are fluctuations. Therefore, $Ev_n = 0$. We assume that the v_n are uncorrelated and have the same standard deviation σ:

$$\operatorname{cov}(v_m, v_n) = \delta_{m,n}\sigma^2, \tag{2.27}$$

where $\delta_{m,n}$ is the Kronecker delta symbol which is equal to one if $m = n$ and equal to zero otherwise. Since the model is monoexponential, the difference equation (2.6) used in Prony's method is described by

$$s_n + \lambda\, s_{n-1} = 0. \tag{2.28}$$

Then, the least squares criterion (2.9) becomes

$$J(\ell) = \sum_n \left(w_n + \ell\, w_{n-1}\right)^2. \tag{2.29}$$

Here and below, the summations are over $n = 2, \ldots, N$. Differentiating $J(\ell)$ with respect to ℓ and equating the result to zero yields the least squares solution $\hat{\ell}$ for ℓ:

$$\hat{\ell} = -\frac{\overline{w_n\, w_{n-1}}}{\overline{w_{n-1}^2}}\,. \tag{2.30}$$

Next, using (2.26), we write out all terms of the numerator and the denominator. The results are

$$-\overline{y_n y_{n-1}} - \overline{y_n v_{n-1}} - \overline{v_n y_{n-1}} - \overline{v_n v_{n-1}} \tag{2.31}$$

and

$$\overline{y_{n-1}^2} + 2\overline{y_{n-1} v_{n-1}} + \overline{v_{n-1}^2}, \tag{2.32}$$

respectively. We will assume that the measurement points $n\Delta$ and their number N are such that all mean lagged products may be approximated by their expectations. Then,

$$\overline{y_n v_{n-1}} = \frac{1}{N-1} \sum_n y_n v_{n-1} \tag{2.33}$$

is approximated by

$$\frac{1}{N-1} \sum_n y_n \, E v_{n-1} = 0 \tag{2.34}$$

where use has been made of the fact that y_n is a constant and $Ev_{n-1} = 0$. Similarly,

$$\overline{v_n y_{n-1}} = \overline{y_{n-1} v_{n-1}} \approx 0. \tag{2.35}$$

Furthermore,

$$\overline{v_n v_{n-1}} \approx 0 \tag{2.36}$$

since the expectations and, by (2.27), the covariance of v_n and v_{n-1} are equal to zero. Finally,

$$\overline{v_{n-1}^2} \approx \sigma^2. \tag{2.37}$$

Substituting (2.33)–(2.37) in (2.30)–(2.32) yields

$$\hat{\ell} \approx -\frac{\overline{y_n y_{n-1}}}{\overline{y_{n-1}^2} + \sigma^2} \tag{2.38}$$

while the result if all $v_n \equiv 0$ would have been

$$\hat{\ell} = \lambda = -\frac{\overline{y_n y_{n-1}}}{\overline{y_{n-1}^2}}. \tag{2.39}$$

■

Expressions (2.38) and (2.39) show that the estimate $\hat{\ell}$ of λ deviates systematically from the true value. This, of course, also produces systematic errors in the estimates of α and β. Similar results may be derived for multiexponential models. Then, complex conjugate pairs of $\hat{\ell}_k$ may occur, resulting in oscillatory solutions. We emphasize that this is an example of a *systematic* error in the estimates caused by fluctuations, that is, by *nonsystematic* errors in the observations. Also, this systematic error could only be traced by including the fluctuations in the model of the observations.

The next example illustrates the computation of the precision of two different estimators of the slope of a straight line through the origin.

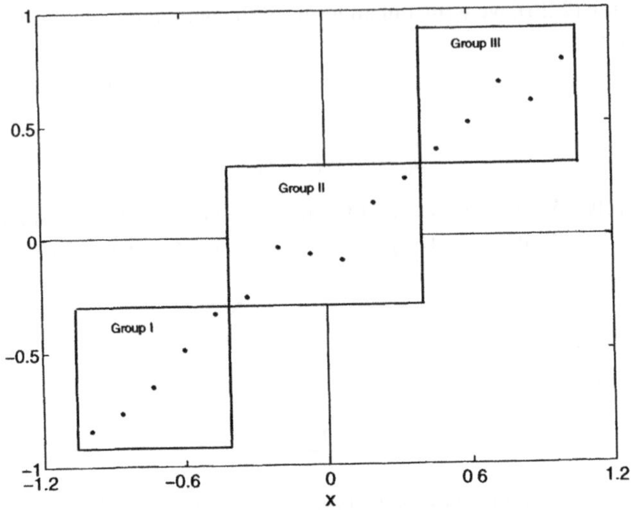

Figure 2.3. Straight-line observations (dots) and their distribution over three groups. (Example 2.10)

■ **EXAMPLE 2.10**

Estimation of the slope of a straight line through the origin

Suppose that the expectation of the observations $w = (w_1 \dots w_N)^T$ is described by

$$Ew_n = g_n(\alpha) = \alpha x_n\,, \tag{2.40}$$

where the x_n are exactly known measurement points and N is even. Furthermore, suppose that the covariance of the observations is

$$\operatorname{cov}(w_m, w_n) = \delta_{m,n}\sigma^2. \tag{2.41}$$

Then, the observations and, therefore, the fluctuations are uncorrelated and have equal variance σ^2. We now compute the variance of two different methods for estimating the parameter α. The first method is the *method of grouping*. In this method, the observations are divided in three groups as follows. The observations in Group I and Group III correspond to the pN smallest and the pN largest values of the x_n, respectively, with $0 < p \leq 0.5$. Group II consists of the observations located between Group I and Group III. However, if $p = 0.5$, all observations are in Group I or Group III, and Group II is empty.

In the method of grouping, the estimate of α is taken as the slope of the straight line connecting the center of gravity of Group I with that of Group III:

$$\breve{a} = \frac{\overline{w}_{III} - \overline{w}_I}{\overline{x}_{III} - \overline{x}_I}\,, \tag{2.42}$$

where $\overline{w}_{III}$ and $\overline{w}_I$ are the averages of the observations in Group III and Group I, respectively, and $\overline{x}_{III}$ and $\overline{x}_I$ are defined accordingly. Observations in Group

II are not used. The numerator of $\breve{a}$ is a linear combination of $2pN$ observations. Since the denominator is constant, $\breve{a}$ is a linear combination of the observations as well. The coefficients of the first pN terms of this linear combination are equal to $1/[(pN(\overline{x}_{III} - \overline{x}_I)]$. Those of the last pN terms are equal to $-1/[(pN(\overline{x}_{III} - \overline{x}_I)]$. The variance of linear combinations of stochastic variables is discussed in Appendix A. Since in the problem at hand the w_n are uncorrelated, the relevant expression is (A.13). Applying this to $\breve{a}$ yields

$$\operatorname{var}\breve{a} = \frac{2\sigma^2}{pN(\overline{x}_{III} - \overline{x}_I)^2}. \tag{2.43}$$

As an example, suppose that observations are made at the measurement points $x_n = -1 + 2(n-1)/15$, $n = 1, \ldots, 16$. Therefore, $N = 16$. Then, if $pN = 5$, $\operatorname{var}\breve{a} = 0.1860\,\sigma^2$. Figure 2.3. shows a realization of such an experiment for $\alpha = 0.75$ and normally distributed fluctuations with $\sigma = 0.1$. Next, we take $pN = 8$, which means that all observations are used and that Group II is empty. Then, it is found that $\operatorname{var}\breve{a} = 0.2197\,\sigma^2$. So, this variance is larger than that for $pN = 5$.

Next, consider estimating the parameter α using the least squares method. Then, the estimate of α is taken as that value of a that minimizes the least squares criterion

$$\sum_n (w_n - a x_n)^2 \tag{2.44}$$

with $n = 1, \ldots, N$. Therefore, the solution is

$$\hat{a} = \arg\min_a \sum_n (w_n - a x_n)^2. \tag{2.45}$$

Differentiating the least squares criterion with respect to a and equating the result to zero yields

$$\hat{a} = \frac{\overline{w_n x_n}}{\overline{x_n^2}}. \tag{2.46}$$

Since this is a linear combination of the observations, the variance may again be computed using the results of Appendix A. The weights are equal to $x_n/(N\overline{x_n^2})$. Since the observations are uncorrelated, the relevant equation is (A.13). Applying this to $\hat{a}$ yields

$$\operatorname{var}\hat{a} = \frac{\sigma^2}{N\overline{x_n^2}}. \tag{2.47}$$

For the experiment described above, this expression yields $\operatorname{var}\hat{a} = 0.1654\,\sigma^2$. ∎

In Example 2.10, we have computed the variance of two different methods for estimating the same parameter from the same observations. The method of grouping was found to be more precise when some of the observations were not used. This result might perhaps not have been predicted on intuitive grounds. Furthermore, the example showed that in the cases considered, the method of grouping is less precise than the least squares method. Thus, the modeling of the observations as stochastic variables enabled us to compare the results of a method for different *experimental designs*, in the sense of different choices of measurement points. It also enabled us to compare the precision of different *methods*.

We continue the list of advantages of statistically modeling observations with:

- **The possibility to find precise estimation methods**

  The possibility to find precise estimation methods and to use a priori knowledge will be discussed in Chapter 5.

- **The possibility to compute attainable precision**

  This is the possibility to compute the precision that may be attained for a particular set of observations. It enables the experimenter to investigate the appropriateness of the set of observations available for the purposes in mind. The computation of the attainable precision will be discussed in Chapter 4.

- **Experimental design**

  This is the possibility to optimize or improve the precision of an estimator through selection of the experimental variables that can be freely chosen.

■ **EXAMPLE 2.11**

Optimal design

Suppose that in Example 2.10 the location of the measurement points x_n may be freely chosen under the restriction that $| x_n | \leq 1$ for all n. Then, (2.47) shows that var $\hat{a}$ is smallest if all x_n are equal to either 1 or -1. For this choice, var $\hat{a} = 0.0625\,\sigma^2$. This is much smaller than the result var $\hat{a} = 0.1654\,\sigma^2$ found earlier for equidistant x_n. ■

A discussion of experimental design will be presented in Section 4.10.

2.5 CONCLUSIONS

In this chapter, the strength of the statistical parametric approach to parameter estimation has been demonstrated. The modeling of the observations in this approach has two main ingredients: a parametric function model for the expectations of the observations and a description of the distribution of the observations around these expectations. The distribution of the observations will be the subject of the next chapter where distributions often used in practice and their most important properties will be discussed.

2.6 COMMENTS AND REFERENCES

In [27], Scharf and Demeure discuss Prony's method and refer to the source material. For a discussion of the method of exponential peeling, see Seber and Wild [28]. The method of grouping has been proposed by Wald [33].

CHAPTER 3

DISTRIBUTIONS OF OBSERVATIONS

3.1 INTRODUCTION

In the preceding chapter, statistical parametric models of observations have been introduced. In these models, the expectations of the observations are values of the parametric function underlying the observations. Joint distributions of the observations around these expectations are the subject of this chapter.

First, in Section 3.2, relevant statistical notions are introduced, the Fisher score vector in particular. Then, in Sections 3.3–3.5, the joint normal, the Poisson, and the multinomial distribution are discussed. These have been chosen primarily because of their practical importance. They are examples of so-called exponential families of distributions. This general class of distributions is the subject of Section 3.6. Next, in Section 3.7, we present a numerical example illustrating statistical properties of the Fisher score vector. Then, in Section 3.8, complex stochastic variables are introduced. The joint normal distribution of real and complex stochastic variables is discussed in Section 3.9.

3.2 EXPECTATION, COVARIANCE, AND FISHER SCORE

Let the vector of observations $w = (w_1 \dots w_N)^T$ be modeled as a vector of stochastic variables. Then, the $N \times 1$ *(vector) expectation* μ of w is defined as

$$\mu = Ew = (Ew_1 \dots Ew_n)^T. \tag{3.1}$$

Thus, the nth element of μ is described by

$$\mu_n = Ew_n. \tag{3.2}$$

If the w_n are *continuous* stochastic variables, the *expectation* Ew_n is defined as

$$Ew_n = \int \omega_n \, p(\omega) \, d\omega, \tag{3.3}$$

where

$$\omega = (\omega_1 \dots \omega_N)^T \tag{3.4}$$

and $p(\omega)$ is the *joint probability density function* of the elements of w. Thus, w_n is a stochastic variable while ω_n is the independent variable corresponding to it. Furthermore,

$$d\omega = d\omega_1 \, d\omega_2 \dots d\omega_N \tag{3.5}$$

and

$$\int \cdot \, d\omega = \int \cdots \int \cdot \; d\omega_1 \, d\omega_2 \dots d\omega_N, \tag{3.6}$$

where the integrations are performed over all possible values the elements of w may assume. Moreover, since $p(\omega)$ is a probability density function, we have

$$\int p(\omega) \, d\omega = 1. \tag{3.7}$$

Throughout the book, the symbol C will be used for the $N \times N$ covariance matrix C of w:

$$C = \text{cov}(w, w) = E\left[(w - \mu)(w - \mu)^T\right] \tag{3.8}$$

with elements

$$c_{mn} = \text{cov}(w_m, w_n) = E\left[(w_m - \mu_m)(w_n - \mu_n)\right] \tag{3.9}$$

so that

$$c_{nn} = \sigma_n^2, \tag{3.10}$$

where σ_n^2 is the variance of w_n.

The covariance matrix of linear combinations of the elements of w is computed as follows. Suppose that the vector of stochastic variables $u = (u_1 \dots u_M)^T$ is obtained from $w = (w_1 \dots w_N)^T$ by

$$u = Dw, \tag{3.11}$$

where D is an $M \times N$ matrix. Then, $u - Eu = D(w - Ew)$ and, therefore,

$$\begin{aligned} \text{cov}(u, u) &= E\left[(u - Eu)(u - Eu)^T\right] \\ &= D\,E\left[(w - Ew)(w - Ew)^T\right] D^T \\ &= D\,C\,D^T. \end{aligned} \tag{3.12}$$

If all observations are uncorrelated, $\text{cov}(w_m, w_n) = 0$ for $m \neq n$. Then, the covariance matrix is diagonal and is described by

$$C = \text{diag}\,(\sigma_1^2 \dots \sigma_N^2). \tag{3.13}$$

Further properties of covariance matrices will be discussed in Section 4.3.

If the w_n are *discrete* stochastic variables, the expectation of the w_n is defined as

$$Ew_n = \sum \omega_n \, p(\omega), \tag{3.14}$$

where $p(\omega)$ is the *joint probability function* of the w_n and the summation is over all possible discrete values the elements of w may assume. Furthermore, since $p(\omega)$ is a probability function, we have

$$\sum p(\omega) = 1. \tag{3.15}$$

In this book, probability (density) functions will usually be parametric in a vector of parameters $\theta = (\theta_1 \ldots \theta_K)^T$. This will be expressed by using $p(\omega;\theta)$ instead of $p(\omega)$. We will also use the *joint log-probability (density) function* defined as

$$\boxed{q(\omega;\theta) = \ln p(\omega;\theta)}\,. \tag{3.16}$$

Then, the $K \times 1$ *Fisher score vector* s_θ is defined as

$$\boxed{s_\theta = \frac{\partial q(w;\theta)}{\partial \theta}}\,. \tag{3.17}$$

The element s_{θ_k} is the *Fisher score* of the stochastic variables w for the parameter θ_k. Note that in (3.17), the vector of independent variables ω has been replaced by the corresponding vector of observations w. So, Fisher scores are stochastic variables since the observations w are stochastic variables.

For Fisher scores, we present the following theorem:

Theorem 3.1 *Under suitable conditions, the expectation of the Fisher score vector is equal to the null vector:*

$$\boxed{Es_\theta = o}\,. \tag{3.18}$$

Proof. By (3.7),

$$\int p(\omega;\theta)\, d\omega = 1. \tag{3.19}$$

Hence, if $p(\omega;\theta)$ is such that differentiating with respect to θ and integrating over ω may be interchanged, we obtain

$$\frac{\partial}{\partial\theta}\int p(\omega;\theta)\, d\omega = \int \frac{\partial p(\omega;\theta)}{\partial\theta}\, d\omega = \int \frac{\partial \ln p(\omega;\theta)}{\partial\theta}\, p(\omega;\theta) d\omega = o \tag{3.20}$$

or

$$Es_\theta = E\frac{\partial q(w;\theta)}{\partial\theta} = o\,, \tag{3.21}$$

where o is the $K \times 1$ null vector. ■

The condition that differentiation and integration may be interchanged is called a *regularity condition.* Regularity conditions are discussed in more detail in Subsection 4.4.1. Theorem 3.1 is also true if the stochastic variables w are discrete. For the proof, (3.15) may be used.

3.3 THE JOINT REAL NORMAL DISTRIBUTION

3.3.1 The joint real normal probability density function

The elements of the real vector of stochastic variables $w = (w_1 \dots w_N)^T$ are said to possess a *joint real normal distribution* if their probability density function is defined as

$$p(\omega) = \frac{1}{(2\pi)^{\frac{N}{2}} (\det C)^{\frac{1}{2}}} \exp\left(-\frac{1}{2}(\omega - \mu)^T C^{-1} (\omega - \mu) \right), \qquad (3.22)$$

where μ is the $N \times 1$ vector expectation and C is the $N \times N$ covariance matrix of w. In (3.22), the elements of the $N \times 1$ vector of independent variables $\omega = (\omega_1 \dots \omega_N)^T$ correspond to those of the vector of stochastic variables $w = (w_1 \dots w_N)^T$. The usual shorthand notation for the joint normal distribution (3.22) is

$$\mathcal{N}(\mu; C). \qquad (3.23)$$

If the w_n are uncorrelated,

$$C = \operatorname{diag}(\sigma_1^2 \dots \sigma_N^2) \qquad (3.24)$$

and

$$p(\omega) = \prod_n \frac{1}{(2\pi)^{\frac{1}{2}} \sigma_n} \exp\left(-\frac{1}{2} \left(\frac{\omega_n - \mu_n}{\sigma_n} \right)^2 \right). \qquad (3.25)$$

This is the product of all marginal probability density functions of the w_n. Therefore, if normally distributed stochastic variables are uncorrelated, they are also independent. If all σ_n are identical and equal to σ, (3.25) becomes

$$p(\omega) = \frac{1}{(2\pi)^{\frac{N}{2}} \sigma^N} \prod_n \exp\left(-\frac{1}{2\sigma^2} (\omega_n - \mu_n)^2 \right). \qquad (3.26)$$

If, in addition, all μ_n are identical, the stochastic variables w_n are called *independent and identically normally distributed (iind)*.

By (3.22), the general expression for $q(\omega) = \ln p(\omega)$ for normally distributed w is

$$q(\omega) = -\frac{N}{2} \ln 2\pi - \frac{1}{2} \ln \det C - \frac{1}{2} (\omega - \mu)^T C^{-1} (\omega - \mu). \qquad (3.27)$$

Expression (3.25) shows that for uncorrelated w_n we have

$$q(\omega) = -\frac{N}{2} \ln 2\pi - \sum_n \ln \sigma_n - \frac{1}{2} \sum_n \left(\frac{\omega_n - \mu_n}{\sigma_n} \right)^2, \qquad (3.28)$$

and for *iind* fluctuations $w_n - \mu_n$ we obtain

$$q(\omega) = -\frac{N}{2} \ln 2\pi - N \ln \sigma - \frac{1}{2\sigma^2} \sum_n (\omega_n - \mu_n)^2. \qquad (3.29)$$

If $\mu_n = Ew_n = g_n(\theta)$ is substituted in (3.22), the resulting expression shows the dependence of the probability density function $p(\omega; \theta)$ on the parameters θ:

$$p(\omega; \theta) = \frac{1}{(2\pi)^{\frac{N}{2}} (\det C)^{\frac{1}{2}}} \exp\left(-\frac{1}{2} [\omega - g(\theta)]^T C^{-1} [\omega - g(\theta)] \right), \qquad (3.30)$$

where $g(\theta) = [g_1(\theta) \ldots g_N(\theta)]^T$ with $g_n(\theta) = g(x_n; \theta)$. The corresponding expression for the log-probability density function $q(\omega; \theta) = \ln p(\omega; \theta)$ is

$$q(\omega;\theta) = -\frac{N}{2}\ln 2\pi - \frac{1}{2}\ln\det C - \frac{1}{2}[\omega - g(\theta)]^T C^{-1}[\omega - g(\theta)]. \tag{3.31}$$

Equation (3.29) shows that for *iind* fluctuations (3.31) simplifies to

$$q(\omega;\theta) = -\frac{N}{2}\ln 2\pi - N\ln\sigma - \frac{1}{2\sigma^2}\sum_n [\omega_n - g_n(\theta)]^2 . \tag{3.32}$$

3.3.2 The Fisher score of normal observations

By definition, the Fisher score vector is the gradient of $q(w;\theta)$ with respect to the parameters θ. Assume that C does not depend on θ. Then, expression (3.31) shows that both first terms of $q(w;\theta)$ do not depend on θ. The last term is a quadratic expression in the fluctuations $d_n(\theta) = w_n - g_n(\theta)$:

$$-\frac{1}{2}\sum_m\sum_n d_m(\theta)\, c_{mn}^{-}\, d_n(\theta), \tag{3.33}$$

where c_{mn}^{-} is the (m,n)th element of C^{-1}. Then, the Fisher score of the normally distributed observations w for parameter θ_k is equal to

$$\begin{aligned} s_{\theta_k} &= \frac{\partial q(w;\theta)}{\partial\theta_k} \\ &= \frac{1}{2}\sum_m\sum_n \frac{\partial g_m(\theta)}{\partial\theta_k}\, c_{mn}^{-}\, d_n(\theta) + d_m(\theta)\, c_{mn}^{-}\,\frac{\partial g_n(\theta)}{\partial\theta_k} \\ &= \sum_m\sum_n \frac{\partial g_m(\theta)}{\partial\theta_k}\, c_{mn}^{-}\, d_n(\theta) = \frac{\partial g^T(\theta)}{\partial\theta_k}\, C^{-1}\, d(\theta), \end{aligned} \tag{3.34}$$

with $d(\theta) = w - g(\theta)$, where use has been made of the fact that C^{-1} is symmetric because C is and, hence, $c_{mn}^{-} = c_{nm}^{-}$. Therefore, the Fisher score vector of normally distributed observations is described by

$$s_\theta = \frac{\partial g^T(\theta)}{\partial\theta}\, C^{-1} d(\theta). \tag{3.35}$$

3.4 THE POISSON DISTRIBUTION

3.4.1 The Poisson probability function

A stochastic variable w_n is said to be Poisson distributed if its probability function is defined as

$$p(\omega_n) = \exp(-\lambda_n)\,\frac{\lambda_n^{\omega_n}}{\omega_n!}\,, \tag{3.36}$$

where $\omega_n = 0, 1, \ldots$ and λ_n is a positive scalar. The Poisson probability function describes the probability of a counting result w_n and is, therefore, a discrete stochastic variable. It can be shown that

$$E w_n = \lambda_n \tag{3.37}$$

and that

$$\operatorname{var} w_n = \lambda_n . \tag{3.38}$$

This implies that the ratio of the standard deviation of a Poisson distributed stochastic variable to its expectation is equal to

$$\frac{1}{\sqrt{\lambda_n}} . \tag{3.39}$$

This shows that the larger the expectation, the smaller the relative fluctuations of a Poisson stochastic variable.

If the stochastic variables $w = (w_1 \dots w_N)^T$ are independent and Poisson distributed, their joint probability function is the product of the marginal probability functions and is equal to

$$\begin{aligned} p(\omega) &= \prod_n \exp(-\lambda_n) \frac{\lambda_n^{\omega_n}}{\omega_n!} \\ &= \exp\left(-\sum_n \lambda_n\right) \prod_n \frac{\lambda_n^{\omega_n}}{\omega_n!} . \end{aligned} \tag{3.40}$$

Then, the corresponding joint log-probability density function is

$$q(\omega) = \sum_n -\lambda_n + \omega_n \ln \lambda_n - \ln \omega_n! \,. \tag{3.41}$$

If $Ew_n = g_n(\theta)$ is substituted for the λ_n in (3.40), the resulting expression shows the dependence of the joint probability function $p(\omega; \theta)$ on the parameters θ:

$$p(\omega; \theta) = \exp\left(-\sum_n g_n(\theta)\right) \prod_n \frac{g_n^{\omega_n}(\theta)}{\omega_n!} . \tag{3.42}$$

The corresponding joint log-probability function is

$$q(\omega; \theta) = \sum_n -g_n(\theta) + \omega_n \ln g_n(\theta) - \ln \omega_n! \,. \tag{3.43}$$

Equation (3.38) shows that the variances of the w_n depend on θ since $\lambda_n = Ew_n = g_n(\theta)$. This is a difference with the variances of normally distributed observations that need not depend on θ. Furthermore, since the w_n are assumed to be independent, they are also uncorrelated. Therefore, the off-diagonal elements of their covariance matrix are equal to zero while the diagonal elements, which are the variances, are equal to the $g_n(\theta)$. Then,

$$C = \operatorname{diag} g(\theta). \tag{3.44}$$

3.4.2 The Fisher score of Poisson observations

The Fisher score of independent and Poisson distributed observations w for the parameter θ_k follows directly from (3.43):

$$s_{\theta_k} = \frac{\partial q(w;\theta)}{\partial \theta_k} = \sum_n \frac{\partial g_n(\theta)}{\partial \theta_k} \frac{d_n(\theta)}{g_n(\theta)} \tag{3.45}$$

with $d_n(\theta) = w_n - g_n(\theta)$. This may be alternatively written

$$s_{\theta_k} = \frac{\partial g^T(\theta)}{\partial \theta_k} \operatorname{diag}\left[\frac{1}{g_1(\theta)} \cdots \frac{1}{g_N(\theta)} \right] d(\theta) \tag{3.46}$$

with $d(\theta) = w - g(\theta)$. Therefore, the Fisher score vector for the parameters θ is described by

$$\begin{aligned} s_\theta &= \frac{\partial g^T(\theta)}{\partial \theta} \left[\operatorname{diag} g(\theta) \right]^{-1} d(\theta) \\ &= \frac{\partial g^T(\theta)}{\partial \theta} C^{-1} d(\theta), \end{aligned} \tag{3.47}$$

where (3.44) has been used. This result is formally similar to (3.35).

3.5 THE MULTINOMIAL DISTRIBUTION

3.5.1 The multinomial probability function

Suppose that the elements of the vector of stochastic variables $(w_1 \ldots w_N)^T$ are counting results obtained by assigning a total of M independent counts to N different counters with probabilities $\varphi_1, \ldots, \varphi_N$, respectively. Therefore,

$$\sum_{n=1}^{N} w_n = M \tag{3.48}$$

and

$$\sum_{n=1}^{N} \varphi_n = 1. \tag{3.49}$$

Then, the joint probability function of the elements of $w = (w_1 \ldots w_{N-1})^T$ may be shown to be

$$p(\omega) = \frac{M!}{\prod_{n=1}^{N} \omega_n!} \varphi_1^{\omega_1} \varphi_2^{\omega_2} \ldots \varphi_N^{\omega_N}, \tag{3.50}$$

where ω_n corresponds to w_n while ω_N and φ_N are defined as

$$\omega_N = M - \sum_{n=1}^{N-1} \omega_n \tag{3.51}$$

and

$$\varphi_N = 1 - \sum_{n=1}^{N-1} \varphi_n\,. \tag{3.52}$$

The probability function (3.50) is called the multinomial probability function. For example, for $N = 3$ and $M = 10$ it is described by

$$p(\omega_1, \omega_2) = \frac{10!}{\omega_1!\, \omega_2!\, \omega_3!} \varphi_1^{\omega_1} \varphi_2^{\omega_2} \varphi_3^{\omega_3} \tag{3.53}$$

with $\omega_3 = 10 - \omega_1 - \omega_2$ and $\varphi_3 = 1 - \varphi_1 - \varphi_2$.

With respect to the expectation and the covariance matrix of multinomial stochastic variables, we have the following theorem:

Theorem 3.2 *Suppose that the elements of* $w = (w_1 \dots w_{N-1})^T$ *are multinomially distributed with probabilities* $\varphi = (\varphi_1 \dots \varphi_{N-1})^T$ *and that* $w_N = M - \sum_{n=1}^{N-1} w_n$ *and* $\varphi_N = 1 - \sum_{n=1}^{N-1} \varphi_n$. *Then,*

$$Ew_p = M\varphi_p \tag{3.54}$$

and

$$\operatorname{cov}(w_p, w_q) = M\varphi_p(\delta_{pq} - \varphi_q) \tag{3.55}$$

where $p, q = 1, \dots, N$ *and* δ_{pq} *is the Kronecker delta symbol.*

Proof. Consider the counting result w_n as the sum of M independent stochastic variables v_{nm} with $m = 1, \dots, M$, which are equal to one with probability φ_n and equal to zero otherwise. Then, $Ev_{nm} = \varphi_n \cdot 1 + (1 - \varphi_n) \cdot 0 = \varphi_n$ and

$$Ew_n = E \sum_{m=1}^{M} v_{nm} = \sum_{m=1}^{M} Ev_{nm} = M\varphi_n. \tag{3.56}$$

The covariance of w_p and w_q is equal to

$$\begin{aligned} \operatorname{cov}(w_p, w_q) &= E[(w_p - M\varphi_p)(w_q - M\varphi_q)] \\ &= E\left[\sum_m (v_{pm} - \varphi_p) \sum_m (v_{qm} - \varphi_q)\right] \\ &= \sum_{m_1} \sum_{m_2} E[(v_{p\,m_1} - \varphi_p)(v_{q\,m_2} - \varphi_q)]. \end{aligned} \tag{3.57}$$

For $m_1 \neq m_2$, the $v_{p\,m_1}$ and $v_{q\,m_2}$ are independent and, therefore, uncorrelated. Hence,

$$\operatorname{cov}(w_p, w_q) = \sum_m E[(v_{pm} - \varphi_p)(v_{qm} - \varphi_q)]. \tag{3.58}$$

In the summand, we have $Ev_{pm}v_{qm} = 0$ for $p \neq q$ since $v_{pm} = 1$ implies $v_{qm} = 0$. For $p = q$, the term is equal to $\varphi_p \cdot 1 + (1 - \varphi_p) \cdot 0 = \varphi_p$. Then, for $p = q$:

$$\begin{aligned} \operatorname{cov}(w_p, w_q) &= \operatorname{var} w_p = \sum_m (\varphi_p - \varphi_p^2 - \varphi_p^2 + \varphi_p^2) \\ &= M\varphi_p(1 - \varphi_p) \end{aligned} \tag{3.59}$$

and for $p \neq q$:

$$\operatorname{cov}(w_p, w_q) = \sum_m (-\varphi_p\varphi_q - \varphi_p\varphi_q + \varphi_p\varphi_q) = -M\varphi_p\varphi_q\,. \tag{3.60}$$

This completes the proof. ■

Theorem 3.2 shows that multinomially distributed observations are negatively correlated. For example, for stochastic variables distributed according to (3.53) with $\varphi_1 = 0.25$ and $\varphi_2 = 0.5$, the covariance matrix of $w = (w_1\ w_2)^T$ is

$$10 \begin{pmatrix} 0.1875 & -0.125 \\ -0.125 & 0.25 \end{pmatrix}. \tag{3.61}$$

Next, suppose that $Ew_n = g_n(\theta)$. Then, (3.54) and (3.55) show that

$$\varphi_p = \frac{g_p(\theta)}{M} \tag{3.62}$$

and

$$c_{pq} = \mathrm{cov}(w_p, w_q) = \frac{1}{M} g_p(\theta) \left[M \delta_{pq} - g_q(\theta) \right] . \tag{3.63}$$

Therefore, the $(N-1) \times (N-1)$ covariance matrix of $w = (w_1 \ldots w_{N-1})^T$ is described by

$$C = \frac{1}{M} \begin{pmatrix} g_1(M - g_1) & -g_1 g_2 & \cdots & -g_1 g_{N-1} \\ -g_1 g_2 & g_2(M - g_2) & \cdots & -g_2 g_{N-1} \\ \vdots & & \ddots & \vdots \\ -g_1 g_{N-1} & \cdots & \cdots & g_{N-1}(M - g_{N-1}) \end{pmatrix} , \tag{3.64}$$

where, for brevity, the argument θ has been left out from the $g_n(\theta)$. Furthermore, (3.52) shows that

$$g_N = M - \sum_{n=1}^{N-1} g_n . \tag{3.65}$$

Substituting (3.62) for the φ_n in (3.50) produces the probability function of the observations $w = (w_1 \ldots w_{N-1})^T$ parametric in θ:

$$p(\omega; \theta) = \frac{M!}{M^M \prod_{n=1}^{N} \omega_n!} g_1^{\omega_1} g_2^{\omega_2} \ldots g_N^{\omega_N} \tag{3.66}$$

with $\omega_N = M - \sum_{n=1}^{N-1} \omega_n$ and $g_N = M - \sum_{n=1}^{N-1} g_n$.

Finally, this expression shows that the multinomial log-probability function is described by

$$q(\omega; \theta) = \ln M! - M \ln M - \sum_{n=1}^{N} \ln \omega_n! + \sum_{n=1}^{N} \omega_n \ln g_n , \tag{3.67}$$

where $\omega_N = M - \sum_{n=1}^{N-1} \omega_n$ and $g_N = M - \sum_{n=1}^{N-1} g_n$.

3.5.2 The Fisher score of multinomial observations

Expression (3.67) shows that the Fisher score vector of the multinomial observations $w = (w_1 \ldots w_{N-1})^T$ for the parameters θ is described by

$$s_\theta = \frac{\partial q(w;\theta)}{\partial \theta} = \sum_{n=1}^{N-1} \frac{w_n}{g_n} \frac{\partial g_n}{\partial \theta} + \frac{w_N}{g_N} \frac{\partial g_N}{\partial \theta} = \sum_{n=1}^{N-1} \left(\frac{w_n}{g_n} - \frac{w_N}{g_N} \right) \frac{\partial g_n}{\partial \theta} \tag{3.68}$$

with $w_N = M - \sum_{n=1}^{N-1} w_n$ and $g_N = M - \sum_{n=1}^{N-1} g_n$. Furthermore, straightforward calculations show that

$$\left(\frac{w_n}{g_n} - \frac{w_N}{g_N} \right) = \frac{1}{g_N} \left[d_1 + \ldots + \left(\frac{g_N}{g_n} + 1 \right) d_n + \ldots + d_{N-1} \right] , \tag{3.69}$$

where $n = 1, \ldots, N-1$ and $d_n = w_n - g_n$. Then, (3.68) may be written

$$s_\theta = \frac{\partial g^T}{\partial \theta} R\, d \tag{3.70}$$

with

$$R = \frac{1}{g_N} \begin{pmatrix} \frac{g_N}{g_1} + 1 & 1 & 1 & \cdots & 1 \\ 1 & \frac{g_N}{g_2} + 1 & 1 & \cdots & 1 \\ \vdots & & \ddots & & \vdots \\ 1 & 1 & 1 & & \frac{g_N}{g_{N-1}} + 1 \end{pmatrix} \tag{3.71}$$

while $d = (d_1 \ldots d_{N-1})^T$ and $g = (g_1 \ldots g_{N-1})^T$. Multiplication shows that the matrix R is the inverse of the $(N-1) \times (N-1)$ covariance matrix C, defined by (3.64). Thus,

$$C^{-1} = R. \tag{3.72}$$

Then, after reintroducing the argument θ, (3.70) becomes

$$s_\theta = \frac{\partial g^T(\theta)}{\partial \theta} C^{-1} d(\theta), \tag{3.73}$$

a result formally similar to (3.35) and (3.47).

3.6 EXPONENTIAL FAMILIES OF DISTRIBUTIONS

3.6.1 Definition and examples

A parametric family of joint probability (density) functions $p(\omega;\theta)$ is said to be an *exponential family* if it is described by

$$\boxed{p(\omega;\theta) = \alpha(\theta)\,\beta(\omega)\,\exp\!\left(\gamma^T(\theta)\,\delta(\omega)\right)}, \tag{3.74}$$

where:

$\alpha(\theta)$ is a scalar function of the elements of θ only,

$\beta(\omega)$ is a scalar function of the elements of ω only,

$\gamma(\theta) = [\,\gamma_1(\theta) \ldots \gamma_L(\theta)\,]^T$ is a vector function of the elements of θ only,

and

$\delta(\omega) = [\,\delta_1(\omega) \ldots \delta_L(\omega)\,]^T$ is a vector function of the elements of ω only.

The parametric family of log-probability (density) functions corresponding to (3.74) is

$$q(\omega;\theta) = \ln \alpha(\theta) + \ln \beta(\omega) + \gamma^T(\theta)\delta(\omega). \tag{3.75}$$

If $\delta(\omega) = \omega$, the corresponding exponential family is said to be *linear* or *regular*. Then, the probability (density) function is described by

$$p(\omega;\theta) = \alpha(\theta)\beta(\omega)\exp\left(\gamma^T(\theta)\,\omega\right) \tag{3.76}$$

and the log-probability (density) function by

$$q(\omega;\theta) = \ln \alpha(\theta) + \ln \beta(\omega) + \gamma^T(\theta)\,\omega. \tag{3.77}$$

We give three examples.

■ **EXAMPLE 3.1**

The joint normal distribution

Suppose that $p(\omega;\theta)$ is the parametric family of joint normal probability density functions (3.30):

$$p(\omega;\theta) = \frac{1}{(2\pi)^{\frac{N}{2}}(\det C)^{\frac{1}{2}}} \exp\left(-\frac{1}{2}\left[\omega - g(\theta)\right]^T C^{-1}\left[\omega - g(\theta)\right]\right). \tag{3.78}$$

The argument of the exponential function in this expression may be written

$$-\frac{1}{2}g^T(\theta)C^{-1}g(\theta) - \frac{1}{2}\omega^T C^{-1}\omega + g^T(\theta)C^{-1}\omega, \tag{3.79}$$

where the symmetry of the covariance matrix C has been used. Then, (3.78) is an exponential family of distributions parametric in θ with

$$\alpha(\theta) = \frac{1}{(2\pi)^{\frac{N}{2}}(\det C)^{\frac{1}{2}}} \exp\left(-\frac{1}{2}g^T(\theta)C^{-1}g(\theta)\right), \tag{3.80}$$

$$\beta(\omega) = \exp\left(-\frac{1}{2}\omega^T C^{-1}\omega\right), \tag{3.81}$$

$$\gamma(\theta) = C^{-1}g(\theta), \tag{3.82}$$

and

$$\delta(\omega) = \omega. \tag{3.83}$$

The last expression shows that the family of joint normal distributions with constant covariance matrix is *linear* exponential. ■

■ **EXAMPLE 3.2**

The Poisson distribution

Suppose that $p(\omega;\theta)$ is the parametric family of Poisson probability functions (3.42). This expression is equivalent to

$$p(\omega;\theta) = \exp\left(-\sum_n g_n(\theta)\right) \frac{\exp(\sum_n \omega_n \ln g_n(\theta))}{\prod_n \omega_n!}. \tag{3.84}$$

Thus, this family of Poisson probability functions parametric in θ is exponential with

$$\alpha(\theta) = \exp\left(-\sum_n g_n(\theta)\right), \tag{3.85}$$

$$\beta(\omega) = \frac{1}{\prod_n \omega_n!}, \tag{3.86}$$

$$\gamma(\theta) = \left[\ln g_1(\theta) \ldots \ln g_N(\theta)\right]^T, \tag{3.87}$$

and

$$\delta(\omega) = \omega. \tag{3.88}$$

The last expression shows that the parametric family of Poisson distributions (3.84) is *linear* exponential. ■

EXAMPLE 3.3

The multinomial distribution

Suppose that $p(\omega;\theta)$ is the parametric family of multinomial probability functions (3.66). This expression is equivalent to

$$p(\omega;\theta) = \frac{M!}{M^M \prod_{n=1}^{N} \omega_n!} \exp\left(\sum_{n=1}^{N} \omega_n \ln g_n(\theta)\right), \tag{3.89}$$

with $\omega_N = M - \sum_{n=1}^{N-1} \omega_n$ and $g_N(\theta) = M - \sum_{n=1}^{N-1} g_n(\theta)$, respectively. Thus,

$$p(\omega;\theta) = \frac{M!}{M^M \prod_{n=1}^{N} \omega_n!} g_N^M(\theta) \exp\left(\sum_{n=1}^{N-1} \omega_n \ln \frac{g_n(\theta)}{g_N(\theta)}\right). \tag{3.90}$$

Then, this family of multinomial distributions parametric in θ is exponential with

$$\alpha(\theta) = g_N^M(\theta), \tag{3.91}$$

$$\beta(\omega) = \frac{M!}{M^M \prod_{n=1}^{N} \omega_n!}, \tag{3.92}$$

$$\gamma(\theta) = \left[\ln \frac{g_1(\theta)}{g_N(\theta)} \ldots \ln \frac{g_{N-1}(\theta)}{g_N(\theta)}\right]^T, \tag{3.93}$$

and

$$\delta(\omega) = (\omega_1 \ldots \omega_{N-1})^T. \tag{3.94}$$

The last expression shows that the family of multinomial probability functions (3.90) is *linear* exponential. ■

Examples 3.1–3.3 show the generality of linear exponential families of distributions. Further examples are the binomial, the negative binomial, the beta, the gamma, and the exponential distribution. The examples also illustrate that the definition of $\alpha(\theta)$, $\beta(\omega)$, $\gamma(\theta)$, and $\delta(\omega)$ is not unique since proportionality constants for these functions may be freely chosen provided the products $\alpha(\theta)\beta(\omega)$ or $\gamma^T(\theta)\delta(\omega)$ remain the same.

3.6.2 Properties of exponential families of distributions

In this section, two general properties of exponential families of distributions will be derived.

Property 1 *Normalization*
The function $\alpha(\theta)$ in the expression for an exponential family of probability density functions

$$\alpha(\theta)\,\beta(\omega)\,\exp(\gamma^T(\theta)\delta(\omega)) \tag{3.95}$$

satisfies

$$\frac{1}{\alpha(\theta)} = \int \beta(\omega) \exp(\gamma^T(\theta)\delta(\omega))\, d\omega. \tag{3.96}$$

Proof. Since $p(\omega;\theta)$ is a probability density function, we have

$$\int p(\omega;\theta)d\omega = \alpha(\theta)\int \beta(\omega)\exp\big(\gamma^T(\theta)\delta(\omega)\big)\,d\omega = 1, \tag{3.97}$$

which implies (3.96). ■

Property 1 shows that the dependence of $p(\omega;\theta)$ on θ is characterized by the vector $\gamma(\theta)$. An analogous proof for discrete stochastic variables is obtained by replacing the integrals in the proof by appropriate summations.

Property 2 (Jennrich and Moore) *A general expression for the Fisher score*
The Fisher score vector of exponential family distributed observations is described by

$$s_\theta = \frac{\partial \mu_\delta^T(\theta)}{\partial\theta} C_{\delta\delta}^{-1}[\delta(w) - \mu_\delta(\theta)], \tag{3.98}$$

where $C_{\delta\delta} = \operatorname{cov}((\delta(w), \delta(w))$ and $\mu_\delta(\theta) = E\delta(w)$. If the exponential family is linear, this becomes

$$\boxed{s_\theta = \frac{\partial g^T(\theta)}{\partial\theta} C^{-1} d(\theta)}, \tag{3.99}$$

where $d(\theta) = w - g(\theta)$ and the elements of $C = \operatorname{cov}(w, w)$ typically depend on θ.

Proof. Differentiating (3.75) with respect to θ and substituting w for ω produces the Fisher score vector of exponential family distributed observations

$$s_\theta = \frac{\partial q(w;\theta)}{\partial\theta} = \frac{1}{\alpha(\theta)}\frac{\partial\alpha(\theta)}{\partial\theta} + \frac{\partial\gamma^T(\theta)}{\partial\theta}\delta(w), \tag{3.100}$$

where

$$\frac{\partial\gamma^T(\theta)}{\partial\theta} = \begin{pmatrix} \frac{\partial\gamma_1}{\partial\theta_1} & \cdots & \frac{\partial\gamma_1}{\partial\theta_K} \\ \vdots & \ddots & \vdots \\ \frac{\partial\gamma_L}{\partial\theta_1} & \cdots & \frac{\partial\gamma_L}{\partial\theta_K} \end{pmatrix}^T . \tag{3.101}$$

In the right-hand member of this expression, for brevity, the argument of the $\gamma_\ell(\theta)$ has been left out. Applying Theorem 3.1 to (3.100) yields

$$Es_\theta = \frac{1}{\alpha(\theta)}\frac{\partial\alpha(\theta)}{\partial\theta} + \frac{\partial\gamma^T(\theta)}{\partial\theta}\mu_\delta(\theta) = o, \tag{3.102}$$

where the $L \times 1$ vector $\mu_\delta(\theta) = E\delta(w)$. Subtracting this from (3.100) shows that

$$s_\theta = \frac{\partial q(w;\theta)}{\partial\theta} = \frac{\partial\gamma^T(\theta)}{\partial\theta}[\delta(w) - \mu_\delta(\theta)], \tag{3.103}$$

where, by definition,

$$\mu_\delta(\theta) = E\delta(w) = \int\cdots\int \delta(\omega)\,p(\omega;\theta)\,d\omega. \tag{3.104}$$

For brevity, we now introduce the notation: $p = p(\omega;\theta)$, $q = q(\omega;\theta) = \ln p(\omega;\theta)$, and $\mu_\delta = \mu_\delta(\theta)$. Then, under suitable regularity conditions, differentiating (3.104) with respect to θ yields

$$\frac{\partial \mu_\delta}{\partial \theta^T} = \int \delta(\omega) \frac{\partial p}{\partial \theta^T}\, d\omega = \int \delta(\omega) \frac{\partial q}{\partial \theta^T}\, p\, d\omega. \tag{3.105}$$

Combining this with (3.103) produces

$$\begin{aligned}\frac{\partial \mu_\delta}{\partial \theta^T} &= \int \delta(\omega)\,[\delta(\omega) - \mu_\delta]^T \frac{\partial \gamma(\theta)}{\partial \theta^T}\, p\, d\omega \\ &= \left(\int [\delta(\omega) - \mu_\delta]\,[\delta(\omega) - \mu_\delta]^T\, p\, d\omega \right) \frac{\partial \gamma(\theta)}{\partial \theta^T} \\ &= C_{\delta\delta} \frac{\partial \gamma(\theta)}{\partial \theta^T}\,. \end{aligned} \tag{3.106}$$

Equations (3.103) and (3.106) show that for nonsingular $C_{\delta\delta}$:

$$s_\theta = \frac{\partial \mu_\delta^T}{\partial \theta} C_{\delta\delta}^{-1} \left[\delta(w) - \mu_\delta\right]. \tag{3.107}$$

This proves (3.98).

If the exponential family is linear, $\delta(w) = w$ and $\mu_\delta(\theta) = Ew = g(\theta)$. Substituting these expressions in (3.103) shows that then

$$s_\theta = \frac{\partial \gamma^T(\theta)}{\partial \theta} d(\theta)\,. \tag{3.108}$$

Also, since in this particular case $C_{\delta\delta} = C$, (3.106) shows that

$$\frac{\partial g(\theta)}{\partial \theta^T} = C \frac{\partial \gamma(\theta)}{\partial \theta^T}\,. \tag{3.109}$$

Combining this with (3.108) yields

$$s_\theta = \frac{\partial g^T(\theta)}{\partial \theta} C^{-1} d(\theta)\,. \tag{3.110}$$

This completes the proof. ■

The following examples illustrate Property 2.

■ **EXAMPLE 3.4**

Fisher score vector of normally, Poisson, and multinomially distributed observations

In the Examples 3.1, 3.2, and 3.3, it has been shown that the joint normal, the Poisson, and the multinomial probability (density) functions are linear exponential families. Therefore, (3.99) must describe the relevant Fisher score vectors. This is confirmed by (3.35), (3.47), and (3.73), respectively. ■

3.7 STATISTICAL PROPERTIES OF FISHER SCORES

The expressions (3.35), (3.47), and (3.73) for the Fisher score vector of normal, Poisson, and multinomial observations have the form

$$s_\theta = A(\theta)d(\theta), \tag{3.111}$$

where

$$A(\theta) = \frac{\partial g^T(\theta)}{\partial \theta} C^{-1} \tag{3.112}$$

is a $K \times N$ matrix with elements depending on θ. Thus, the elements of s_θ are linear combinations of the fluctuations $d_n(\theta)$ of the observations around their expectations.

In addition, let the w_n be independent. Then, their covariance matrix is described by

$$C = \text{diag}\,(c_{11} \ldots c_{NN}), \tag{3.113}$$

where $c_{nn} = \text{var}\, w_n$, and the kth element of s_θ is equal to

$$s_{\theta_k} = \sum_n \frac{1}{c_{nn}} \frac{\partial g_n(\theta)}{\partial \theta_k} d_n(\theta)\,. \tag{3.114}$$

In Appendix A, the variance of a weighted sum of uncorrelated stochastic variables is shown to be equal to the quadratically weighted sum of their individual variances. Consequently, the variance of the kth element of s_θ is described by

$$\begin{aligned} \text{var}\, s_{\theta_k} &= \sum_n \frac{1}{c_{nn}^2} \left(\frac{\partial g_n(\theta)}{\partial \theta_k} \right)^2 \text{var}\, d_n(\theta) \\ &= \sum_n \frac{1}{c_{nn}} \left(\frac{\partial g_n(\theta)}{\partial \theta_k} \right)^2, \end{aligned} \tag{3.115}$$

where (2.25) has been used. This implies that, typically, with each additional observation the variance of the Fisher score increases. This is illustrated in the following example.

■ EXAMPLE 3.5

The variance of the Fisher score of Poisson distributed observations for the height and the location of a Lorentz line

Suppose that observations are made with as expectation model a spectral line model, called Lorentz line:

$$g(x; \theta) = \frac{\theta_1}{1 + (x - \theta_2)^2}, \tag{3.116}$$

where the vector of unknown parameters is $\theta = (\theta_1\ \theta_2)^T$. The parameters θ_1 and θ_2 are the *height* and the *location* of the line, respectively. The line attains its maximum value θ_1 at the location $x = \theta_2$. The solid line in Fig. 3.1.(a) and 3.1.(b) shows a Lorentz line with $\theta_1 = 2500$ and $\theta_2 = 0.125$. In Fig. 3.1.(a) the following measurement points, called Design 1, have been chosen:

$$x_n = -2.5 + 0.5(n - 1), \quad n = 1, \ldots, 11, \tag{3.117}$$

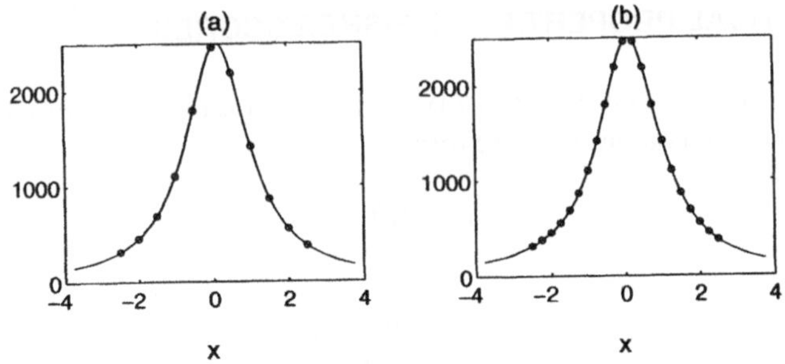

Figure 3.1. Lorentz line expectation model (solid line) and its values at the measurement points (circles). In (a), 11 observations are used; in (b), 21 observations are used. (Example 3.5)

while Design 2, shown in Fig. 3.1.(b), is

$$x_n = -2.5 + 0.25(n-1), \quad n = 1, \ldots, 21. \tag{3.118}$$

Thus, in Design 2, additional measurement points have been inserted between those of Design 1. Suppose that the observations are independent and Poisson distributed. Then, by (3.44),

$$c_{nn} = g_n(\theta). \tag{3.119}$$

Furthermore,

$$g_n(\theta) = \frac{\theta_1}{1 + (x_n - \theta_2)^2} \tag{3.120}$$

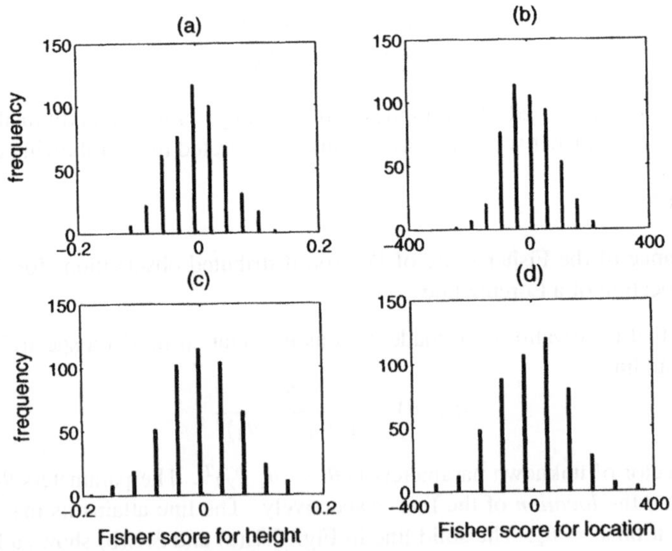

Figure 3.2. Histograms of the Fisher scores for height and location of a Lorentz line. In (a) and (b), 11 observations are used; in (c) and (d), 21 observations are used. (Example 3.5)

and hence,

$$\frac{\partial g_n(\theta)}{\partial \theta_1} = \frac{1}{\theta_1}\, g_n(\theta)\,, \tag{3.121}$$

and

$$\frac{\partial g_n(\theta)}{\partial \theta_2} = \frac{2}{\theta_1}(x_n - \theta_2)\, g_n^2(\theta). \tag{3.122}$$

Then, by (3.115), the variances of the elements s_{θ_1} and s_{θ_2} of the Fisher score vector are

$$\text{var}\, s_{\theta_1} = \frac{1}{\theta_1^2} \sum_n g_n(\theta) \tag{3.123}$$

and

$$\text{var}\, s_{\theta_2} = \frac{4}{\theta_1^2} \sum_n (x_n - \theta_2)^2\, g_n^3(\theta). \tag{3.124}$$

For Design 1, the variances (3.123) and (3.124) are equal to 0.0020 and 7.34×10^3 and, for Design 2, to 0.0039 and 1.45×10^4, respectively. These results show that the variances with Design 2 are nearly twice as large as those with Design 1, as could have been expected. For either design, 500 sets of Poisson observations with expectations (3.120) have been simulated next. From each of these sets, the Fisher scores (3.45) have been computed for θ_1 and θ_2. Histograms of these Fisher scores are shown in Fig. 3.2. As predicted, they are wider as the number of observations increases while their clustering around the origin agrees with (3.18). ■

In Chapter 4, we will return to this increase of the variance of the Fisher score with the number of observations.

3.8 COMPLEX STOCHASTIC VARIABLES

In this section, scalar and vector complex stochastic variables are introduced and their most important properties are discussed. Next, the same properties of vectors of both real and complex stochastic variables are addressed.

3.8.1 Scalar complex stochastic variables

Suppose that u and v are jointly distributed real scalar stochastic variables. Next, define the scalar complex stochastic variable z and its complex conjugate z^* by

$$z = u + jv \tag{3.125}$$

and

$$z^* = u - jv \tag{3.126}$$

where the superscript $*$ denotes complex conjugation and $j^2 = -1$. Then, the relation of the real stochastic variables u and v and the complex stochastic variables z and z^* is described by

$$\begin{pmatrix} z \\ z^* \end{pmatrix} = \begin{pmatrix} 1 & j \\ 1 & -j \end{pmatrix} \begin{pmatrix} u \\ v \end{pmatrix}. \tag{3.127}$$

The expectations of z and z^* are defined as

$$Ez = Eu + jEv \tag{3.128}$$

and

$$Ez^* = Eu - jEv\,. \tag{3.129}$$

Therefore,

$$Ez^* = (Ez)^*. \tag{3.130}$$

The variance of a complex stochastic variable is defined as

$$\operatorname{var} z = E\left[(z - Ez)\,(z^* - Ez^*)\right]. \tag{3.131}$$

Then,

$$\operatorname{var} z = E\,|z - Ez|^2 = E[(u - Eu)^2 + (v - Ev)^2] = \operatorname{var} u + \operatorname{var} v\,. \tag{3.132}$$

Finally, the covariance of the scalar complex stochastic variables z_1 and z_2 is defined as

$$\operatorname{cov}(z_1, z_2) = E\left[(z_1 - Ez_1)\,(z_2^* - Ez_2^*)\right]. \tag{3.133}$$

This definition implies (3.131).

3.8.2 Vectors of complex stochastic variables

Next, let u and v be the real vectors

$$u = (u_1 \ldots u_L)^T \tag{3.134}$$

and

$$v = (v_1 \ldots v_L)^T \tag{3.135}$$

with elements that are jointly distributed stochastic variables and define

$$t = \begin{pmatrix} u \\ v \end{pmatrix}. \tag{3.136}$$

Furthermore, define

$$z = (z_1 \ldots z_L)^T \tag{3.137}$$

with

$$z_\ell = u_\ell + jv_\ell \tag{3.138}$$

and let the elements of the vector z^* be the complex conjugates of the corresponding elements of z. Next, define

$$s = \begin{pmatrix} z \\ z^* \end{pmatrix}. \tag{3.139}$$

Then,

$$s = At, \tag{3.140}$$

where the complex $2L \times 2L$ matrix A is defined as

$$A = \begin{pmatrix} I & jI \\ I & -jI \end{pmatrix} \tag{3.141}$$

with I the identity matrix of order L. Thus, the expression (3.140) is consistent with (3.127).

Generally, the covariance matrix $\text{cov}(z, z)$ of a vector of complex stochastic variables z is defined by its (m, n)th element

$$\text{cov}(z_m, z_n) = E\left[(z_m - Ez_m)\,(z_n^* - Ez_n^*)\right]. \tag{3.142}$$

Therefore,

$$\text{cov}(z_m, z_n) = (\text{cov}(z_n, z_m))^* \,. \tag{3.143}$$

This shows that the covariance matrix of a complex vector of stochastic variables is Hermitian, that is, equal to its complex conjugate transpose. Furthermore, the definition (3.142) is equivalent to

$$\text{cov}(z, z) = E\left[(z - Ez)(z - Ez)^H\right] \tag{3.144}$$

for any vector of complex stochastic variables z, where the superscript H indicates complex conjugate transposition. This expression shows that the covariance matrix of the vector s as defined by (3.139) is described by

$$\text{cov}(s, s) = \begin{pmatrix} \text{cov}(z, z) & \text{cov}(z, z^*) \\ \text{cov}(z^*, z) & \text{cov}(z^*, z^*) \end{pmatrix}. \tag{3.145}$$

Definition (3.144) may also be instrumental in the proof that the covariance matrix of s as defined by (3.140) is described by

$$\text{cov}(s, s) = A\,\text{cov}(t, t)\,A^H\,. \tag{3.146}$$

3.8.3 Vectors of real and complex stochastic variables

Suppose that the elements of the vector s of complex stochastic variables z and z^* are jointly distributed with the elements of the real vector

$$r = (r_1 \ldots r_K)^T. \tag{3.147}$$

Then, the combined vector of real and complex stochastic variables is

$$w = \begin{pmatrix} r \\ s \end{pmatrix} = \text{diag}\,(I\ A) \begin{pmatrix} r \\ t \end{pmatrix} \tag{3.148}$$

where, as before,

$$s = \begin{pmatrix} z \\ z^* \end{pmatrix} \tag{3.149}$$

and

$$t = \begin{pmatrix} u \\ v \end{pmatrix}, \tag{3.150}$$

while I is the identity matrix of order K and the $2L \times 2L$ matrix A is defined by (3.141). The block diagonal notation $\text{diag}\,(I\ A)$ is defined by Definition B.15. Thus,

$$w = (r_1 \ldots r_K\ z_1 \ldots z_L\ z_1^* \ldots\ z_L^*)^T = (r^T\ z^T\ z^H)^T. \tag{3.151}$$

Next, define

$$y = \begin{pmatrix} r \\ t \end{pmatrix}. \tag{3.152}$$

Then,

$$w = By, \tag{3.153}$$

where the $(K + 2L) \times (K + 2L)$ complex matrix B is defined as

$$B = \mathrm{diag}(I \;\; A). \tag{3.154}$$

The covariance matrix C of the vector w as defined by (3.151) is described by

$$C = \mathrm{cov}(w, w) = \begin{pmatrix} \mathrm{cov}(r, r) & \mathrm{cov}(r, z) & \mathrm{cov}(r, z^*) \\ \mathrm{cov}(z, r) & \mathrm{cov}(z, z) & \mathrm{cov}(z, z^*) \\ \mathrm{cov}(z^*, r) & \mathrm{cov}(z^*, z) & \mathrm{cov}(z^*, z^*) \end{pmatrix}. \tag{3.155}$$

Also, using the definition (3.144), we may show that the covariance matrix of w as defined by (3.153) is described by

$$C = BYB^H, \tag{3.156}$$

where $Y = \mathrm{cov}(y, y)$. Therefore,

$$Y = B^{-1}C\,B^{-H} \tag{3.157}$$

where B^{-H} is defined as $(B^H)^{-1}$. In the next section, this relation of the covariance matrices Y and C will be used in the derivation of the joint real–complex normal distribution.

3.9 THE JOINT REAL–COMPLEX NORMAL DISTRIBUTION

Suppose that the elements of the real $(K + 2L) \times 1$ vector y defined by (3.152) are jointly normally distributed. Then, by (3.22), their probability density function is

$$p(\eta) = \frac{1}{(2\pi)^{\frac{K+2L}{2}} (\det Y)^{\frac{1}{2}}} \exp\Big(-\frac{1}{2}(\eta - \mu_y)^T Y^{-1}(\eta - \mu_y) \Big), \tag{3.158}$$

where the elements of the $(K + 2L) \times 1$ vector of independent variables η correspond to those of y, and Y and μ_y are the covariance matrix and the vector expectation of y, respectively. Then, by (3.153), $w = By$. Furthermore, (3.157) shows that

$$Y^{-1} = B^H C^{-1} B. \tag{3.159}$$

Therefore,

$$\begin{aligned} (y - \mu_y)^T Y^{-1} (y - \mu_y) &= (y - \mu_y)^T B^H C^{-1} B (y - \mu_y) \\ &= (w - \mu_w)^H C^{-1} (w - \mu_w) \end{aligned} \tag{3.160}$$

since $\mu_w = Ew = B\mu_y$. Also, from (3.156),

$$\det C = \det B \det Y \det B^H \tag{3.161}$$

since, by Theorem B.7, the determinant of the product of square matrices is the product of the determinants of the matrices. Then, (3.154) and Definition B.19 show that

$$\det B = \det A. \tag{3.162}$$

Next, (B.34) is used for the computation of the determinant of the partitioned matrix A:

$$\det A = \det I \, \det\left(-jI - jI\,I^{-1}I\right) = \det(-2jI) = (-2j)^L. \tag{3.163}$$

Similarly, $\det B^H = (2j)^L$. Combining these results with (3.161) shows that

$$\det Y = 4^{-L} \det C. \tag{3.164}$$

Substituting this result and (3.160) in (3.158) yields the *joint real–complex normal probability density function*

$$\frac{1}{2^{\frac{K}{2}} \pi^{\frac{K+2L}{2}} (\det C)^{\frac{1}{2}}} \exp\left(-\frac{1}{2}\,(\omega - \mu_w)^H \, C^{-1}\,(\omega - \mu_w)\right), \tag{3.165}$$

where the elements of the $(K + 2L) \times 1$ vector of variables ω correspond to those of w.

Next, we consider special cases of the probability density function (3.165). First, suppose that all elements of w are real. Then, $2L = 0$ and (3.165) simplifies to the real joint normal distribution (3.22). Next, suppose that all elements of w are complex. Then, $K = 0$ and (3.165) becomes

$$\frac{1}{\pi^L (\det S)^{\frac{1}{2}}} \exp\left(-\frac{1}{2}\,(\omega - \mu_s)^H \, S^{-1}\,(\omega - \mu_s)\right), \tag{3.166}$$

where $S = \text{cov}(s, s)$ and s is defined by (3.149). In applications mentioned in the literature, often

$$E[(z_m - \mu_{z_m})\,(z_n - \mu_{z_n})] = E[(z_m - \mu_{z_m})^*\,(z_n - \mu_{z_n})^*] = 0, \tag{3.167}$$

that is, $\text{cov}(z, z^*)$ and $\text{cov}(z^*, z)$ are $L \times L$ null matrices. Then, the complex stochastic variables z are called *circularly complex* and

$$S = \begin{pmatrix} Z & O \\ O & Z^* \end{pmatrix}, \tag{3.168}$$

where $Z = \text{cov}(z, z)$ with $z = (z_1 \ldots z_L)^T$ and O is the $L \times L$ null matrix. Substituting (3.168) in (3.166) yields the joint circularly complex normal distribution

$$\frac{1}{\pi^L \det Z} \exp\left(-\,(\zeta - \mu_z)^H \, Z^{-1}\,(\zeta - \mu_z)\right), \tag{3.169}$$

where the elements of the vector ζ correspond to those of z while Theorem B.8 and $Z^* = Z^T$ have been used to show that $\det S = (\det Z)^2$.

3.10 COMMENTS AND REFERENCES

Useful references to the distributions discussed in Sections 3.3–3.5 and to many others are [21], [19], and [29]. There exists specialized literature on exponential families of distributions discussed in Section 3.6, for example, Barndorff-Nielsen's book [3]. However, for our purposes, relatively limited introductions to these distributions such as given by Jennrich [17] and by Lehmann and Casella [22] are sufficient. The expression for the Fisher score vector of exponential family distributed observations presented in Subsection

3.6.2 is a slightly less general version of the expression derived by Jennrich and Moore [18]. It will be used in the rest of the book to generalize results and methods. Historically, the Jennrich–Moore result has generalized a variety of earlier results specialized to particular distributions or to independent observations. The main motive to include Section 3.7 has been to illustrate that Fisher scores are stochastic variables with an expectation zero and a variance increasing with the number of observations. Finally, complex stochastic variables discussed in Section 3.8 have been included since they arise naturally in many disciplines of applied science, in signal processing in particular. The derivation of the expression for the joint real–complex normal distribution in Section 3.9 is a somewhat modified version of the derivation by the author in [32].

3.11 PROBLEMS

3.1 Show for discrete stochastic variables that the expectation of the Fisher score vector is equal to the null vector.

3.2 The scalar stochastic variable u is Poisson distributed with parameter λ. Show that both Eu and $\operatorname{var} u$ are equal to λ.

3.3 A scalar discrete stochastic variable u has a *binomial* distribution if its probability function is described by

$$p(v) = \binom{M}{v} \varphi^v (1-\varphi)^{M-v}, \qquad 0 \leq \varphi \leq 1.$$

In this expression, φ is the probability of success of an experiment with two possible outcomes: 1 ("success") or 0 ("failure"), M is the total number of independent experiments, and v, such that $0 \leq v \leq M$, is the number of successes occurring in the M experiments.

(a) Show that $Eu = M\varphi$ and that $\operatorname{var} u = M\varphi(1-\varphi)$.

(b) Let the elements of $w = (w_1 \dots w_N)^T$ be N independent, binomially distributed stochastic variables with $Ew_n = M\varphi_n$. Derive a parametric expression in θ for the joint probability function and the joint log-probability function of the elements of w if $Ew_n = g_n(\theta)$.

(c) Derive an expression for the covariance matrix of w.

(d) Use the expression for the joint log-probability function derived under (b) to find an expression for the Fisher score vector.

3.4

(a) Show that the binomial distribution of Problem 3.3 is a linear exponential family.

(b) Derive an expression for the Fisher score vector of the binomially distributed stochastic variables of Problem 3.3 from the general expression for the Fisher score vector of linear exponential family distributed stochastic variables. Verify that this expression is identical to that found in Problem 3.3(d).

3.5 Suppose that the elements of $w = (w_1 \ldots w_N)^T$ are independent and binomially distributed stochastic variables with a number of M_n independent experiments for w_n and $Ew_n = g_n(\theta)$.

(a) Derive a parametric expression in θ for the joint probability function and the joint log-probability function of w.

(b) Derive an expression for the covariance matrix of w.

(c) Use the expression for the joint log-probability function derived under (a) to find an expression for the Fisher score vector of w.

3.6

(a) Show that the binomial distribution of Problem 3.5 is a linear exponential family.

(b) Derive an expression for the Fisher score vector of the binomial stochastic variables of Problem 3.5 from the general expression for the Fisher score vector of linear exponential family distributed stochastic variables. Verify that this expression is identical to that found in Problem 3.5(c).

3.7 A scalar continuous stochastic variable u has an *exponential* distribution if its probability density function is described by

$$p(v) = \lambda \exp(-\lambda v), \qquad v \geq 0$$

and is equal to zero elsewhere. The scalar λ is positive.

(a) Show that $Eu = 1/\lambda$ and that $\operatorname{var} u = 1/\lambda^2$.

(b) Let the elements of $w = (w_1 \ldots w_N)^T$ be N independent, exponentially distributed stochastic variables with $Ew_n = 1/\lambda_n$. Derive a parametric expression in θ for the joint probability density function and the joint log-probability density function of the elements of w if $Ew_n = g_n(\theta)$.

(c) Derive an expression for the covariance matrix of w.

(d) Use the expression for the log-probability density function derived under (b) to find an expression for the Fisher score vector of w.

3.8

(a) Show that the exponential distribution of Problem 3.7 is a linear exponential family.

(b) Derive an expression for the Fisher score vector of the exponentially distributed stochastic variables of Problem 3.7 from the general expression for the Fisher score vector of linear exponential family distributed stochastic variables. Verify that this expression is identical to that found in Problem 3.7(d).

3.9 A scalar continuous stochastic variable u has a *Maxwell* distribution if its probability density function is described by

$$p(v) = \left(\frac{2}{\pi}\right)^{\frac{1}{2}} \frac{1}{\alpha^3} v^2 \exp\left(-\frac{v^2}{2\alpha^2}\right), \qquad v \geq 0$$

and is equal to zero elsewhere. The scalar α is positive. It can be shown that for this distribution $Eu = 2\,(2/\pi)^{\frac{1}{2}}\,\alpha$. Let the elements of $w = (w_1 \ldots w_N)^T$ be N independent stochastic variables having a Maxwell distribution with $Ew_n = g_n(\theta)$.

(a) Derive a parametric expression in θ for the joint probability density function and the joint log-probability density function of the elements of w if $Ew_n = g_n(\theta)$.

(b) Is this distribution a linear exponential family of distributions?

3.10 The observations $w = (w_1 \ldots w_N)^T$ are normally distributed with $Ew = g(\theta)$. The elements of the covariance matrix C of w also depend on the elements of θ. Is this distribution an exponential family of distributions?

3.11 Prove (3.69).

3.12 Show that the matrix R defined by (3.71) is the inverse of the covariance matrix C defined by (3.64).

3.13 The stochastic variables $w = (w_1 \ldots w_{N-1})^T$ are multinomially distributed with $w_N = M - \sum_{n=1}^{N-1} w_n$. Show that the $N \times N$ covariance matrix of $(w_1 \ldots w_N)^T$ is singular.

3.14 The stochastic variables $w = (w_1 \ldots w_{N-1})^T$ are multinomially distributed with $w_N = M - \sum_{n=1}^{N-1} w_n$ and expectations $Ew_n = g_n(\theta)$. Show that the kth element of the Fisher score vector s_θ is described by

$$s_{\theta_k} = \sum_n \frac{1}{g_n(\theta)} \frac{\partial g_n(\theta}{\partial \theta_k} d_n(\theta) + \frac{1}{g_N(\theta)} \left(\sum_n \frac{\partial g_n(\theta)}{\partial \theta_k} \right) \left(\sum_n d_n(\theta) \right)$$

with $n = 1, \ldots, N-1$ and $d_n(\theta) = w_n - g_n(\theta)$.

3.15 Let $z = u + jv$ be a scalar complex stochastic variable. Show that the variance of z does not depend on $\operatorname{cov}(u, v)$.

3.16 Let $z = u + jv$ be a scalar complex stochastic variable and let r be a scalar real stochastic variable. Show that z and r are uncorrelated if and only if u and v are uncorrelated with r.

3.17 Let $z = u + jv$ be a scalar complex stochastic variable. Express $E\,(z - Ez)^2$ in $\operatorname{var} u$, $\operatorname{var} v$, and $\operatorname{cov}(u, v)$.

3.18 The joint distribution of the elements of a vector of complex stochastic variables z is circularly complex. What are then the conditions to be met by the real parts u_m and the imaginary parts v_m of the elements z_m of z?

3.19 Suppose that the $(K + 2L) \times 1$ vector of real–complex normally distributed stochastic variables w is described by $(r^T\ t^T)^T$ with $r = (r_1 \ldots r_K)^T$ real and $t = (z^T\ z^H)^T$ complex, and let R and T be the covariance matrices of r and t, respectively. Derive an expression for the joint real–complex normal probability density function of the elements of w if the elements of r are not correlated with those of t.

CHAPTER 4

PRECISION AND ACCURACY

4.1 INTRODUCTION

This chapter deals with precision and accuracy of parameter estimates computed from statistical observations. The parameters are those of the expectation model introduced in Chapter 2. The statistical properties of the observations are defined by their distribution—for example, one of the distributions described in Chapter 3.

First, in Section 4.2, standard deviation and bias of estimators are introduced as measures of their precision and accuracy. Then, after a description of properties of covariance matrices in Section 4.3, the Fisher information matrix, defined as the covariance matrix of the Fisher score vector, is introduced in Section 4.4. Fisher information is a key notion in the approach to parameter estimation followed in this book. It is closely connected with the Cramér–Rao lower bound on the variance of unbiased estimators described in Section 4.5. In Section 4.6, important practical properties of the Cramér–Rao lower bound are discussed. Then, in Section 4.7, the Cramér–Rao lower bound for real parameters discussed thus far is generalized to complex parameters. A general expression for the Cramér–Rao lower bound for exponential families of distributions is derived in Section 4.8. Then, in Section 4.9, the Cramér–Rao lower bound is used to define identifiability of parameters. Finally, a procedure for minimizing the Cramér–Rao lower bound in an appropriate sense with respect to free experimental variables, called optimal experimental design, is presented in Section 4.10.

4.2 PROPERTIES OF ESTIMATORS

This chapter addresses precision and accuracy in measurement. In our terminology, these are the precision and accuracy of the estimator used. An *estimator* is a scalar or vector function of the observations intended to measure the value of one or more parameters. The estimator is not a function of any of the parameters to be estimated. The value assumed by the estimator for a particular set of observations is called the *estimate*. The *estimand* is the quantity to be estimated. In this book, this will always be the hypothetical true value of the parameter.

Example 2.10, concerned with estimating the slope of a straight line, shows that different estimators may be devised for the same parameter from the same observations. The example also shows that these different estimators have a different precision. Therefore, the question is justified which of all possible estimators is *most* precise. This question will be discussed in Chapter 5. A related question is if this ultimate precision can be computed. This chapter is mainly concerned with the latter question, but first we will discuss how precision and accuracy are defined in this book.

Suppose that

$$t = (t_1 \dots t_K)^T \tag{4.1}$$

is an estimator of the vector of parameters

$$\theta = (\theta_1 \dots \theta_K)^T. \tag{4.2}$$

Any of the elements t_k is a function of the observations and is a stochastic variable since the observations are stochastic variables. Then, the *bias* β_{t_k} of the estimator t_k of the parameter θ_k is defined as

$$\boxed{\beta_{t_k} = Et_k - \theta_k}, \tag{4.3}$$

where the expectation is taken with respect to the probability (density) function of the observations. The bias is specific of a particular estimator. If the bias is equal to zero, the estimator concerned is called *unbiased*. An estimator is called *asymptotically unbiased* if its bias vanishes when the number of observations tends to infinity. Bias defines and quantifies the *accuracy* or, equivalently, the systematic error of an estimator. An estimator is said to be more accurate as its bias is smaller.

Bias may have different sources:

- Bias may be a systematic deviation of the estimator caused by the fluctuations of the observations and not vanishing if the number of observations tends to infinity. An example is the systematic error in the results of Prony's method discussed in Example 2.9.
- Bias may be a systematic deviation of the estimator caused by modeling errors which are inadequacies of the parametric model used to represent the expectations of the observations. An example is the absence of drift in the model used in Example 2.8.
- Bias may be a systematic deviation of an asymptotically unbiased estimator as a result of the use of an insufficient number of observations.

For the first type of bias, there is no cure except for taking a different estimator. The difficulty with this type of bias is that it may go undetected. If the number of observations used is increased, the parameter estimates may converge to a particular value that the experimenter is tempted to take as the true value of the parameter. These considerations show the importance of applying a chosen estimator first to numerically simulated observations since, unfortunately, in most cases an expression for this type of bias is difficult to derive.

The second type of bias, caused by modeling errors, is so serious since it will corrupt the results of any estimator used. To avoid this, the experimenter is tempted to extend the expectation model so as to include *all* possible contributions. This may increase the number of nuisance parameters substantially. Generally, estimating these additional nuisance parameters causes the standard deviations of the estimates of the target parameters to increase. This decrease of precision will be addressed in Section 4.6.5.

The last type of bias depends on the number of observations used. Since expressions for this dependence are, typically, not available, this bias may best be computed by applying the estimator to simulated observations.

The standard deviation of the kth element t_k of the vector estimator t defines and quantifies the *precision* of t_k or, equivalently, its nonsystematic error. This standard deviation is defined as the square root of

$$\boxed{\operatorname{var} t_k = E\,(t_k - Et_k)^2}, \tag{4.4}$$

where the expectations are taken with respect to the probability (density) function of the observations. The standard deviation is specific of a particular estimator. An estimator is said to be more precise as its standard deviation is smaller. Generally, we will study the precision of a vector estimator using its covariance matrix

$$\operatorname{cov}(t,t) = \begin{pmatrix} \operatorname{var} t_1 & \operatorname{cov}(t_1,t_2) & \dots & \operatorname{cov}(t_1,t_K) \\ \operatorname{cov}(t_2,t_1) & \operatorname{var} t_2 & & \vdots \\ \vdots & & \ddots & \vdots \\ \operatorname{cov}(t_K,t_1) & \dots & \dots & \operatorname{var} t_K \end{pmatrix}, \tag{4.5}$$

where, as usual,

$$\operatorname{cov}(t_k,t_\ell) = E[(t_k - Et_k)\,(t_\ell - Et_\ell)]\,. \tag{4.6}$$

Then, (4.6) implies that

$$\operatorname{cov}(t_k,t_\ell) = E[t_k t_\ell] - Et_k\,Et_\ell. \tag{4.7}$$

Since $\operatorname{cov}(t_k,t_\ell) = \operatorname{cov}(t_\ell,t_k)$ for real t_k and t_ℓ, the covariance matrix $\operatorname{cov}(t,t)$ is symmetric. Further properties of the covariance matrix will be dealt with in Section 4.3.

The definition of the variance of t_k shows that the variance and, therefore, the standard deviation are measured in terms of deviations of the estimator from its expectation. Consequently, it measures the fluctuations of the estimates from experiment to experiment as a result of the fluctuations in the observations and does so relative to the mean value of the estimator. We have seen earlier in this section that this mean value—that is, the expectation of the estimator—need not be equal to the true value of the parameter. Generally, in measurement we want to know the deviation of our estimator t_k from the true value θ_k of the parameter. This leads to the definition of the *mean squared error (mse)*:

$$\boxed{\operatorname{mse} t_k = E\,(t_k - \theta_k)^2}. \tag{4.8}$$

This definition shows that

$$
\begin{aligned}
\operatorname{mse} t_k &= E\left(t_k - Et_k + Et_k - \theta_k\right)^2 \\
&= E\left(t_k - Et_k\right)^2 + 2E\left[\left(t_k - Et_k\right)\left(Et_k - \theta_k\right)\right] + E\left(Et_k - \theta_k\right)^2 \\
&= E\left(t_k - Et_k\right)^2 + 2E\left(t_k - Et_k\right)\left(Et_k - \theta_k\right) + \left(Et_k - \theta_k\right)^2 \\
&= E\left(t_k - Et_k\right)^2 + 0 + \left(Et_k - \theta_k\right)^2 \\
&= \sigma_{t_k}^2 + \beta_{t_k}^2
\end{aligned}
\tag{4.9}
$$

since $E\left(t_k - Et_k\right) = 0$, and Et_k and θ_k are not functions of the observations with respect to which the expectations are taken. The conclusion is that the mean squared error is equal to the sum of the variance and the square of the bias.

Generally, estimators should approach the exact value of the parameter in some sense as the number of observations increases. This convergence process may take place in various ways. Of these, we briefly describe the two most important ones for our purposes. An estimator t is defined as *convergent in quadratic mean* if its mean squared error vanishes asymptotically. Expression (4.9) shows that convergence in quadratic mean occurs if and only if the elements of t are asymptotically unbiased and their variances vanish asymptotically. An estimator is defined as *consistent* if the probability that it deviates less than a specified amount from its exact value may be made arbitrarily close to one if the number of observations is taken sufficiently large. That is, for any $\varepsilon, \delta > 0$, a value N' for the number of observations N may be found such that for all $N > N'$ we have

$$\Pr(\mid t_k - \theta_k \mid < \varepsilon) > 1 - \delta\,, \tag{4.10}$$

where $\Pr(A)$ denotes the probability of the event A. Consistency does not imply asymptotic unbiasedness. It is a statement about the concentration of mass of the probability density function of the estimator, not about its moments. On the other hand, convergence in quadratic mean is a statement about moments of t_k. It may be shown that convergence in quadratic mean implies consistency but that the converse is not true.

4.3 PROPERTIES OF COVARIANCE MATRICES

4.3.1 Real covariance matrices

Suppose that u and v are an $M \times 1$ and an $N \times 1$ vector of real stochastic variables, respectively. Then, the *covariance matrix* of u and v is defined as follows:

Definition 4.1 *The $M \times N$ covariance matrix of the $M \times 1$ vector u and the $N \times 1$ vector v is defined as*

$$\operatorname{cov}(u, v) = E[(u - Eu)(v - Ev)^T] \tag{4.11}$$

with elements

$$\operatorname{cov}(u_m, v_n) = E[(u_m - Eu_m)(v_n - Ev_n)], \tag{4.12}$$

where the expectations are taken with respect to the joint probability (density) function of the elements of u and v.

This definition shows that

$$\mathrm{cov}(v, u) = (\mathrm{cov}(u, v))^T. \tag{4.13}$$

The expressions (4.11) and (4.12) also imply that

$$\mathrm{cov}(u, v) = E[uv^T] - Eu\, Ev^T \tag{4.14}$$

or

$$\mathrm{cov}(u_m, v_n) = E[u_m v_n] - Eu_m Ev_n. \tag{4.15}$$

Definition 4.2 *The $N \times N$ covariance matrix of the $N \times 1$ vector of stochastic variables u is defined as*

$$\mathrm{cov}(u, u) = E[(u - Eu)(u - Eu)^T] \tag{4.16}$$

with elements

$$\mathrm{cov}(u_m, u_n) = E[(u_m - Eu_m)(u_n - Eu_n)]. \tag{4.17}$$

Theorem 4.1 *The covariance matrix of a vector of real stochastic variables is symmetric and positive semidefinite. It is positive definite if and only if the stochastic variables are linearly independent.*

Proof. Equation (4.17) shows that $\mathrm{cov}(u_m, u_n) = \mathrm{cov}(u_n, u_m)$. Therefore, the covariance matrix $\mathrm{cov}(u, u)$ is symmetric.

Next, let the scalar stochastic variable y be any real linear combination of the elements of u:

$$y = a_1 u_1 + \cdots + a_N u_N = a^T u, \tag{4.18}$$

with $a = (a_1 \ldots a_N)^T$. Since, by definition, the variance of a stochastic variable is the expectation of a quadratic quantity, we have

$$\mathrm{var}\, y \geq 0. \tag{4.19}$$

By expression (A.12), the variance of the linear combination of stochastic variables y is equal to

$$\mathrm{var}\, y = a^T \mathrm{cov}(u, u) a. \tag{4.20}$$

Equations (4.19) and (4.20) show that

$$\mathrm{var}\, y = a^T \mathrm{cov}(u, u)\, a \geq 0. \tag{4.21}$$

This result is true for any a. Therefore, $\mathrm{cov}(u, u)$ is positive semidefinite. However,

$$\mathrm{var}\, y = a^T \mathrm{cov}(u, u)\, a > 0 \tag{4.22}$$

for $a \neq o$ if and only if the elements of u are linearly independent stochastic variables. Then, cov (u, u) is positive definite. This completes the proof. ■

For an explanation of the concepts linear dependence and independence of stochastic variables, see Section A.1. Real definite matrices are discussed in Section C.1.

4.3.2 Complex covariance matrices

In Section (4.3.1), properties of covariance matrices of vectors of *real* stochastic variables have been discussed. In this subsection, we discuss the corresponding properties if at least one of the vectors is complex in the sense that it has one or more complex elements. Suppose that the $M \times 1$ vector u and the $N \times 1$ vector v are such vectors. Then, the *covariance matrix* of u and v is defined as follows:

Definition 4.3 *The $M \times N$ covariance matrix of the complex $M \times 1$ vector u and the complex $N \times 1$ vector v is defined as*

$$\text{cov}(u, v) = E[(u - Eu)(v - Ev)^H] \tag{4.23}$$

with elements

$$\text{cov}(u_m, v_n) = E[(u_m - Eu_m)(v_n - Ev_n)^*]. \tag{4.24}$$

This definition shows that

$$\text{cov}(v, u) = (\text{cov}(u, v))^H . \tag{4.25}$$

The expressions (4.23) and (4.24) also imply that

$$\text{cov}(u, v) = E[uv^H] - Eu\, Ev^H \tag{4.26}$$

or

$$\text{cov}(u_m, v_n) = E[u_m\, v_n^*] - Eu_m Ev_n^* . \tag{4.27}$$

Definition 4.4 *The $N \times N$ covariance matrix of the complex $N \times 1$ vector of stochastic variables u is defined as*

$$\text{cov}(u, u) = E[(u - Eu)(u - Eu)^H] \tag{4.28}$$

with elements

$$\text{cov}(u_m, u_n) = E[(u_m - Eu_m)(u_n - Eu_n)^*] . \tag{4.29}$$

This definition shows that the diagonal elements $\text{cov}(u_m, u_n)$ with $m = n$ are equal to

$$\text{var}\, u_m = E \mid u_m - Eu_m \mid^2 \tag{4.30}$$

and are, therefore, real and nonnegative.

Theorem 4.2 *The covariance matrix of a vector of complex stochastic variables is Hermitian and positive semidefinite. It is positive definite if and only if the stochastic variables are linearly independent.*

Proof. Equation (4.29) shows that

$$\text{cov}(u_m, u_n) = \text{cov}(u_n^*, u_m^*) = (\text{cov}(u_n, u_m))^* . \tag{4.31}$$

Therefore, the covariance matrix $\text{cov}(u, u)$ is equal to its conjugate transpose, that is, it is Hermitian. Furthermore, let the complex scalar stochastic variable y be any complex linear combination of the elements of u:

$$y = a_1^* u_1 + \cdots + a_N^* u_N = a^H u , \tag{4.32}$$

where a is a complex vector. Since, by definition, the variance of a stochastic variable is the expectation of a quadratic quantity, it is nonnegative. That is,

$$\operatorname{var} y = E\left[(y - Ey)(y - Ey)^*\right] \geq 0\,. \tag{4.33}$$

Substituting (4.32) for y in this expression yields

$$\operatorname{var} y = a^H \operatorname{cov}(u, u)\, a \geq 0\,. \tag{4.34}$$

Therefore, $\operatorname{cov}(u, u)$ is positive semidefinite. This result is true for any a. Analogous to the variance of real y, $\operatorname{var} y > 0$ for $a \neq o$ if and only if the elements of u are linearly independent stochastic variables. This completes the proof. ■

Complex definite matrices are discussed in Section C.2.

4.4 FISHER INFORMATION

4.4.1 Definition of the Fisher information matrix

The results in this section are derived under *regularity conditions*. A probability density function $p(\omega; \theta)$ will be called *regular* if:

- The range of the ω_n is independent of θ, and
- The partial derivatives of $p(\omega; \theta)$ with respect to θ up to third order exist and are bounded by integrable functions of ω.

In Section 3.2, we defined the $K \times 1$ Fisher score vector s_θ of a set of stochastic variables $w = (w_1 \ldots w_N)^T$ representing observations as

$$s_\theta = \frac{\partial q(w; \theta)}{\partial \theta}, \tag{4.35}$$

where $\theta = (\theta_1 \ldots \theta_K)^T$ is the $K \times 1$ vector of parameters and $q(\omega; \theta) = \ln p(\omega; \theta)$ is the log-probability density function of the observations. In the same section, a proof was presented that under suitable regularity conditions

$$E s_\theta = o. \tag{4.36}$$

The regularity conditions concerned may be shown to be implied by the regularity conditions mentioned at the beginning of this subsection. If (4.36) is true, the $K \times K$ covariance matrix of s_θ is described by

$$\operatorname{cov}(s_\theta, s_\theta) = E\left[(s_\theta - E s_\theta)(s_\theta - E s_\theta)^T\right] = E\left[s_\theta s_\theta^T\right]. \tag{4.37}$$

The $K \times K$ matrix

$$\boxed{F_\theta = E\left[s_\theta s_\theta^T\right] = E\left[\frac{\partial q(w; \theta)}{\partial \theta}\frac{\partial q(w; \theta)}{\partial \theta^T}\right]} \tag{4.38}$$

is called the *Fisher information matrix* of the observations w. Since the Fisher information matrix is a covariance matrix, it is positive semidefinite. It is positive definite if and only if the elements of the Fisher score vector are linearly independent stochastic variables. See Theorem 4.1.

■ **EXAMPLE 4.1**

The Fisher information matrix of normally distributed observations

Suppose that the observations $w = (w_1 \ldots w_N)^T$ are jointly normally distributed. Then, the Fisher score vector is described by (3.35):

$$s_\theta = \frac{\partial g^T(\theta)}{\partial \theta} C^{-1} d(\theta)\,, \tag{4.39}$$

where $d(\theta) = w - g(\theta)$. This shows that

$$\begin{aligned} F_\theta &= E\left[s_\theta\, s_\theta^T\right] \\ &= E\left[\frac{\partial g^T(\theta)}{\partial \theta} C^{-1} d(\theta)\, d^T(\theta) C^{-1} \frac{\partial g(\theta)}{\partial \theta^T}\right] \\ &= \frac{\partial g^T(\theta)}{\partial \theta} C^{-1} E\left[d(\theta)\, d^T(\theta)\right] C^{-1} \frac{\partial g(\theta)}{\partial \theta^T} \\ &= \frac{\partial g^T(\theta)}{\partial \theta} C^{-1} C C^{-1} \frac{\partial g(\theta)}{\partial \theta^T} \\ &= \frac{\partial g^T(\theta)}{\partial \theta} C^{-1} \frac{\partial g(\theta)}{\partial \theta^T}. \end{aligned} \tag{4.40}$$

■

■ **EXAMPLE 4.2**

The Fisher information matrix of Poisson distributed observations

Suppose that the observations are independent and jointly Poisson distributed. Then, the Fisher score vector is described by (3.47):

$$s_\theta = \frac{\partial g^T(\theta)}{\partial \theta} C^{-1}\, d(\theta). \tag{4.41}$$

This implies that the *form* of the Fisher information matrix is the same as that of normally distributed observations:

$$F_\theta = \frac{\partial g^T(\theta)}{\partial \theta} C^{-1} \frac{\partial g(\theta)}{\partial \theta^T}. \tag{4.42}$$

However, the covariance matrix of independent Poisson distributed observations is described by (3.44):

$$C = \operatorname{diag} g(\theta). \tag{4.43}$$

Therefore, different from the covariance matrix of normally distributed observations, it depends on θ. The equations (4.42) and (4.43) show that for independent, Poisson distributed observations the (p, q)th element f_{pq} of F_θ is described by

$$f_{pq} = \sum_{n=1}^{N} \frac{1}{g_n(\theta)} \frac{\partial g_n(\theta)}{\partial \theta_p} \frac{\partial g_n(\theta)}{\partial \theta_q}\,. \tag{4.44}$$

■

■ **EXAMPLE 4.3**

The Fisher information matrix of multinomially distributed observations

Suppose that the observations $w = (w_1 \ldots w_{N-1})^T$ are multinomially distributed. Then, the Fisher score vector is described by (3.73):

$$s_\theta = \frac{\partial g^T(\theta)}{\partial \theta} C^{-1} d(\theta). \tag{4.45}$$

Therefore, the Fisher information matrix is described by

$$F_\theta = \frac{\partial g^T(\theta)}{\partial \theta} C^{-1} \frac{\partial g(\theta)}{\partial \theta^T}, \tag{4.46}$$

where the $(N-1) \times (N-1)$ covariance matrix C, defined by (3.64), depends on θ. Thus, the form of this Fisher information matrix is the same as that of normally or Poisson distributed observations. ■

If θ is scalar, the Fisher information matrix reduces to a scalar called *Fisher information* that is described by

$$\boxed{F_\theta = E s_\theta^2 = E\left(\frac{\partial q(w;\theta)}{\partial \theta}\right)^2}. \tag{4.47}$$

An alternative expression for the Fisher information matrix (4.38) is obtained as follows. Equation (4.36) shows that the expectation of the kth element of the Fisher score vector satisfies

$$\int \frac{\partial q(\omega;\theta)}{\partial \theta_k} p(\omega;\theta)\, d\omega = 0. \tag{4.48}$$

Then, under the regularity conditions, differentiating both members of this equation with respect to the parameter θ_ℓ yields

$$\begin{aligned}
&\int \frac{\partial^2 q(\omega;\theta)}{\partial \theta_k \partial \theta_\ell} p(\omega;\theta)\, d\omega + \int \frac{\partial q(\omega;\theta)}{\partial \theta_k} \frac{\partial p(\omega;\theta)}{\partial \theta_\ell}\, d\omega \\
&= \int \frac{\partial^2 q(\omega;\theta)}{\partial \theta_k \partial \theta_\ell} p(\omega;\theta)\, d\omega + \int \frac{\partial q(\omega;\theta)}{\partial \theta_k} \frac{\partial q(\omega;\theta)}{\partial \theta_\ell} p(\omega;\theta)\, d\omega \\
&= 0.
\end{aligned} \tag{4.49}$$

Hence,

$$E\left[\frac{\partial q(w;\theta)}{\partial \theta_k} \frac{\partial q(w;\theta)}{\partial \theta_\ell}\right] = -E \frac{\partial^2 q(w;\theta)}{\partial \theta_k \partial \theta_\ell}. \tag{4.50}$$

Therefore, an alternative to (4.38) is

$$\boxed{F_\theta = -E \frac{\partial^2 q(w;\theta)}{\partial \theta\, \partial \theta^T}} \tag{4.51}$$

or

$$F_\theta = -E \frac{\partial s_\theta}{\partial \theta^T}. \tag{4.52}$$

The corresponding form for the Fisher information for a scalar parameter θ is

$$\boxed{F_\theta = Es_\theta^2 = -E\frac{\partial^2 q(w;\theta)}{\partial\theta^2}} \tag{4.53}$$

or

$$F_\theta = -E\frac{\partial s_\theta}{\partial\theta}\,. \tag{4.54}$$

The corresponding alternative expression for the Fisher information matrix of discrete observations is the same and is derived analogously. Using the alternative form (4.51), we now derive the Fisher information matrix of normally, Poisson, and multinomially distributed observations that we earlier derived using the original expression (4.38) in the Examples 4.1–4.3.

■ EXAMPLE 4.4

The alternative form of the Fisher information matrix of normally distributed observations

For normally distributed observations, the kth element of the Fisher score vector is described by (3.35):

$$\frac{\partial q(w;\theta)}{\partial\theta_k} = \frac{\partial g^T(\theta)}{\partial\theta_k}C^{-1}d(\theta)\,, \tag{4.55}$$

where the covariance matrix C is supposed to be independent of the parameters θ. Then,

$$\frac{\partial^2 q(w;\theta)}{\partial\theta_k\partial\theta_\ell} = \frac{\partial^2 g^T(\theta)}{\partial\theta_k\partial\theta_\ell}C^{-1}d(\theta) - \frac{\partial g^T(\theta)}{\partial\theta_k}C^{-1}\frac{\partial g(\theta)}{\partial\theta_\ell}. \tag{4.56}$$

Therefore, the (k,ℓ)th element of the Fisher information matrix F_θ is equal to

$$\begin{aligned} -E\frac{\partial^2 q(w;\theta)}{\partial\theta_k\partial\theta_\ell} &= -E\left[\frac{\partial^2 g^T(\theta)}{\partial\theta_k\partial\theta_\ell}C^{-1}d(\theta)\right] + \frac{\partial g^T(\theta)}{\partial\theta_k}C^{-1}\frac{\partial g(\theta)}{\partial\theta_\ell} \\ &= \frac{\partial g^T(\theta)}{\partial\theta_k}C^{-1}\frac{\partial g(\theta)}{\partial\theta_\ell} \end{aligned} \tag{4.57}$$

since $Ed(\theta) = E[w - g(\theta)] = o$. Therefore,

$$F_\theta = \frac{\partial g^T(\theta)}{\partial\theta}C^{-1}\frac{\partial g(\theta)}{\partial\theta^T}. \tag{4.58}$$

This result agrees with (4.40). ■

■ EXAMPLE 4.5

The alternative form of the Fisher information matrix of Poisson distributed observations

The kth element of the Fisher score vector of Poisson distributed observations follows from (3.47):

$$\frac{\partial q(w;\theta)}{\partial\theta_k} = \frac{\partial g^T(\theta)}{\partial\theta_k}C^{-1}d(\theta) \tag{4.59}$$

with

$$C = \operatorname{diag} g(\theta). \tag{4.60}$$

Then,

$$\frac{\partial^2 q(w;\theta)}{\partial\theta_k\,\partial\theta_\ell} = \frac{\partial}{\partial\theta_\ell}\left(\frac{\partial g^T(\theta)}{\partial\theta_k}C^{-1}\right)d(\theta) - \frac{\partial g^T(\theta)}{\partial\theta_k}C^{-1}\frac{\partial g(\theta)}{\partial\theta_\ell}. \tag{4.61}$$

Since $Ed(\theta) = o$, this implies that

$$E\frac{\partial^2 q(w;\theta)}{\partial\theta_k\,\partial\theta_\ell} = -\frac{\partial g^T(\theta)}{\partial\theta_k}C^{-1}\frac{\partial g(\theta)}{\partial\theta_\ell} \tag{4.62}$$

and, therefore,

$$F_\theta = \frac{\partial g^T(\theta)}{\partial\theta}C^{-1}\frac{\partial g(\theta)}{\partial\theta^T}, \tag{4.63}$$

a result identical to (4.42). ■

■ **EXAMPLE 4.6**

The alternative form of the Fisher information matrix of multinomially distributed observations

Equation (3.73) shows that the kth element of the Fisher score vector of multinomially distributed observations is described by

$$\frac{\partial q(w;\theta)}{\partial\theta_k} = \frac{\partial g^T(\theta)}{\partial\theta_k}C^{-1}d(\theta) \tag{4.64}$$

with C the $(N-1)\times(N-1)$ covariance matrix of the observations defined by (3.64). As in the previous example, this implies that

$$F_\theta = \frac{\partial g^T(\theta)}{\partial\theta}C^{-1}\frac{\partial g(\theta)}{\partial\theta^T}, \tag{4.65}$$

a result identical to (4.46). ■

4.4.2 The Fisher information matrix for exponential families of distributions

Expression (3.98) describes the Fisher score vector for exponential family distributed observations

$$s_\theta = \frac{\partial\mu_\delta^T}{\partial\theta}C_{\delta\delta}^{-1}\left[\delta(w) - \mu_\delta\right]. \tag{4.66}$$

Then,

$$
\begin{aligned}
F_\theta &= E\left[s_\theta s_\theta^T\right] \\
&= E\left[\frac{\partial \mu_\delta^T}{\partial \theta} C_{\delta\delta}^{-1} \{\delta(w) - \mu_\delta\} \{\delta(w) - \mu_\delta\}^T C_{\delta\delta}^{-1} \frac{\partial \mu_\delta}{\partial \theta^T}\right] \\
&= \frac{\partial \mu_\delta^T}{\partial \theta} C_{\delta\delta}^{-1} E\left[\{\delta(w) - \mu_\delta\} \{\delta(w) - \mu_\delta\}^T\right] C_{\delta\delta}^{-1} \frac{\partial \mu_\delta}{\partial \theta^T} \\
&= \frac{\partial \mu_\delta^T}{\partial \theta} C_{\delta\delta}^{-1} C_{\delta\delta} C_{\delta\delta}^{-1} \frac{\partial \mu_\delta}{\partial \theta^T} \\
&= \frac{\partial \mu_\delta^T}{\partial \theta} C_{\delta\delta}^{-1} \frac{\partial \mu_\delta}{\partial \theta^T}\,. \qquad (4.67)
\end{aligned}
$$

The Fisher information matrix of *linear* exponential family distributed observations is of particular practical importance. The relevant expression for the Fisher score vector (3.99) shows that in this special case

$$
\boxed{F_\theta = \frac{\partial g^T(\theta)}{\partial \theta} C^{-1} \frac{\partial g(\theta)}{\partial \theta^T}}\,. \qquad (4.68)
$$

This result also follows from (4.67) by substituting $\mu_\delta = E\delta(w) = Ew = g(\theta)$ and $C_{\delta\delta} = C$.

We present three examples.

■ **EXAMPLE 4.7**

The Fisher information matrix for normal, Poisson, and multinomial families of distributions

In Examples 3.1–3.3, it has been shown that joint normal, Poisson, and multinomial probability (density) functions are linear exponential families. Therefore, (4.68) must describe the relevant Fisher information matrices. This is confirmed by (4.40), (4.42), and (4.46). ■

4.4.3 Inflow of Fisher information

4.4.3.1 *Inflow of Fisher information for independent observations* If the observations $w = (w_1 \ldots w_N)^T$ are independent, their joint probability (density) function is, by definition, equal to the product of the marginal probability (density) functions of all observations

$$
p(\omega;\theta) = p_1(\omega_1;\theta)p_2(\omega_2;\theta)\ldots p_N(\omega_N;\theta). \qquad (4.69)
$$

Therefore,

$$
q(\omega;\theta) = \sum_n q_n(\omega_n;\theta). \qquad (4.70)
$$

Then, the kth element of the Fisher score vector is equal to

$$
s_{\theta_k} = \frac{\partial q(w;\theta)}{\partial \theta_k} = \sum_n \frac{\partial q_n(w_n;\theta)}{\partial \theta_k}\,. \qquad (4.71)
$$

Since the terms of this sum are Fisher scores, their expectations are equal to zero. Furthermore, they are independent stochastic variables since the w_n are. Therefore,

$$\operatorname{cov}(s_{\theta_k}, s_{\theta_\ell}) = \sum_n E\left[\frac{\partial q_n(w_n;\theta)}{\partial \theta_k}\frac{\partial q_n(w_n;\theta)}{\partial \theta_\ell}\right]. \tag{4.72}$$

This expression shows that the Fisher information matrix $F_{(N)}$ of $w = (w_1 \ldots w_N)^T$ for the parameter vector θ is described by

$$F_{(N)} = \sum_n E\left[\frac{\partial q_n(w_n;\theta)}{\partial \theta}\frac{\partial q_n(w_n;\theta)}{\partial \theta^T}\right] \tag{4.73}$$

with $n = 1, \ldots, N$. Then, the difference of the Fisher information matrix $F_{(N+1)}$ of the observations $(w_1 \ldots w_{N+1})^T$ and the Fisher information matrix $F_{(N)}$ of the observations $(w_1 \ldots w_N)^T$ is the matrix:

$$E\left[\frac{\partial q_{N+1}(w_{N+1};\theta)}{\partial \theta}\frac{\partial q_{N+1}(w_{N+1};\theta)}{\partial \theta^T}\right]. \tag{4.74}$$

Since this is a covariance matrix, it is positive semidefinite and, therefore,

$$F_{(N+1)} \succeq F_{(N)}, \tag{4.75}$$

expressing in this book positive semidefiniteness of the difference $F_{(N+1)} - F_{(N)}$. See Section C.1. A property of positive semidefinite matrices is that their diagonal elements are nonnegative. So, as a result of any additional independent observation, the diagonal elements of the Fisher information matrix either increase or remain the same. The inflow of Fisher information—that is, the increase of the diagonal elements—may, however, be quite different from measurement point to measurement point as the following example shows.

■ **EXAMPLE 4.8**

Contribution of Poisson distributed observations to the Fisher information for exponential decay parameters

Suppose that independent, Poisson distributed observations $w = (w_1 \ldots w_N)^T$ are available with expectations

$$Ew_n = g_n(\theta) = \theta_1 \exp(-\theta_2 x_n), \tag{4.76}$$

where $\theta = (\theta_1\ \theta_2)^T$ are the parameters to be estimated. Then, (4.44) shows that the Fisher information for the parameter θ_k is equal to

$$f_{kk} = \sum_{n=1}^{N} \frac{1}{g_n(\theta)}\left(\frac{\partial g_n(\theta)}{\partial \theta_k}\right)^2 \tag{4.77}$$

with $k = 1, 2$. The partial derivatives in this expression are $\partial g_n(\theta)/\partial\theta_1 = (1/\theta_1)g_n(\theta)$ and $\partial g_n(\theta)/\partial\theta_2 = -x_n g_n(\theta)$. Equation (4.77) shows that the contribution of the measurement point x_n to the Fisher information for the parameter θ_k is

$$\frac{1}{g_n(\theta)}\left(\frac{\partial g_n(\theta)}{\partial \theta_k}\right)^2. \tag{4.78}$$

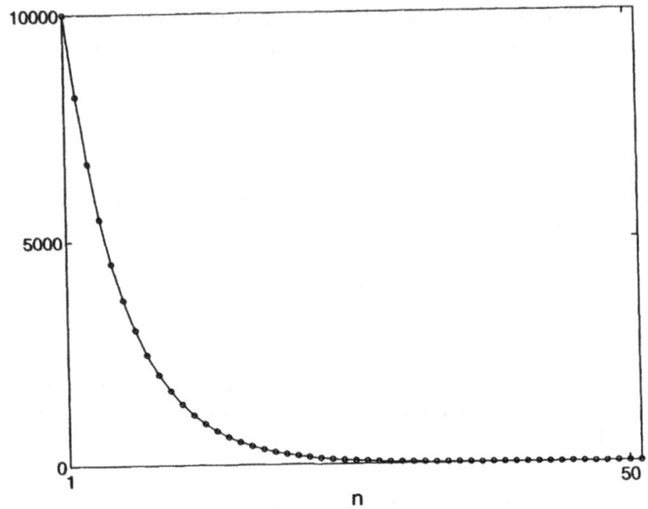

Figure 4.1. Monoexponential expectation model (solid line) and its values at the measurement points (circles). (Example 4.8)

Suppose that $\theta = (10000 \;\; 1)^T$ and that the measurement points are $x_n = (n-1) \times 0.2$, $n = 1, \ldots, 51$. Figure 4.1. shows the expectation model and its values at the measurement points. Figures 4.2.(a) and 4.2.(b) show the contribution of each measurement point to the diagonal elements f_{11} and f_{22} of the Fisher information matrix.

These represent the Fisher information for the amplitude and the decay constant, respectively. Figures 4.2.(c) and 4.2.(d) show the elements f_{11} and f_{22} themselves as a function of the first N observations. ■

From Example 4.8, the following conclusions may be drawn. In the first place, Figs. 4.2.(a) and 4.2.(b) show that the contribution to the Fisher information varies substantially from point to point. The inflow of Fisher information for the amplitude parameter θ_1 is maximum for $n = 1$ and that for the decay constant θ_2 is maximum for $n = 11$. Points with $n > 25$ do not substantially contribute anymore to f_{11} but those for $25 < n < 50$ still do, to some extent, to f_{22}. In any case, it is not worthwhile to include observations for $n > 50$ since, as Figs. 4.2.(c) and 4.2.(d) show, these points hardly contribute to f_{11} or to f_{22}. We will return to this behavior in Section 4.6.3.

4.4.3.2 Inflow of Fisher information for linear exponential families of distributions The results discussed in Subsection 4.4.3.1 apply to independent observations. If the distribution of the observations is a linear exponential family, the assumption of independence may be dropped. Then, for N observations $w = (w_1 \ldots w_N)^T$ with expectations $g(\theta) = [g_1(\theta) \ldots g_N(\theta)]^T$ and covariance matrix C, the Fisher information matrix is described by (4.68):

$$F_{(N)} = \frac{\partial g^T(\theta)}{\partial \theta} C^{-1} \frac{\partial g(\theta)}{\partial \theta^T}. \tag{4.79}$$

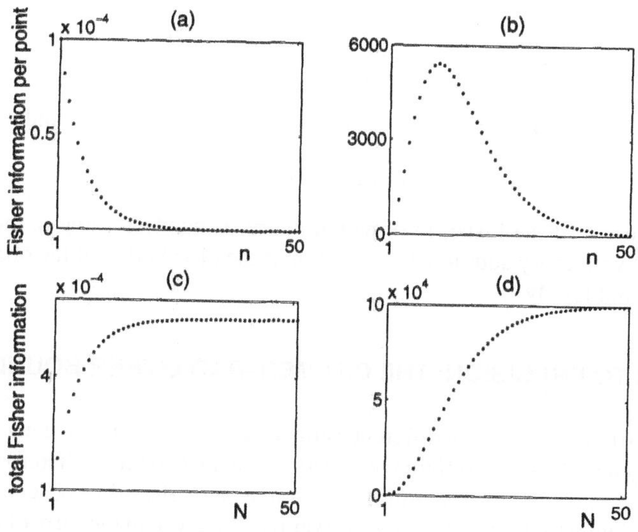

Figure 4.2. (a,b) Contribution of each measurement point to the Fisher information for the amplitude and the decay constant, respectively. (c,d) Joint contribution of the first N measurement points to the Fisher information for the same parameters. (Example 4.8)

Next, suppose that the observation w_{N+1} is added. Then, the corresponding Fisher information matrix becomes

$$F_{(N+1)} = \begin{pmatrix} \dfrac{\partial g^T(\theta)}{\partial \theta} & \dfrac{\partial g_{N+1}(\theta)}{\partial \theta} \end{pmatrix} \begin{pmatrix} C & c \\ c^T & d \end{pmatrix}^{-1} \begin{pmatrix} \dfrac{\partial g(\theta)}{\partial \theta^T} \\ \dfrac{\partial g_{N+1}(\theta)}{\partial \theta^T} \end{pmatrix}, \tag{4.80}$$

where

$$\begin{pmatrix} C & c \\ c^T & d \end{pmatrix} \tag{4.81}$$

is the covariance matrix of $(w_1 \ldots w_{N+1})^T$ with $c = (c_{1,N+1} \ldots c_{N,N+1})^T$ and $d = c_{N+1,N+1}$. Using the special form of Frobenius formula (B.37), we may show that the inverse of this covariance matrix is equal to

$$\begin{pmatrix} C^{-1} + \dfrac{C^{-1}cc^T C^{-1}}{d - c^T C^{-1} c} & \dfrac{-C^{-1}c}{d - c^T C^{-1} c} \\ \dfrac{-c^T C^{-1}}{d - c^T C^{-1} c} & \dfrac{1}{d - c^T C^{-1} c} \end{pmatrix}. \tag{4.82}$$

In this expression,

$$\frac{1}{d - c^T C^{-1} c} > 0 \tag{4.83}$$

since it is a diagonal element of the inverse of a symmetric and positive definite matrix. Substituting (4.82) in (4.80) and rearranging shows that

$$F_{(N+1)} = F_{(N)} + P^T P \tag{4.84}$$

with

$$P = \frac{1}{(d - c^T C^{-1} c)^{\frac{1}{2}}} \left(c^T C^{-1} \frac{\partial g(\theta)}{\partial \theta^T} - \frac{\partial g_{N+1}(\theta)}{\partial \theta^T} \right) . \tag{4.85}$$

Therefore, since any matrix product $P^T P$ is symmetric and positive semidefinite, we have

$$F_{(N+1)} \succeq F_{(N)} . \tag{4.86}$$

The conclusion is that the Fisher information matrix is nondecreasing in the sense of inequality (4.86) with every additional observation if the distribution of the observations is a linear exponential family.

4.5 LIMITS TO PRECISION: THE CRAMÉR–RAO LOWER BOUND

In Example 2.10, we showed that different estimators of the same parameters from the same observations generally have a different precision. The question may, therefore, be posed as to what precision may be achieved ultimately from a particular set of observations. For the general class of unbiased estimators, the answer may be given in the form of a lower bound on their variance, the so-called *Cramér–Rao lower bound.*

4.5.1 The Cramér–Rao lower bound for scalar functions of scalar parameters

In this subsection, we will discuss the simplest Cramér–Rao lower bound: that for a scalar function of a scalar parameter. We then have the following theorem:

Theorem 4.3 *Suppose that observations $w = (w_1 \dots w_N)^T$ are available with joint probability (density) function $p(\omega; \theta)$ where θ is an unknown scalar parameter. Furthermore, suppose that $\rho(\theta)$ is a scalar function of θ and that $r(w)$ is an unbiased estimator of $\rho(\theta)$. Then, under regularity conditions,*

$$\boxed{\operatorname{var} r(w) \geq \frac{1}{F_\theta} \left(\frac{d\rho}{d\theta} \right)^2} , \tag{4.87}$$

where F_θ is the Fisher information $E s_\theta^2$ and $\rho = \rho(\theta)$.

The expression (4.87) is called the Cramér–Rao inequality. The scalar quantity

$$\frac{1}{F_\theta} \left(\frac{d\rho}{d\theta} \right)^2 \tag{4.88}$$

is the Cramér–Rao lower bound on the variance of unbiased estimators of $\rho(\theta)$.

Proof. By assumption,

$$E r(w) = \rho \tag{4.89}$$

with

$$E r(w) = \int r(\omega)\, p(\omega; \theta)\, d\omega . \tag{4.90}$$

Thus,

$$\frac{d\rho}{d\theta} = \frac{d}{d\theta} \int r(\omega)\, p(\omega; \theta)\, d\omega . \tag{4.91}$$

Under regularity conditions, differentiating and integrating in this expression may be interchanged. Then,

$$\frac{d\rho}{d\theta} = \int r(\omega)\,\frac{dq(\omega;\theta)}{d\theta}\,p(\omega;\theta)\,d\omega = E\left[r(w)\,s_\theta\right], \tag{4.92}$$

where s_θ is the Fisher score. However, since $Es_\theta = 0$, (4.92) is equal to

$$\frac{d\rho}{d\theta} = E\left[\{r(w) - \rho\}\, s_\theta\right] = \mathrm{cov}\left(r(w), s_\theta\right). \tag{4.93}$$

Then, by the Cauchy–Schwarz inequality (A.19) we obtain

$$\left(\frac{d\rho}{d\theta}\right)^2 \leq \mathrm{var}\, r(w)\ \mathrm{var}\, s_\theta = \mathrm{var}\, r(w)\ F_\theta\,. \tag{4.94}$$

This completes the proof for continuous w. The proof for discrete w is analogous. ■

By Corollary A.1, equality in (4.94) occurs if and only if $r(w) - Er(w)$ is proportional to $s_\theta - Es_\theta$ with probability one. Since $Es_\theta = 0$ and $Er(w) = \rho$, this implies proportionality of $r(w) - \rho$ to s_θ. Then, (4.93) shows that the proportionality constant is equal to $(d\rho/\,d\theta)\,F_\theta^{-1}$. Therefore, a necessary and sufficient condition for the unbiased estimator $r(w)$ to attain the Cramér–Rao lower bound is

$$\boxed{\frac{d\rho(\theta)}{d\theta}F_\theta^{-1}s_\theta = r(w) - \rho(\theta)}\,, \tag{4.95}$$

where the argument θ has been reintroduced. Equation (4.95) is also a sufficient condition for $r(w)$ to be unbiased since $Es_\theta = o$. Thus, (4.95) is a necessary and sufficient condition for $r(w)$ to be unbiased for $\rho(\theta)$ and to attain the Cramér–Rao lower bound. If $\rho(\theta) = \theta$ and, therefore, $r(w) = t(w)$, the necessary and sufficient condition for $t(w)$ to be unbiased and to attain the Cramér–Rao lower bound is described by

$$\boxed{F_\theta^{-1}s_\theta = t(w) - \theta}\,. \tag{4.96}$$

■ EXAMPLE 4.9

Cramér–Rao lower bound for estimating the slope of a straight line from Poisson distributed observations

Suppose that the parameter θ of the expectations $Ew_n = g_n(\theta) = \theta x_n$ is estimated from independent, Poisson distributed observations $w = (w_1 \ldots w_N)^T$. The Fisher information of such observations is described by (4.42):

$$F_\theta = \frac{\partial g^T(\theta)}{\partial\theta}C^{-1}\frac{\partial g(\theta)}{\partial\theta^T}. \tag{4.97}$$

Since the observations are independent and Poisson distributed, $C = \mathrm{diag}\, g(\theta)$. See (3.44). Then,

$$F_\theta = x^T\,\mathrm{diag}\,(\frac{1}{g_1(\theta)} \cdots \frac{1}{g_N(\theta)})\,x = \frac{1}{\theta}\sum_n x_n \tag{4.98}$$

with $x = (x_1 \ldots x_N)^T$. Therefore, the Cramér–Rao lower bound on the variance of unbiased estimators of θ is

$$\frac{\theta}{\sum_n x_n}. \tag{4.99}$$

The Fisher score of independent and Poisson distributed observations is described by (3.47):

$$s_\theta = \frac{\partial g^T}{\partial \theta} C^{-1} [w - g(\theta)]. \tag{4.100}$$

Hence, in this example,

$$s_\theta = \frac{1}{\theta} \sum_n (w_n - \theta x_n). \tag{4.101}$$

Equations (4.96), (4.98), and (4.101) show after some rearrangements that the necessary and sufficient condition for an estimator $t(w)$ of θ to be unbiased and to attain the Cramér–Rao lower bound is

$$t(w) = \frac{\sum_n w_n}{\sum_n x_n}. \tag{4.102}$$

Thus, the condition for attaining the Cramér–Rao lower bound has produced a closed-form expression for the estimator doing so. ■

■ **EXAMPLE 4.10**

Cramér–Rao lower bound for estimating an exponent from normally distributed observations

Suppose that the parameter θ of the expectations $Ew_n = g_n(\theta) = x_n^\theta$, $n = 1, \ldots, N$ with $x_n > 0$ is estimated from uncorrelated, normally distributed observations $w = (w_1 \ldots w_N)^T$ with equal variance σ^2. The Fisher information of such observations is described by (4.40):

$$F_\theta = \frac{\partial g^T(\theta)}{\partial \theta} C^{-1} \frac{\partial g(\theta)}{\partial \theta^T} \tag{4.103}$$

with $g_n(\theta) = x_n^\theta$ and $C = \sigma^2 I$. Since $\partial g(\theta)/\partial \theta = (x_1^\theta \ln x_1 \ldots x_N^\theta \ln x_N)^T$,

$$F_\theta = \frac{1}{\sigma^2} \sum_n (x_n^\theta \ln x_n)^2. \tag{4.104}$$

The Cramér–Rao lower bound is the reciprocal of this quantity.

The Fisher score of normally distributed observations is described by (3.35):

$$s_\theta = \frac{\partial g^T(\theta)}{\partial \theta} C^{-1} [w - g(\theta)]. \tag{4.105}$$

Hence, in this example,

$$s_\theta = \frac{1}{\sigma^2} \sum_n x_n^\theta (w_n - x_n^\theta) \ln x_n. \tag{4.106}$$

Equations (4.96), (4.104), and (4.106) show after some rearrangements that the necessary and sufficient condition for an estimator $t(w)$ of θ to be unbiased and to attain the Cramér–Rao lower bound is

$$t(w) = \theta + \frac{\sum_n x_n^\theta (w_n - x_n^\theta) \ln x_n}{\sum_n (x_n^\theta \ln x_n)^2}. \tag{4.107}$$

This expression shows that there exists no unbiased estimator for θ attaining the Cramér–Rao lower bound since, as stated in Section 4.2, an estimator may not be a function of parameters to be estimated. ■

4.5.2 The Cramér–Rao lower bound for vector functions of vector parameters

Theorem 4.4 *Suppose that observations $w = (w_1 \ldots w_N)^T$ are available with joint probability (density) function $p(\omega; \theta)$ where $\theta = (\theta_1 \ldots \theta_K)^T$ is a vector of unknown parameters. Furthermore, suppose that $\rho(\theta) = [\rho_1(\theta) \ldots \rho_L(\theta)]^T$ is a vector of functions of the parameters θ and that $r(w) = [r_1(w) \ldots r_L(w)]^T$ is an unbiased estimator of $\rho(\theta)$. Then, under suitable regularity conditions,*

$$\boxed{\text{cov}\,(r(w), r(w)) \succeq \frac{\partial \rho}{\partial \theta^T} F_\theta^{-1} \frac{\partial \rho^T}{\partial \theta}} \tag{4.108}$$

or

$$\text{cov}\,(r(w), r(w)) - \frac{\partial \rho}{\partial \theta^T} F_\theta^{-1} \frac{\partial \rho^T}{\partial \theta} \succeq O\,, \tag{4.109}$$

where:

- $\rho = \rho(\theta)$
- $\text{cov}\,(r(w), r(w))$ *is the $L \times L$ covariance matrix of the estimator $r(w)$,*
- $\partial \rho / \partial \theta^T$ *is the $L \times K$ Jacobian matrix of ρ with respect to θ,*
- F_θ *is the Fisher information matrix $E\left[s_\theta\, s_\theta^T\right]$,*
- O *is the $L \times L$ null matrix, and*
- $A \succeq B$ *expresses that the difference $A - B$ of the real symmetric matrices A and B is positive semidefinite. See Section C.1.*

Expression (4.108) is called Cramér–Rao inequality. The matrix

$$\boxed{\frac{\partial \rho}{\partial \theta^T} F_\theta^{-1} \frac{\partial \rho^T}{\partial \theta}} \tag{4.110}$$

is called Cramér–Rao lower bound matrix. Its diagonal elements are the Cramér–Rao lower bounds on the variances of unbiased estimators of the elements of ρ.

Proof. By assumption,

$$Er(w) = \rho \tag{4.111}$$

with

$$Er(w) = \int r(\omega)\, p(\omega; \theta)\, d\omega. \tag{4.112}$$

Therefore,

$$\frac{\partial \rho}{\partial \theta^T} = \frac{\partial}{\partial \theta^T} \int r(\omega) p(\omega; \theta)\, d\omega. \tag{4.113}$$

Under regularity conditions, differentiating and integrating in this expression may be interchanged. Then,

$$\frac{\partial \rho}{\partial \theta^T} = \int r(\omega)\frac{\partial q(\omega;\theta)}{\partial \theta^T} p(\omega;\theta)\, d\omega = E[r(w)\, s_\theta^T]\,. \tag{4.114}$$

Since $Es_\theta = o$, this may be written

$$\frac{\partial \rho}{\partial \theta^T} = \operatorname{cov}\left(r(w), s_\theta\right), \tag{4.115}$$

which is the $L \times K$ covariance matrix of $r(w)$ and s_θ. Next, consider the $(L+K) \times 1$ vector

$$\begin{pmatrix} r(w) \\ s_\theta \end{pmatrix}. \tag{4.116}$$

Then, the covariance matrix of this vector,

$$\begin{pmatrix} \operatorname{cov}(r(w), r(w)) & \operatorname{cov}(r(w), s_\theta) \\ \operatorname{cov}(s_\theta, r(w)) & \operatorname{cov}(s_\theta, s_\theta) \end{pmatrix}, \tag{4.117}$$

is positive semidefinite and equal to

$$\begin{pmatrix} \operatorname{cov}(r(w), r(w)) & \dfrac{\partial \rho}{\partial \theta^T} \\ \dfrac{\partial \rho^T}{\partial \theta} & F_\theta \end{pmatrix}. \tag{4.118}$$

Next, suppose that F_θ is nonsingular. Then, F_θ^{-1} exists. Consider the $(L+K) \times L$ matrix

$$\begin{pmatrix} I \\ -F_\theta^{-1}\dfrac{\partial \rho^T}{\partial \theta} \end{pmatrix}, \tag{4.119}$$

where I is the identity matrix of order L. Then, by Theorem C.5, the product matrix

$$\begin{pmatrix} I & -\dfrac{\partial \rho}{\partial \theta^T} F_\theta^{-1} \end{pmatrix} \begin{pmatrix} \operatorname{cov}(r(w), r(w)) & \dfrac{\partial \rho}{\partial \theta^T} \\ \dfrac{\partial \rho^T}{\partial \theta} & F_\theta \end{pmatrix} \begin{pmatrix} I \\ -F_\theta^{-1}\dfrac{\partial \rho^T}{\partial \theta} \end{pmatrix} \tag{4.120}$$

is positive semidefinite since it is of the form $P^T V P$ with V positive semidefinite. Partitioned multiplication in (4.120) produces the matrix

$$\operatorname{cov}(r(w), r(w)) - \frac{\partial \rho}{\partial \theta^T} F_\theta^{-1} \frac{\partial \rho^T}{\partial \theta} \tag{4.121}$$

and, since this is positive semidefinite, this completes the proof. ■

In (4.110), F_θ is the Fisher information matrix (4.38). Thus,

$$F_\theta^{-1} = \left(E\left[\frac{\partial q(w;\theta)}{\partial \theta}\frac{\partial q(w;\theta)}{\partial \theta^T}\right]\right)^{-1}. \tag{4.122}$$

Then, if (4.51) is true, this matrix is also equal to

$$F_\theta^{-1} = -\left(E\frac{\partial^2 q(w;\theta)}{\partial \theta\, \partial \theta^T}\right)^{-1}. \tag{4.123}$$

Therefore, (4.110) with F_θ^{-1} defined by (4.122) and the same expression with F_θ^{-1} defined by (4.123) are alternative forms of the Cramér–Rao lower bound matrix.

■ **EXAMPLE 4.11**

Functions to be estimated are the parameters themselves

If the vector function $\rho = \rho(\theta)$ is the $K \times 1$ vector of parameters θ,

$$\frac{\partial \rho}{\partial \theta^T} = I \,, \tag{4.124}$$

where I is the identity matrix of order K. Then, the Cramér–Rao inequality (4.108) becomes

$$\boxed{\operatorname{cov}\left(t(w), t(w)\right) \succeq F_\theta^{-1}} \,, \tag{4.125}$$

where $t(w)$ is an unbiased estimator of θ. In words: the Cramér–Rao lower bound matrix for the parameters is equal to the inverse of the Fisher information matrix. If θ is scalar, this becomes

$$\operatorname{var} t(w) \geq \frac{1}{F_\theta} \,. \tag{4.126}$$

■

■ **EXAMPLE 4.12**

Cramér–Rao lower bound for sum and difference of parameters

Suppose that $\theta = (\theta_1 \; \theta_2)^T$ and that $\rho = \rho(\theta)$ is the 2×1 vector

$$\rho = \begin{pmatrix} \theta_1 + \theta_2 \\ \theta_1 - \theta_2 \end{pmatrix} . \tag{4.127}$$

Then,

$$\frac{\partial \rho}{\partial \theta^T} = \begin{pmatrix} 1 & 1 \\ 1 & -1 \end{pmatrix} \tag{4.128}$$

and hence, if the 2×1 vector $r(w)$ is an unbiased estimator of ρ,

$$\operatorname{cov}\left(r(w), r(w)\right) \succeq \begin{pmatrix} 1 & 1 \\ 1 & -1 \end{pmatrix} F_\theta^{-1} \begin{pmatrix} 1 & 1 \\ 1 & -1 \end{pmatrix} . \tag{4.129}$$

■

■ **EXAMPLE 4.13**

Cramér–Rao lower bound for location and area of Lorentz line

Suppose that the expectation of the observations $w = (w_1 \ldots w_N)^T$ is described by

$$Ew_n = g(x_n; \theta) = \frac{\theta_1}{1 + \left(\dfrac{x_n - \theta_2}{\theta_3}\right)^2} \tag{4.130}$$

with parameters $\theta = (\theta_1 \; \theta_2 \; \theta_3)^T$ where θ_1 is the height of the line, θ_2 is its location, and θ_3 is its half-width. The expectation model $g(x; \theta)$ corresponding to (4.130) is called *Lorentz line*.

In practical measurement, the quantities of interest are usually the area and the location of the line. Simple calculations show that the area is equal to $\pi\theta_1\theta_3$. Then,

$$\rho = \begin{pmatrix} \pi\theta_1\theta_3 \\ \theta_2 \end{pmatrix} \tag{4.131}$$

and, therefore,

$$\frac{\partial \rho}{\partial \theta^T} = \begin{pmatrix} \pi\theta_3 & 0 & \pi\theta_1 \\ 0 & 1 & 0 \end{pmatrix}. \tag{4.132}$$

Next, suppose that the observations have a Poisson distribution as in Example 4.2. Then, the (p, q)th element of F_θ is described by (4.44):

$$f_{pq} = \sum_n \frac{1}{g_n(\theta)} \frac{\partial g_n(\theta)}{\partial \theta_p} \frac{\partial g_n(\theta)}{\partial \theta_q} \tag{4.133}$$

with

$$g_n(\theta) = \frac{\theta_1}{1+\xi_n^2}, \tag{4.134}$$

$$\frac{\partial g_n(\theta)}{\partial \theta_1} = \frac{1}{\theta_1} g_n(\theta), \tag{4.135}$$

$$\frac{\partial g_n(\theta)}{\partial \theta_2} = \frac{2}{\theta_1\theta_3} \xi_n g_n^2(\theta), \tag{4.136}$$

and

$$\frac{\partial g_n(\theta)}{\partial \theta_3} = \frac{2}{\theta_1\theta_3} \xi_n^2 g_n^2(\theta), \tag{4.137}$$

where $\xi = (x_n - \theta_2)/\theta_3$. The equations (4.132)–(4.137) define the 2×2 Cramér–Rao lower bound matrix for estimating of area and location:

$$\frac{\partial \rho}{\partial \theta^T} F_\theta^{-1} \frac{\partial \rho^T}{\partial \theta}. \tag{4.138}$$

■

4.6 PROPERTIES OF THE CRAMÉR–RAO LOWER BOUND

4.6.1 Interpretation of the expression for the Cramér–Rao lower bound

The Cramér–Rao inequality (4.108) states that the difference of cov $(r(w), r(w))$, which is the covariance matrix of any unbiased estimator $r(w)$ of the functions $\rho(\theta)$, and the Cramér–Rao lower bound matrix is a positive semidefinite matrix. We have seen that the diagonal elements of such a matrix are nonnegative. This implies that any diagonal element of cov $(r(w), r(w))$ is larger than or equal to the corresponding diagonal element of the Cramér–Rao lower bound matrix. However, the diagonal elements of the covariance matrix are the variances of the estimators $r_\ell(w)$, $\ell = 1, \ldots, L$. *Therefore, any of these variances is larger than or equal to the corresponding diagonal element of the Cramér–Rao lower bound matrix.* This implies that no unbiased estimator can be found that is more precise than a hypothetical unbiased estimator that attains the Cramér–Rao lower bound, that is, has variances equal to the diagonal elements of the Cramér–Rao lower bound matrix. In what follows, these diagonal elements will be called *Cramér–Rao variances*. Their square roots will be called *Cramér–Rao standard deviations*. The Cramér–Rao inequality does not imply that the off-diagonal elements of the covariance matrix of any unbiased estimator are necessarily larger than or equal to the corresponding elements of the Cramér–Rao lower bound matrix.

Expression (4.108) also shows that an expression for the Cramér–Rao lower bound matrix can be derived only if the probability (density) function of the observations w and its dependence on the parameters θ are specified. Otherwise, the Fisher information matrix cannot be computed. Furthermore, the Cramér–Rao lower bound matrix can, usually, be numerically computed only if numerical values for the parameters θ are supplied.

4.6.2 The Cramér–Rao lower bound as a measure of efficiency of estimation

The Cramér–Rao lower bound would be of theoretical interest only if no estimators would exist that actually attain or approach it. The following theorem establishes necessary and sufficient conditions for the existence of unbiased estimators that actually attain the Cramér–Rao lower bound, that is, have variances equal to the corresponding Cramér–Rao variances.

Theorem 4.5 *Let $w = (w_1 \ldots w_N)^T$ be the vector of observations and θ the $K \times 1$ vector of unknown parameters. Then, under regularity conditions, a necessary and sufficient condition for an estimator $r(w)$ of the $L \times 1$ vector function $\rho = \rho(\theta)$ to be unbiased and to attain the Cramér–Rao lower bound is*

$$\boxed{\frac{\partial \rho}{\partial \theta^T} F_\theta^{-1} s_\theta = r(w) - \rho}, \tag{4.139}$$

where $\partial \rho / \partial \theta^T$ is the $L \times K$ Jacobian matrix of ρ with respect to θ, F_θ is the Fisher information matrix, and s_θ is the Fisher score vector.

Proof. First, assume that the estimator satisfies (4.139). Then,

$$E[r(w) - \rho] = E\left[\frac{\partial \rho}{\partial \theta^T} F_\theta^{-1} s_\theta\right] = \frac{\partial \rho}{\partial \theta^T} F_\theta^{-1} E s_\theta = o \tag{4.140}$$

since the expectation of the elements of s_θ is equal to zero. Thus,

$$Er(w) = \rho. \tag{4.141}$$

Therefore, $r(w)$ is an unbiased estimator of ρ. Furthermore, (4.139) and (4.141) show that

$$\begin{aligned}
\operatorname{cov}(r(w), r(w)) &= E\left[\{r(w) - Er(w)\}\{r(w) - Er(w)\}^T\right] \\
&= E\left[\{r(w) - \rho\}\{r(w) - \rho\}^T\right] \\
&= E\left[\frac{\partial \rho}{\partial \theta^T} F_\theta^{-1} s_\theta\, s_\theta^T F_\theta^{-1} \frac{\partial \rho^T}{\partial \theta}\right] \\
&= \frac{\partial \rho}{\partial \theta^T} F_\theta^{-1} E\left[s_\theta\, s_\theta^T\right] F_\theta^{-1} \frac{\partial \rho^T}{\partial \theta} \\
&= \frac{\partial \rho}{\partial \theta^T} F_\theta^{-1} \frac{\partial \rho^T}{\partial \theta},
\end{aligned} \tag{4.142}$$

where use has been made of the symmetry of F_θ and of the fact that $F_\theta = E[s_\theta\, s_\theta^T]$. Since, by Theorem 4.4,

$$\frac{\partial \rho}{\partial \theta^T} F_\theta^{-1} \frac{\partial \rho^T}{\partial \theta} \tag{4.143}$$

is the Cramér–Rao lower bound matrix, the conclusion is that an estimator $r(w)$ satisfying (4.139) is unbiased and attains the Cramér-Rao lower bound.

Next, assume that $r(w)$ is unbiased and attains the Cramér–Rao lower bound. That is, $Er(w) = \rho$ and $\operatorname{cov}(r(w), r(w))$ is equal to (4.143). Then, returning to the proof of the Cramér–Rao inequality in Section 4.5.2, we see that here the matrix (4.118) is equal to

$$\begin{pmatrix} \dfrac{\partial \rho}{\partial \theta^T} F_\theta^{-1} \dfrac{\partial \rho^T}{\partial \theta} & \dfrac{\partial \rho}{\partial \theta^T} \\ \dfrac{\partial \rho^T}{\partial \theta} & F_\theta \end{pmatrix}. \tag{4.144}$$

The next step of the proof, postmultiplying by (4.119) and premultiplying by its transpose, produces an $L \times L$ null matrix when applied to (4.144). However, since (4.144) is the $(L + K) \times (L + K)$ covariance matrix (4.117) of the vector

$$\begin{pmatrix} r(w) \\ s_\theta \end{pmatrix}, \tag{4.145}$$

this null matrix is the covariance matrix of the vector

$$\begin{pmatrix} I & -\dfrac{\partial \rho}{\partial \theta^T} F_\theta^{-1} \end{pmatrix} \begin{pmatrix} r(w) \\ s_\theta \end{pmatrix} = r(w) - \frac{\partial \rho}{\partial \theta^T} F_\theta^{-1} s_\theta. \tag{4.146}$$

This can be true for all $\rho(\theta)$ and F_θ only if the elements of the vector

$$\begin{aligned}
r(w) &- \frac{\partial \rho}{\partial \theta^T} F_\theta^{-1} s_\theta - E\left[r(w) - \frac{\partial \rho}{\partial \theta^T} F_\theta^{-1} s_\theta\right] \\
&= r(w) - \frac{\partial \rho}{\partial \theta^T} F_\theta^{-1} s_\theta - \rho
\end{aligned} \tag{4.147}$$

are constant and equal to zero with probability one. See Section A.1. Then,

$$\frac{\partial \rho}{\partial \theta^T} F_\theta^{-1} s_\theta = r(w) - \rho. \tag{4.148}$$

This completes the proof. ■

Corollary 4.1 *If in Theorem 4.5 the vector function $\rho = \rho(\theta)$ is the vector of parameters θ itself, then the necessary and sufficient condition for an estimator $t(w)$ to be unbiased and to attain the Cramér–Rao lower bound is*

$$\boxed{F_\theta^{-1} s_\theta = t(w) - \theta}. \tag{4.149}$$

Proof. If $\rho = \theta$, $\partial\rho/\partial\theta^T = I$. Substituting this in (4.139) yields (4.149). ■

The condition (4.149) is mentioned in the literature as

$$s_\theta = F_\theta \left[t(w) - \theta\right], \tag{4.150}$$

which is the same. Furthermore, the conditions (4.139) and (4.149) are, respectively, consistent with (4.95) and (4.96) for scalar parameters.

Unbiased estimators attaining the Cramér–Rao lower bound are called *efficient unbiased*. Although often no efficient unbiased estimator exists, an estimator can be usually found that attains the Cramér–Rao lower bound *asymptotically* and is, in addition, asymptotically unbiased. The term "asymptotically" means that such estimators, discussed in Chapter 5, possess these properties for large numbers of observations. They are, therefore, *asymptotically efficient unbiased*. Moreover, as will be demonstrated in Chapter 5, asymptotically efficient unbiased estimators may already behave asymptotically for unexpectedly small numbers of observations. This often justifies the use of the Cramér–Rao lower bound as a reference to which the performance of estimators may be compared. We define the *efficiency* of an estimator as the ratio of the relevant Cramér–Rao variance to the mean squared error of the estimator. This implies that the efficiency of an efficient unbiased estimator is equal to one or, equivalently, 100%. The efficiency of an asymptotically efficient unbiased estimator is asymptotically equal to one. If estimators have an efficiency less than one, they either have a variance exceeding the Cramér–Rao variance or are biased, or both. Strictly, the proposed definition allows of efficiencies exceeding one. This is because biased estimators might have a variance sufficiently smaller than Cramér–Rao variance to more than offset the bias contribution to the mean squared error. If serious inefficiency occurs, this may be a reason to reconsider the choice of estimator or experimental design—for example, the number of observations and the measurement points.

A further application of the Cramér–Rao lower bound is the study of the feasibility of observations for the purposes of the experimenter. Before any actual measurement, the Cramér–Rao variances of unbiased estimators of the parameters may be computed from the assumed statistical properties of the observations, the expectation model, nominal values of the parameters, and the experimental design chosen. If the computed Cramér–Rao standard deviation is not sufficient for the purposes of the experimenter, a different experimental design has to be chosen since this precision is the highest that might be realized by the (asymptotically) unbiased estimators often used in practice and discussed in Chapter 5.

4.6.3 Monotonicity with the number of observations

In Section 4.4.3, the inflow of Fisher information was studied. The Fisher information was shown to either increase or remain the same with every additional observation if the observations are independent or if their distribution is a linear exponential family. More specifically, under these conditions, the difference of the Fisher information matrix $F_{(N+1)}$ of $N + 1$ observations from the Fisher information matrix $F_{(N)}$ of N observations was found to be positive semidefinite:

$$F_{(N+1)} - F_{(N)} \succeq O. \tag{4.151}$$

This result will now be used to show that under the same conditions the Cramér–Rao lower bound decreases typically with every additional observation.

Since $F_{(N+1)}$ and $F_{(N)}$ are the covariance matrices of the corresponding Fisher score vectors, they are positive semidefinite. Here, they will also be assumed to be positive definite and, therefore, nonsingular. So, their inverses $F_{(N+1)}^{-1}$ and $F_{(N)}^{-1}$ exist. Then, Theorem C.7 shows that

$$F_{(N+1)} \succeq F_{(N)} \tag{4.152}$$

implies that

$$F_{(N)}^{-1} \succeq F_{(N+1)}^{-1} \; . \tag{4.153}$$

As a result, the Cramér–Rao variances decrease or remain the same with an increasing number of observations. Therefore, as a rule, increasing the number of observations improves the attainable precision. However, the degree of improvement may strongly differ from measurement point to measurement point.

4.6.4 Propagation of standard deviation

The Cramér–Rao lower bound is also illustrative of what in measurement is often called *error propagation*. We will explain this using a simple example. Suppose that $\rho = \rho(\theta) = (\rho_1 \; \rho_2)^T$ with $\rho_\ell = \rho_\ell(\theta)$, $\ell = 1, 2$, and $\theta = (\theta_1 \; \theta_2)^T$. Then, the Cramér–Rao lower bound matrix (4.110) becomes

$$\begin{pmatrix} \dfrac{\partial \rho_1}{\partial \theta_1} & \dfrac{\partial \rho_1}{\partial \theta_2} \\ \\ \dfrac{\partial \rho_2}{\partial \theta_1} & \dfrac{\partial \rho_2}{\partial \theta_2} \end{pmatrix} F_\theta^{-1} \begin{pmatrix} \dfrac{\partial \rho_1}{\partial \theta_1} & \dfrac{\partial \rho_2}{\partial \theta_1} \\ \\ \dfrac{\partial \rho_1}{\partial \theta_2} & \dfrac{\partial \rho_2}{\partial \theta_2} \end{pmatrix} . \tag{4.154}$$

Next, consider F_θ^{-1} as the covariance matrix of the hypothetical efficient unbiased estimator $\check{t}$ for θ. Then, F_θ^{-1} may be written

$$F_\theta^{-1} = \begin{pmatrix} \operatorname{var} \check{t}_1 & \operatorname{cov}(\check{t}_1, \check{t}_2) \\ \operatorname{cov}(\check{t}_1, \check{t}_2) & \operatorname{var} \check{t}_2 \end{pmatrix} . \tag{4.155}$$

Equations (4.154) and (4.155) show that the Cramér–Rao variances for unbiasedly estimating the elements of $\rho(\theta)$ are described by

$$\operatorname{var} \check{r}_\ell = \left(\frac{\partial \rho_\ell}{\partial \theta_1} \right)^2 \operatorname{var} \check{t}_1 + 2 \frac{\partial \rho_\ell}{\partial \theta_1} \frac{\partial \rho_\ell}{\partial \theta_2} \operatorname{cov}(\check{t}_1, \check{t}_2) + \left(\frac{\partial \rho_\ell}{\partial \theta_2} \right)^2 \operatorname{var} \check{t}_2 \tag{4.156}$$

with $\ell = 1, 2$. This expression looks like popular error propagation laws, but different from these it is not an approximation. Furthermore, the middle term accounts for the covariance of $\check{t}_1$ and $\check{t}_2$.

4.6.5 Influence of estimation of additional parameters

Suppose that in a parameter estimation problem the parameters $\theta_{(K)} = (\theta_1 \ldots \theta_K)^T$ are estimated and that the $K \times K$ Fisher information matrix concerned is

$$F_{(K)} = \begin{pmatrix} f_{11} & f_{12} & \cdots & f_{1K} \\ f_{12} & f_{22} & & \vdots \\ \vdots & & \ddots & \vdots \\ f_{1K} & \cdots & \cdots & f_{KK} \end{pmatrix}, \tag{4.157}$$

where $f_{k\ell} = \text{cov}(s_k, s_\ell) = E[s_k s_\ell]$ with s_k the Fisher score for parameter θ_k. Suppose that $F_{(K)}$ is nonsingular. Since, in addition, $F_{(K)}$ is a covariance matrix and, therefore, positive semidefinite, it is also positive definite. See Theorem C.2. Then, the corresponding Cramér–Rao lower bound matrix $\Psi_{(K)}$ is equal to

$$\Psi_{(K)} = F_{(K)}^{-1} . \tag{4.158}$$

Next, suppose that, in addition to the elements of $\theta_{(K)}$, a parameter θ_{K+1} of the expectation model is estimated from the same observations. Then, the Fisher information matrix $F_{(K+1)}$ for the parameter vector $\theta_{(K+1)} = (\theta_1 \ldots \theta_K \; \theta_{K+1})^T$ becomes

$$F_{(K+1)} = \begin{pmatrix} F_{(K)} & f_{K+1} \\ f_{K+1}^T & f_{K+1,K+1} \end{pmatrix}, \tag{4.159}$$

where $K \times 1$ vector f_{K+1} is defined as

$$f_{K+1} = (f_{1,K+1} \ldots f_{K,K+1})^T \tag{4.160}$$

and the Cramér–Rao lower bound matrix is

$$\Psi_{(K+1)} = F_{(K+1)}^{-1}. \tag{4.161}$$

Then, Corollary B.4 shows that

$$\Psi_{(K+1)} = \begin{pmatrix} \Psi_{(K)} + \frac{1}{f}\, \Psi_{(K)}\, f_{K+1}\, f_{K+1}^T\, \Psi_{(K)} & -\frac{1}{f}\, \Psi_{(K)}\, f_{K+1} \\ -\frac{1}{f}\, f_{K+1}^T\, \Psi_{(K)} & \frac{1}{f} \end{pmatrix} \tag{4.162}$$

with

$$f = f_{K+1,K+1} - f_{K+1}^T \Psi_{(K)} f_{K+1}. \tag{4.163}$$

Theorem C.8 shows that the first K diagonal elements of $\Psi_{(K+1)}$ are larger than or equal to the corresponding diagonal elements of $\Psi_{(K)}$ if $f_{K+1} \neq o$. If $f_{K+1} = o$, these diagonal elements are equal. Then, the Fisher scores for the parameters $\theta_{(K)}$ are not correlated with the Fisher score for the parameter θ_{K+1}. Furthermore, the $(K+1)$th diagonal element of $\Psi_{(K+1)}$ is larger than $1/f_{K+1,K+1}$ if $f_{K+1} \neq o$. If $f_{K+1} = o$, they are equal.

This result may be interpreted as follows. Estimating the additional parameter θ_{K+1} causes the Cramér–Rao variances for the parameters $\theta_{(K)}$ to increase or, at best, to remain the same. They remain the same if the Fisher score for θ_{K+1} is not correlated with the

Fisher scores for the parameters $\theta_{(K)}$. The Cramér–Rao variance for the parameter θ_{K+1} alone is equal to $1/f_{K+1,K+1}$ but, if the parameters $\theta_{(K+1)}$ are estimated jointly, it is larger than $1/f_{K+1,K+1}$ unless the Fisher score for θ_{K+1} is uncorrelated with the Fisher scores for the parameters $\theta_{(K)}$. Therefore, if an additional parameter is estimated, the Cramér–Rao variances for the parameters already present increase or, at best, remain the same. Also, typically, the Cramér–Rao variance for a single parameter is smaller than the Cramér–Rao variance for the same parameter if it is estimated together with other parameters. An illustrative numerical example is the following.

■ EXAMPLE 4.14

Estimating exponential decay parameters with and without an unknown background parameter

Suppose that Poisson distributed observations $w = (w_1 \ldots w_N)^T$ are available with expectations

$$Ew_n = g_n(\theta) = \kappa \left[\theta_1 \exp(-\theta_2 x_n) + \theta_3\right] , \tag{4.164}$$

where the scale factor κ is known. The amplitude θ_1 and the decay constant θ_2 are target parameters. The background θ_3 is, typically, a nuisance parameter.

In a particular experiment, $x_n = n \times 0.2$, $n = 1, \ldots, 20$, $\kappa = 1600$, and the true values of the parameters are $\theta = 1$, $\theta_2 = 1$, and $\theta_3 = 0$. Under these conditions, two different cases are studied. In Case 1, the absence of the background is unknown or ignored, and θ_3 is estimated along with θ_1 and θ_2. In Case 2, the background is assumed to be absent and only θ_1 and θ_2 are estimated.

Subsequently, for both cases the Fisher information matrices and the corresponding Cramér–Rao lower bound matrices are computed from (4.42) and (4.43). The elements of the relevant Jacobian matrices $\partial g(\theta)/\partial \theta^T$ are: $\partial g_n(\theta)/\partial \theta_1 = (1/\theta_1) g_n(\theta)$, $\partial g_n(\theta)/\partial \theta_2 = -x_n g_n(\theta)$, and $\partial g_n(\theta)/\partial \theta_3 = \kappa$.

The results are as follows. In Case 1, the Cramér–Rao variances are : 4.4×10^{-4}, 7.5×10^{-4}, and 1.1×10^{-5} for θ_1, θ_2, and θ_3, respectively. In Case 2, the Cramér–Rao variances are: 3.6×10^{-4} and 2.0×10^{-4} for θ_1 and θ_2, respectively. These results show how including the parameter θ_3 increases the variances for θ_1 and θ_2, that for θ_2 in particular.

■

4.6.6 The Cramér–Rao lower bound for biased estimators

The Cramér–Rao lower bound discussed thus far applies to unbiased estimators only. In this subsection, an extension to biased estimators is discussed that is found in the literature. The purpose of this discussion is describing the conditions under which this extension is valid.

Suppose that t is a biased estimator of θ. In particular, suppose that the expectation of t is described by

$$Et = \theta + \beta(\theta) , \tag{4.165}$$

where the bias vector function $\beta(\theta)$ represents the bias of the elements of t. Then, t is an unbiased estimator of

$$\rho = \theta + \beta(\theta) . \tag{4.166}$$

The Cramér–Rao lower bound for vector functions ρ is described by (4.110):

$$\frac{\partial \rho}{\partial \theta^T} F_\theta^{-1} \frac{\partial \rho^T}{\partial \theta}, \tag{4.167}$$

where F_θ is the Fisher information matrix. Applying (4.167) to (4.166) yields for the Cramér–Rao lower bound on the variance of unbiased estimators of $\theta + \beta(\theta)$:

$$\left(I + \frac{\partial \beta(\theta)}{\partial \theta^T}\right) F_\theta^{-1} \left(I + \frac{\partial \beta^T(\theta)}{\partial \theta}\right). \tag{4.168}$$

In this expression, I is the identity matrix of order K, and $\partial\beta(\theta)/\partial\theta^T$ is the $K \times K$ Jacobian matrix of the $K \times 1$ bias vector function $\beta(\theta)$ with respect to the $K \times 1$ vector θ.

Expression (4.168) is *not* the Cramér–Rao lower bound matrix for all biased estimators of θ. *It is the Cramér–Rao lower bound matrix for all biased estimators of θ that have a bias vector function with the same Jacobian matrix.* This is clearly a subclass of the class of all biased estimators.

■ **EXAMPLE 4.15**

A biased estimator of the variance

Suppose that the observations $w = (w_1 \dots w_N)^T$ are independent and identically normally distributed with unknown equal expectation μ and variance σ^2. Then, the maximum likelihood estimators of these parameters, to be discussed in Chapter 5, are described by, respectively,

$$\bar{w} = \frac{1}{N} \sum_n w_n \tag{4.169}$$

and

$$s^2 = \frac{1}{N} \sum_n (w_n - \bar{w})^2. \tag{4.170}$$

Elementary calculations show that

$$E\bar{w} = \mu \tag{4.171}$$

and

$$Es^2 = \left(1 - \frac{1}{N}\right)\sigma^2. \tag{4.172}$$

Therefore, the estimator $\bar{w}$ is unbiased but the estimator s^2 has a bias

$$-\frac{\sigma^2}{N}. \tag{4.173}$$

The log-probability density function of the observations is described by (3.29):

$$q(\omega; \varphi) = -\frac{N}{2} \ln 2\pi - N \ln \sigma - \frac{1}{2\sigma^2} \sum_n (\omega_n - \mu)^2, \tag{4.174}$$

where $\varphi = (\mu\ \sigma)^T$. Computing the Fisher information matrix using this expression and, from it, the Cramér–Rao lower bound matrix is straightforward. The result is

$$F_\varphi^{-1} = \frac{\sigma^2}{N} \begin{pmatrix} 1 & 0 \\ 0 & \frac{1}{2} \end{pmatrix}. \tag{4.175}$$

Therefore, the Cramér–Rao lower bound matrix for unbiased estimation of $\theta = (\mu\,\sigma^2)^T$ is

$$\begin{aligned}\frac{\partial\theta}{\partial\varphi^T}F_\varphi^{-1}\frac{\partial\theta^T}{\partial\varphi} &= \begin{pmatrix}1 & 0\\ 0 & 2\sigma\end{pmatrix}\frac{\sigma^2}{N}\begin{pmatrix}1 & 0\\ 0 & \frac{1}{2}\end{pmatrix}\begin{pmatrix}1 & 0\\ 0 & 2\sigma\end{pmatrix}\\ &= \frac{\sigma^2}{N}\begin{pmatrix}1 & 0\\ 0 & 2\sigma^2\end{pmatrix}.\end{aligned} \tag{4.176}$$

Then, the Cramér–Rao lower bound matrix for unbiased estimation of $\rho = \rho(\mu, \sigma^2) = [\mu\ \ (1-\frac{1}{N})\sigma^2]^T$ is

$$\frac{\partial\rho}{\partial\theta^T}\frac{\sigma^2}{N}\begin{pmatrix}1 & 0\\ 0 & 2\sigma^2\end{pmatrix}\frac{\partial\rho^T}{\partial\theta} = \frac{\sigma^2}{N}\begin{pmatrix}1 & 0\\ 0 & 2(1-\frac{1}{N})^2\sigma^2\end{pmatrix} \tag{4.177}$$

since

$$\frac{\partial\rho}{\partial\theta^T} = \begin{pmatrix}1 & 0\\ 0 & 1-\dfrac{1}{N}\end{pmatrix}. \tag{4.178}$$

■

The results of Example 4.15 are characteristic of the behavior of the Cramér–Rao lower bound on the variance of biased estimators.

First, the example shows that the Cramér–Rao lower bound matrix computed is specific of a restricted class of biased estimators of the parameters μ and σ^2. It only applies to estimators that have a Jacobian matrix described by (4.178).

Furthermore, the example shows that the Cramér–Rao variance for such a biased estimator of the parameter σ^2 is

$$2(1-\frac{1}{N})^2\,\frac{\sigma^4}{N} \tag{4.179}$$

whereas that for an unbiased estimator is

$$2\frac{\sigma^4}{N}. \tag{4.180}$$

Clearly, depending on the magnitude of N, the effect of the bias may be negligible.

Characteristic of bias in general, the bias (4.173) is of the order N^{-1} whereas the Cramér–Rao standard deviations corresponding to (4.179) and (4.180) are of the order $N^{-\frac{1}{2}}$.

4.7 THE CRAMÉR–RAO LOWER BOUND FOR COMPLEX PARAMETERS OR FUNCTIONS OF PARAMETERS

4.7.1 Introduction

In many applications, some or all parameters of the expectation model are complex by nature. In these applications, estimating the complex parameters directly may be preferred to estimating their real and imaginary parts as separate real parameters. In this section, the Cramér–Rao lower bound matrix is presented for such direct complex estimation. First, the Cramér–Rao lower bound matrix for vectors of real and complex functions of real parameters is derived. Subsequently, this result is extended to include the Cramér–Rao lower bound matrix for vectors of real and complex functions of real and complex parameters. If the functions of the real and complex parameters are the parameters themselves, the result specializes to the Cramér–Rao lower bound matrix for a vector of real and complex parameters. This important special case is also addressed.

4.7.2 The Cramér–Rao lower bound for vectors of real and complex functions of real parameters

Suppose that

$$\varphi = (\varphi_1 \dots \varphi_K)^T \tag{4.181}$$

is a vector of real parameters and that

$$\upsilon = (\upsilon_1 \dots \upsilon_L)^T \tag{4.182}$$

is an $L \times 1$ vector of real functions $\upsilon_\ell = \upsilon_\ell(\varphi)$ of the elements of φ. Then, the Cramér–Rao inequality for unbiased estimators u of υ is described by (4.109):

$$\text{cov}(u, u) - \frac{\partial \upsilon}{\partial \varphi^T} F_\varphi^{-1} \frac{\partial \upsilon^T}{\partial \varphi} \succeq O\,, \tag{4.183}$$

where F_φ is the $K \times K$ Fisher information matrix for the parameters φ.

Next, suppose that ρ is an $L \times 1$ vector of real and complex functions related to the $L \times 1$ vector of real functions υ by

$$\rho = B\upsilon\,, \tag{4.184}$$

where

$$B = \text{diag}\,(I \;\; A) \tag{4.185}$$

with I the identity matrix of order L_1 and A the $2L_2 \times 2L_2$ matrix defined by

$$A = \begin{pmatrix} I & jI \\ I & -jI \end{pmatrix} \tag{4.186}$$

with I the identity matrix of order L_2, and $L_1 + 2L_2 = L$. The relation (4.184) has been introduced in Subsection 3.8.3. Then, the first L_1 elements of ρ are the real functions $\upsilon_1, \dots, \upsilon_{L_1}$. The next L_2 elements are the complex functions $\upsilon_{L_1+1} + j\upsilon_{L_1+L_2+1}, \dots, \upsilon_{L_1+L_2} + j\upsilon_{L_1+2L_2}$ while the last L_2 elements are the complex conjugates of these.

Then, the Cramér–Rao lower bound for unbiased estimators $r = Bu$ of the vector of real and complex functions $\rho = B\upsilon$ is obtained from (4.183) as follows. The left-hand member of (4.183) is positive semidefinite. Therefore, by Theorem C.12,

$$B\left(\text{cov}(u, u) - \frac{\partial \upsilon}{\partial \varphi^T} F_\varphi^{-1} \frac{\partial \upsilon^T}{\partial \varphi}\right) B^H \succeq O. \tag{4.187}$$

In this expression,

$$B\,\text{cov}(u, u)\,B^H = \text{cov}(r, r) \tag{4.188}$$

since $r = Bu$, and, by (4.184),

$$B\frac{\partial \upsilon}{\partial \varphi^T} F_\varphi^{-1} \frac{\partial \upsilon^T}{\partial \varphi} B^H = \frac{\partial \rho}{\partial \varphi^T} F_\varphi^{-1} \frac{\partial \rho^H}{\partial \varphi}. \tag{4.189}$$

Therefore,

$$\operatorname{cov}(r,r) - \frac{\partial \rho}{\partial \varphi^T} F_\varphi^{-1} \frac{\partial \rho^H}{\partial \varphi} \succeq O \tag{4.190}$$

or, equivalently,

$$\boxed{\operatorname{cov}(r,r) \succeq \frac{\partial \rho}{\partial \varphi^T} F_\varphi^{-1} \frac{\partial \rho^H}{\partial \varphi}} . \tag{4.191}$$

The right-hand member of the latter expression is the Cramér–Rao lower bound matrix for unbiased estimators of vectors of real and complex functions of real parameters.

4.7.3 The Cramér–Rao lower bound for vectors of real and complex functions of real and complex parameters

In this Subsection, the results of Subsection 4.7.2 are generalized to the Cramér–Rao lower bound for vectors of real and complex functions of real and complex parameters. Consider the real parameter vector

$$\varphi = \begin{pmatrix} \chi \\ \xi \\ \eta \end{pmatrix} , \tag{4.192}$$

where χ is a $P \times 1$ vector and both ξ and η are $Q \times 1$ vectors. Suppose that the elements of χ are real by nature but that the elements ξ_q and η_q are the real and imaginary parts of the q-th element of the vector of complex parameters

$$\zeta = (\zeta_1 \ldots \zeta_Q)^T \tag{4.193}$$

with

$$\zeta_q = \xi_q + j\eta_q \, . \tag{4.194}$$

Then, the vector of real and complex parameters θ corresponding to the vector φ is defined as

$$\theta = \begin{pmatrix} \chi \\ \zeta \\ \zeta^* \end{pmatrix} = B\varphi , \tag{4.195}$$

where

$$B = \operatorname{diag}(I \;\; A). \tag{4.196}$$

In this expression, I is the identity matrix of order P and

$$A = \begin{pmatrix} I & jI \\ I & -jI \end{pmatrix} , \tag{4.197}$$

where I is the identity matrix of order Q.

Suppose that $\rho(\varphi)$ is an $L \times 1$ vector of real and complex functions of the elements of the real parameter vector φ defined by (4.192) and define the vector $\rho(\theta)$ as the vector $\rho(\varphi)$ after substitution of $B^{-1}\theta$ for φ.

Then, the vector of total differentials of $\rho(\theta)$ with respect to θ is

$$d\rho(\theta) = \frac{\partial \rho(\theta)}{\partial \theta^T} d\theta \, . \tag{4.198}$$

Hence,

$$\frac{\partial \rho(\varphi)}{\partial \varphi^T} = \frac{\partial \rho(\theta)}{\partial \theta^T}\frac{\partial \theta}{\partial \varphi^T} = \frac{\partial \rho(\theta)}{\partial \theta^T} B. \tag{4.199}$$

Substituting this in the right-hand member of (4.191) yields

$$\frac{\partial \rho}{\partial \theta^T} B F_\varphi^{-1} B^H \left(\frac{\partial \rho}{\partial \theta^T}\right)^H = \frac{\partial \rho}{\partial \theta^T}\left(B^{-H} F_\varphi B^{-1}\right)^{-1}\left(\frac{\partial \rho}{\partial \theta^T}\right)^H, \tag{4.200}$$

where, for brevity, the argument θ of ρ has been omitted. In this expression,

$$F_\varphi = E\left[\frac{\partial q(w;\varphi)}{\partial \varphi}\frac{\partial q(w;\varphi)}{\partial \varphi^T}\right] \tag{4.201}$$

is the Fisher information matrix for the real parameters φ. Using the total differential of $q = q(w;\theta)$ with respect to θ in the same way as in (4.198) and (4.199) yields

$$\frac{\partial q}{\partial \varphi} = B^T \frac{\partial q}{\partial \theta} = B^H \frac{\partial q}{\partial \theta^*}, \tag{4.202}$$

where, in the last step, use has been made of q and φ being real. Substituting (4.202) in (4.201) and, next, the result of this substitution in (4.200) yields

$$\frac{\partial \rho}{\partial \theta^T} F_\theta^{-1}\left(\frac{\partial \rho}{\partial \theta^T}\right)^H, \tag{4.203}$$

where

$$\boxed{F_\theta = E\left[\frac{\partial q}{\partial \theta^*}\left(\frac{\partial q}{\partial \theta^*}\right)^H\right]} \tag{4.204}$$

is defined as the *complex Fisher information matrix*. Then, the Cramér–Rao inequality for unbiased estimators r of the vector of real and complex functions ρ of real and complex parameters θ is described by

$$\operatorname{cov}(r,r) - \frac{\partial \rho}{\partial \theta^T} F_\theta^{-1}\left(\frac{\partial \rho}{\partial \theta^T}\right)^H \succeq O \tag{4.205}$$

or, equivalently, by

$$\boxed{\operatorname{cov}(r,r) \succeq \frac{\partial \rho}{\partial \theta^T} F_\theta^{-1}\left(\frac{\partial \rho}{\partial \theta^T}\right)^H}. \tag{4.206}$$

The matrix

$$\boxed{\frac{\partial \rho}{\partial \theta^T} F_\theta^{-1}\left(\frac{\partial \rho}{\partial \theta^T}\right)^H} \tag{4.207}$$

is the Cramér–Rao lower bound matrix for unbiased estimators of vectors of real and complex functions of real and complex parameters.

The most important special case is that the vector of functions $\rho(\theta)$ coincides with the vector of parameters θ. Then,

$$\frac{\partial \rho}{\partial \theta^T} = I, \tag{4.208}$$

where I is the identity matrix of order $P + 2Q$, and (4.205) and (4.206) become

$$\text{cov}(t, t) - F_\theta^{-1} \succeq O \tag{4.209}$$

and

$$\boxed{\text{cov}(t, t) \succeq F_\theta^{-1}}, \tag{4.210}$$

respectively.

The Fisher score vector s_θ of a set of observations w for a vector of real parameters was defined by (3.17). Subsequently, the Fisher information matrix was defined as the covariance matrix (4.38) of s_θ. On the analogy of these definitions for real parameters, (4.204) suggests as definition of the Fisher score vector for complex parameters

$$\boxed{s_\theta = \frac{\partial q}{\partial \theta^*}}. \tag{4.211}$$

By (4.195), the gradient of a function with respect to θ^* and that with respect to θ consist of the same elements but in a different order.

4.8 THE CRAMÉR–RAO LOWER BOUND FOR EXPONENTIAL FAMILIES OF DISTRIBUTIONS

The Fisher information matrix of exponential family distributed observations is described by (4.67):

$$F_\theta = \frac{\partial \mu_\delta^T}{\partial \theta} C_{\delta\delta}^{-1} \frac{\partial \mu_\delta}{\partial \theta^T}. \tag{4.212}$$

In this expression, $C_{\delta\delta}$ is the $L \times L$ covariance matrix of the $L \times 1$ vector of functions $\delta(w)$ of the observations w. This vector is present in expression (3.74) for exponential families of distributions:

$$p(\omega; \theta) = \alpha(\theta)\beta(\omega) \exp\left(\gamma^T(\theta)\delta(\omega)\right) \tag{4.213}$$

while $\mu_\delta = E\delta(w)$. We see that

$$F_\theta^{-1} = \left(\frac{\partial \mu_\delta^T}{\partial \theta} C_{\delta\delta}^{-1} \frac{\partial \mu_\delta}{\partial \theta^T}\right)^{-1} \tag{4.214}$$

is the Cramér–Rao lower bound matrix for all distributions that are exponential families. For practice, the most important special case is that the exponential family is linear. Then,

$$\delta(\omega) = \omega, \tag{4.215}$$

where $\omega = (\omega_1 \ldots \omega_N)^T$. The relevant Cramér–Rao lower bound matrix for estimating the parameter vector θ follows directly from (4.68) and is described by

$$\boxed{F_\theta^{-1} = \left(\frac{\partial g^T(\theta)}{\partial \theta} C^{-1} \frac{\partial g(\theta)}{\partial \theta^T}\right)^{-1}}, \tag{4.216}$$

where C is the covariance matrix of w, and $g(\theta) = [g_1(\theta) \ldots g_N(\theta)]^T$ with $g_n(\theta) = g(x_n; \theta)$. Generally, also the elements of C are functions of the measurement points x_n. The resulting dependence of the elements of the Cramér–Rao lower bound matrix F_θ^{-1} on the x_n is illustrated in the following example.

■ **EXAMPLE 4.16**

Dependence of the Cramér–Rao lower bound on the measurement points

Suppose that the observations $w = (w_1 \dots w_N)^T$ are independent and Poisson distributed with expectations $Ew = g(\theta)$. Then, (4.44) describes the (p,q)th element of the information matrix F_θ:

$$f_{pq} = \sum_n \frac{1}{g_n(\theta)} \frac{\partial g_n(\theta)}{\partial \theta_p} \frac{\partial g_n(\theta)}{\partial \theta_q}. \tag{4.217}$$

Since $g_n(\theta) = g(x_n; \theta)$, this expression shows how the elements of the Fisher information matrix F_θ and, therefore, the elements of the Cramér–Rao lower bound matrix F_θ^{-1} depend on the measurement points x_n. ■

The dependence of the elements of F_θ on the x_n will be used in Section 4.10 for experimental design.

4.9 THE CRAMÉR–RAO LOWER BOUND AND IDENTIFIABILITY

In this section, the concept *identifiability* is introduced. We will show that this concept is often closely connected with nonsingularity of the Jacobian matrix $\partial g(\theta)/\partial \theta^T$ in the sense of Definition B.17. Therefore, this matrix will be studied first. We present two examples of conditions under which $\partial g(\theta)/\partial \theta^T$ is singular. The following theorem describes the first example:

Theorem 4.6 *Suppose that the expectations $g_n(\theta)$ with $\theta = (\theta_1 \dots \theta_K)^T$ may be reparameterized by less than K parameters. Then, $\partial g(\theta)/\partial \theta^T$ is singular.*

Proof. Suppose that the expectations $g_n(\theta)$ may be reparameterized by the parameter vector $\psi(\theta) = [\psi_1(\theta) \dots \psi_{K'}(\theta)]^T$ with $K' < K$. Then,

$$dh_n(\psi) = \sum_{k'=1}^{K'} \frac{\partial h_n(\psi)}{\partial \psi_{k'}} d\psi_{k'}, \tag{4.218}$$

where $h_n(\psi)$ is the reparameterized version of $g_n(\theta)$ and $\psi = \psi(\theta)$. Furthermore,

$$d\psi_{k'} = \sum_{k=1}^{K} \frac{\partial \psi_{k'}}{\partial \theta_k} d\theta_k, \tag{4.219}$$

where $k' = 1, \dots, K'$. By (4.218) and (4.219),

$$\frac{\partial h_n(\psi)}{\partial \theta_k} = \frac{\partial g_n(\theta)}{\partial \theta_k} = \sum_{k'=1}^{K'} \frac{\partial h_n(\psi)}{\partial \psi_{k'}} \frac{\partial \psi_{k'}}{\partial \theta_k} \tag{4.220}$$

for $k = 1, \dots, K$. This expression shows that any of the K columns of the $N \times K$ matrix $\partial g(\theta)/\partial \theta^T$ is a linear combination of the $K' < K$ columns of the $N \times K'$ matrix $\partial h(\psi)/\partial \psi^T$. Then, $\partial g(\theta)/\partial \theta^T$ has linearly dependent columns and is, therefore, singular.
■

The following example is an illustration of Theorem 4.6.

■ **EXAMPLE 4.17**

Parameterization of an exponential model

Suppose that the expectations of the observations $w = (w_1 \ldots w_N)^T$ are described by

$$Ew_n = g_n(\theta) = \theta_1 \exp[-\theta_3(x_n - \theta_2)], \tag{4.221}$$

where $\theta = (\theta_1\, \theta_2\, \theta_3)^T$ are the unknown parameters and $x_n \geq \theta_2$. Thus,

$$g_n(\theta) = \theta_1 \exp(\theta_2\theta_3) \exp(-\theta_3 x_n). \tag{4.222}$$

Then, $g_n(\theta)$ may be reparameterized as

$$h_n(\psi) = \psi_1 \exp(-\psi_2 x_n) \tag{4.223}$$

with $\psi_1 = \theta_1 \exp(\theta_2\theta_3)$ and $\psi_2 = \theta_3$. The elements of the three columns of $\partial g(\theta)/\partial\theta^T$ are

$$\frac{\partial g_n(\theta)}{\partial \theta_1} = \frac{1}{\theta_1} g_n(\theta), \tag{4.224}$$

$$\frac{\partial g_n(\theta)}{\partial \theta_2} = \theta_3\, g_n(\theta), \tag{4.225}$$

and

$$\frac{\partial g_n(\theta)}{\partial \theta_3} = -(x_n - \theta_2)\, g_n(\theta). \tag{4.226}$$

Therefore,

$$\theta_1\theta_3 \frac{\partial g(\theta)}{\partial \theta_1} - \frac{\partial g(\theta)}{\partial \theta_2} = o, \tag{4.227}$$

with o the $N \times 1$ null vector. This shows that both first columns of $\partial g(\theta)/\partial\theta^T$ are linearly dependent and, hence, that this matrix is singular. Note that this singularity is related to the expectation model only and not to statistical properties of the observations. ■

A second, almost trivial, example of a condition under which $\partial g(\theta)/\partial\theta^T$ is singular is $N < K$; that is, the number of observations is smaller than the number of parameters. Then, $\partial g(\theta)/\partial\theta^T$ is singular since the number of its columns exceeds that of its rows.

We now define the concept identifiability. A vector of parameters is said to be *identifiable* from a set of observations if the relevant Fisher information matrix is nonsingular. If, on the other hand, the Fisher information matrix is singular, the vector of parameters is said to be *nonidentifiable* and the Cramér–Rao lower bound matrix does not exist.

■ **EXAMPLE 4.18**

Identifiability from observations with a normal, Poisson, or multinomial distribution

Suppose that the observations have a normal, Poisson, or multinomial distribution. Then, the Fisher information matrix is described by (4.40), (4.42), or (4.46), respectively:

$$\frac{\partial g^T(\theta)}{\partial \theta} C^{-1} \frac{\partial g(\theta)}{\partial \theta^T}, \tag{4.228}$$

where C is the covariance matrix of the observations. Since, apparently, its inverse exists, C is nonsingular. Therefore, because it is a covariance matrix, it is also positive definite and so is C^{-1}. Then, Theorem C.4 shows that the Fisher information matrix (4.228) is positive definite and, therefore, nonsingular if and only if the Jacobian matrix $\partial g(\theta)/\partial \theta^T$ is nonsingular. In conclusion, if the observations have a normal, Poisson, or multinomial distribution, the parameters of the expectation model are identifiable if and only if both the covariance matrix of the observations and the Jacobian matrix $\partial g(\theta)/\partial \theta^T$ are nonsingular. ■

■ **EXAMPLE 4.19**

Identifiability from exponential family distributed observations

The normal, Poisson, and multinomial families of distributions parametric in θ are linear exponential families. It is, therefore, not surprising that the general expression (4.68) for the Fisher information matrix of linear exponential family distributed observations is formally the same as that for the normal, Poisson, and multinomial observations. As a result, the necessary and sufficient condition for identifiability is the same as that in Example 4.18: Both the covariance matrix C and the Jacobian matrix $\partial g(\theta)/\partial \theta^T$ should be nonsingular. ■

4.10 THE CRAMÉR–RAO LOWER BOUND AND EXPERIMENTAL DESIGN

4.10.1 Introduction

Expression (4.216) describes the Cramér–Rao lower bound matrix for estimating parameters of expectation models from linear exponential family distributed observations. It shows that, for this general class of distributions and models, the measurement points enter the expression for the Cramér–Rao lower bound via the expectation model and the covariance matrix of the observations.

For *any* probability density function $p(\omega;\theta)$, the expectation of the nth observation is defined by (3.3):

$$E w_n = \int \omega_n \, p(\omega;\theta)\, d\omega. \tag{4.229}$$

If, in addition, the elements of θ are the parameters of the expectation model of the observations, it is also true that

$$E w_n = g(x_n;\theta). \tag{4.230}$$

Hence,

$$\int \omega_n \, p(\omega;\theta)\, d\omega = g(x_n;\theta) \tag{4.231}$$

for $n = 1, \ldots, N$. These N relations show that $p(\omega;\theta)$ depends on all measurement points x_n. Therefore, the same applies to the Fisher information matrix

$$F_\theta = E\left[\frac{\partial q(w;\theta)}{\partial \theta}\frac{\partial q(w;\theta)}{\partial \theta^T}\right] \tag{4.232}$$

and the Cramér–Rao lower bound matrix F_θ^{-1}.

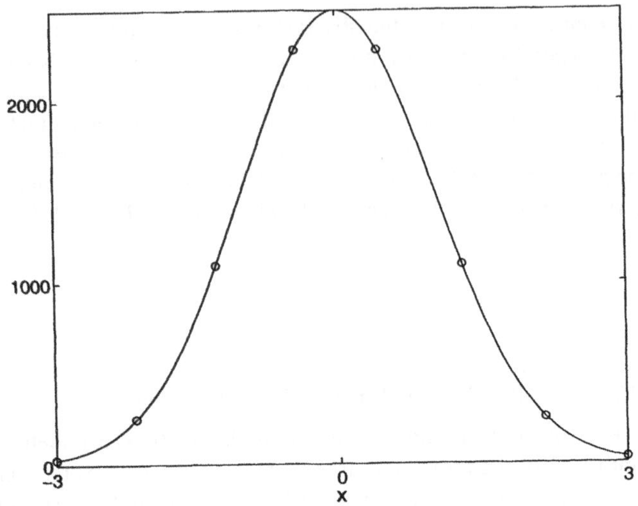

Figure 4.3. Gaussian peak expectation model (solid line) and its values at the measurement points (circles). (Example 4.20)

Often, experimenters can, within certain bounds but otherwise freely, choose the numerical values of some of the experimental variables. A particular choice of these numerical values is called an *experimental design*. Suppose that the problem at hand is estimating the parameters θ of the expectation model $g(x;\theta)$. Furthermore, suppose that N, the number of measurement points, is fixed but that their values x_n may within certain bounds be freely chosen. Then, the choice of the free variables $x = (x_1 \ldots x_N)^T$ is the experimental design. This design may be used to manipulate the Cramér–Rao variances and even to minimize them in a chosen sense. This minimization will be discussed in this subsection and in Subsection 4.10.2. The guiding principle will be the assumption that an estimator is available that attains the Cramér–Rao lower bound. Then, the Cramér–Rao variances are the variances of this hypothetical estimator. The discussion if such estimators exist and, if so, how to find them is postponed until Chapter 5. To illustrate the idea of an optimal design, we present the following relatively simple example.

■ **EXAMPLE 4.20**

Design for estimating a parameter of a Gaussian peak from Poisson distributed observations

Suppose that independent, Poisson distributed observations $w = (w_1 \ldots w_N)^T$ are available with expectations

$$Ew_n = g_n(\mu) = 2500\,\alpha \exp\left(-\frac{1}{2}\left(\frac{x_n - \mu}{\sigma}\right)^2\right), \tag{4.233}$$

where the height parameter α and the half-width σ are supposed to be known whereas the location μ is the target parameter. The design x that minimizes the relevant Cramér–Rao variance may be computed as follows. Expression (4.77) describes the Fisher information

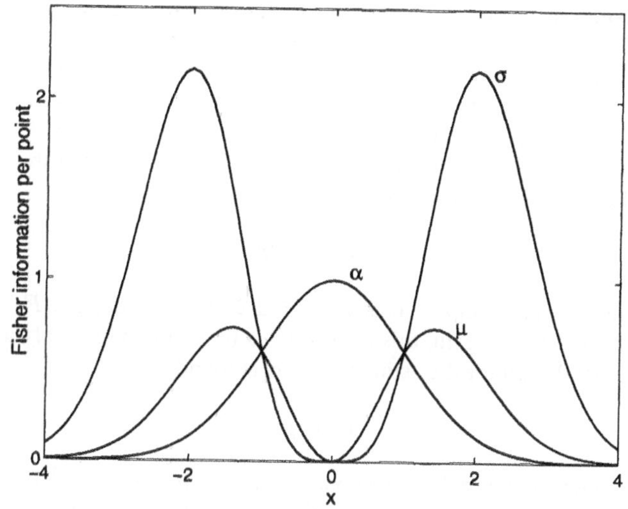

Figure 4.4. Fisher information for the height α, the width σ, and the location μ per point of the Gaussian peak expectation model shown in Fig. 4.3. The observations are Poisson distributed. (Example 4.20)

for μ:

$$F_\mu = \sum_n \frac{1}{g_n(\mu)} \left(\frac{\partial g_n(\mu)}{\partial \mu} \right)^2. \tag{4.234}$$

From (4.233),

$$\frac{\partial g_n(\mu)}{\partial \mu} = \frac{1}{\sigma} \left(\frac{x_n - \mu}{\sigma} \right) g_n(\mu). \tag{4.235}$$

Then, the summand in (4.234) is proportional to

$$\xi^2 \exp\left(-\frac{1}{2}\xi^2 \right) \tag{4.236}$$

with

$$\xi = \frac{x_n - \mu}{\sigma}. \tag{4.237}$$

Elementary calculus shows that the function (4.236) is maximum for $\xi = \pm\sqrt{2}$. Then, taking all measurement points as

$$x_n = \mu \pm \sigma\sqrt{2} \tag{4.238}$$

maximizes F_μ and, therefore, minimizes the Cramér–Rao variance F_μ^{-1}. Thus, this is an *optimal experimental design*. For example, in (4.233) let $\alpha = 1$, $\mu = 0$, and $\sigma = 1$. Suppose first that there are eight measurement points described by $x_n = -3 + \frac{6}{7} \times (n-1)$ for $n = 1, \ldots, 8$. This means that they are equidistant on the interval $[-3, 3]$ as shown in Fig. 4.3. By (4.234), for this design, $F_\mu = 0.73 \times 10^4$. This corresponds to a Cramér–Rao variance $F_\mu^{-1} = 1.37 \times 10^{-4}$. If, on the other hand, all measurement points would be chosen as $x_n = \mu \pm \sigma\sqrt{2} = \pm\sqrt{2}$, then $F_\mu = 1.47 \times 10^4$ and $F_\mu^{-1} = 0.68 \times 10^{-4}$.

The optimal design, therefore, reduces the Cramér–Rao variance by slightly more than a factor of two. In many applications, substantially increasing the number of observations only may reduce the Cramér–Rao variance to such an extent. Similar optimal experimental designs may be computed for unknown σ but known α and μ, as well as for unknown α but known μ and σ. The results are as follows. For the equidistant design, $F_\sigma = 2.12 \times 10^4$ and $F_\sigma^{-1} = 0.47 \times 10^{-4}$, respectively. Simple calculations show that the optimal design implies here taking all measurement points as

$$x_n = \mu \pm 2\sigma\,, \tag{4.239}$$

which, in this numerical example, means $x_n = \pm 2$. For this design, $F_\sigma = 4.33 \times 10^4$ and $F_\sigma^{-1} = 0.23 \times 10^{-4}$. Again, this constitutes a reduction by roughly a factor of two. Finally, the optimal design for the estimation of α is found to be

$$x_n = \mu\,, \tag{4.240}$$

which, in this numerical example, means $x_n = 0$ for all n. Here, F_α^{-1} for the equidistant design and for the optimal design are, respectively, equal to 1.37×10^{-4} and 0.50×10^{-4}, which implies a reduction by a factor of 2.7. Figure 4.4. shows the contribution of each measurement point x_n on interval $[-4, 4]$ to F_α, F_μ, and F_σ, respectively. Note that the origin contributes most to F_α, but nothing at all to F_σ or F_μ. On the other hand, the points that contribute substantially to F_σ or F_μ do not do so to F_α. ■

In Example 4.20, the Cramér–Rao variance has been taken as optimality criterion for the design. If more than one parameter is unknown, the Cramér–Rao variance is replaced by the Cramér–Rao lower bound matrix. Then, the optimality criterion becomes a scalar function of the elements of the this matrix. Optimizing such criteria is the subject of the next subsection.

4.10.2 Experimental design for nonlinear vector parameters

The optimality criterion used in experimental design for vector parameters is chosen by and reflects the purposes of the experimenter. For example, the criterion may be chosen so that the precision of estimates of target parameters is emphasized, possibly at the expense of the precision of estimates of nuisance parameters.

Generally, the optimality criterion Q is a scalar function of the $\frac{1}{2}K(K+1)$ different elements of the Cramér–Rao lower bound:

$$Q = Q(f_{\overline{11}}, f_{\overline{12}}, \ldots, f_{\overline{1K}}, f_{\overline{22}}, f_{\overline{23}}, \ldots, f_{\overline{2K}}, \ldots, f_{\overline{KK}}) \tag{4.241}$$

where $f_{\overline{k\ell}}$ is the (k, ℓ)th element of the Cramér–Rao lower bound matrix F_θ^{-1} and K is the number of parameters. Suppose that $x = (x_1 \ldots x_N)^T$ is the vector of free experimental variables. In the Examples given in this subsection, these free variables will be the measurement points, but they may be other variables as well. The criterion Q is chosen so that the design x *minimizing* it optimizes the precision. Generally, a *necessary* condition for a point x to be the location of a relative or absolute minimum is that it is a *stationary point*. This is defined as a point where the gradient of the function vanishes. Therefore, the minimum sought is among the stationary points and the gradient is used to find it. Since the elements $f_{\overline{pq}}$ are typically nonquadratic, nonlinear functions of the elements of x, so is the criterion Q. Therefore, the elements of the gradient of Q are nonlinear functions of the

x. Consequently, the system of equations obtained by equating all elements of the gradient to zero is a system of nonlinear equations in x that cannot be solved in closed form. This is the reason why Q has to be minimized by iterative numerical methods. Iterative numerical minimization is a subject of Chapter 6. The relevant numerical minimization methods described there require in the first place an expression for the gradient of the function to be minimized. One of these methods, the Newton method, also requires an expression for its Hessian matrix. Therefore, in this subsection, we will derive expressions both for the gradient and for the Hessian matrix of the criterion Q with respect to the vector x.

The total differential of Q at the point $(f_{\overline{11}}, f_{\overline{12}}, \ldots, f_{\overline{1K}}, f_{\overline{22}}, f_{\overline{23}}, \ldots, f_{\overline{2K}}, \ldots, f_{\overline{KK}})$ is equal to

$$dQ = \sum_{p \leq q} \frac{\partial Q}{\partial f_{\overline{pq}}} df_{\overline{pq}}, \tag{4.242}$$

where $q = 1, \ldots, K$. Then, the derivative of Q with respect to any of the x_n is described by

$$\frac{\partial Q}{\partial x_n} = \sum_{p \leq q} \frac{\partial Q}{\partial f_{\overline{pq}}} \frac{\partial f_{\overline{pq}}}{\partial x_n}. \tag{4.243}$$

The partial derivatives $\partial Q / \partial f_{\overline{pq}}$ in this expression follow from the definition of Q whereas the partial derivatives $\partial f_{\overline{pq}} / \partial x_n$ follow from (D.13):

$$\frac{\partial F_\theta^{-1}}{\partial x_n} = -F_\theta^{-1} \frac{\partial F_\theta}{\partial x_n} F_\theta^{-1}. \tag{4.244}$$

Differentiating (4.243) with respect to x_m, we find

$$\frac{\partial^2 Q}{\partial x_m \partial x_n} = \sum_{p \leq q} \sum_{r \leq s} \frac{\partial^2 Q}{\partial f_{\overline{pq}} \partial f_{\overline{rs}}} \frac{\partial f_{\overline{rs}}}{\partial x_m} \frac{\partial f_{\overline{pq}}}{\partial x_n} + \sum_{p \leq q} \frac{\partial Q}{\partial f_{\overline{pq}}} \frac{\partial^2 f_{\overline{pq}}}{\partial x_m \partial x_n}, \tag{4.245}$$

where $s = 1, \ldots, K$. In this expression, the partial derivatives $\partial^2 Q / \partial f_{\overline{pq}} \partial f_{\overline{rs}}$ follow from the definition of Q whereas the partial derivatives $\partial^2 f_{\overline{pq}} / \partial x_m \partial x_n$ follow from (D.14):

$$\frac{\partial^2 F_\theta^{-1}}{\partial x_m \partial x_n} = F_\theta^{-1} \left(\frac{\partial F_\theta}{\partial x_m} F_\theta^{-1} \frac{\partial F_\theta}{\partial x_n} - \frac{\partial^2 F_\theta}{\partial x_m \partial x_n} + \frac{\partial F_\theta}{\partial x_n} F_\theta^{-1} \frac{\partial F_\theta}{\partial x_m} \right) F_\theta^{-1}. \tag{4.246}$$

The equations (4.244) and (4.246) express the derivatives of the elements $f_{\overline{pq}}$ of the Cramér–Rao lower bound matrix directly in the elements of the Fisher information matrix and their derivatives, which are relatively easy to compute.

A relatively simple optimality criterion is the linear combination of the Cramér–Rao variances

$$Q = \lambda_1 f_{\overline{11}} + \cdots + \lambda_K f_{\overline{KK}} \tag{4.247}$$

with $0 \leq \lambda_k \leq 1$ and $\sum_k \lambda_k = 1$. The choice of the weights λ_k depends on the purposes of the experiment. For example, if an equal relative variance for all parameters is pursued, the λ_k could be chosen as

$$\lambda_k = \frac{1/\theta_k^2}{\sum_k 1/\theta_k^2}, \tag{4.248}$$

since thus the quantities

$$\frac{f_{kk}^{-}}{\theta_k^2} \tag{4.249}$$

are uniformly weighted in the criterion.

For the criterion (4.247), the partial derivatives (4.243) and (4.245) simplify to

$$\frac{\partial Q}{\partial x_n} = \sum_{p=1}^{K} \lambda_p \frac{\partial f_{pp}^{-}}{\partial x_n} \tag{4.250}$$

and

$$\frac{\partial^2 Q}{\partial x_m \partial x_n} = \sum_{p=1}^{K} \lambda_p \frac{\partial^2 f_{pp}^{-}}{\partial x_m \partial x_n}. \tag{4.251}$$

The following example is an illustration of the use of criterion (4.247) and its derivatives (4.250) and (4.251).

■ EXAMPLE 4.21

Design for simultaneous estimation of all parameters of a Gaussian peak from Poisson distributed observations

As in Example 4.20, suppose that independent and Poisson distributed observations are available with expectations

$$E w_n = g_n(\theta) = \kappa\, \alpha \exp\left(-\frac{1}{2}\left(\frac{x_n - \mu}{\sigma}\right)^2 \right) \tag{4.252}$$

with $\kappa = 2500$, where, different from Example 4.20, all three elements of the parameter vector $\theta = (\alpha\ \sigma\ \mu)^T$ are unknown. Then, by (4.44), the elements of the Fisher information matrix are equal to

$$f_{pq} = \sum_n \frac{1}{g_n(\theta)} \frac{\partial g_n(\theta)}{\partial \theta_p} \frac{\partial g_n(\theta)}{\partial \theta_q}. \tag{4.253}$$

Simple calculations show that for the model (4.252) we have

$$\begin{aligned} \frac{\partial g_n(\theta)}{\partial \alpha} &= \frac{1}{\alpha} g_n(\theta), \\ \frac{\partial g_n(\theta)}{\partial \sigma} &= \frac{1}{\sigma} \xi_n^2 g_n(\theta), \end{aligned} \tag{4.254}$$

and

$$\frac{\partial g_n(\theta)}{\partial \mu} = \frac{1}{\sigma} \xi_n g_n(\theta)$$

with $\xi_n = (x_n - \mu)/\sigma$.

In this example, $\alpha = 1$, $\sigma = 1$, and $\mu = 0$. This choice implies that expected Poisson count at the point $\xi_n = 0$ is equal to 2500. Furthermore, assume that there are eight measurement points. This number is the constraint under which the design will be optimized. As optimality criterion for the design, (4.247) is chosen:

$$Q = \lambda_1 f_{11}^{-} + \lambda_2 f_{22}^{-} + \lambda_3 f_{33}^{-} \tag{4.255}$$

with $0 \leq \lambda_k \leq 1$ and $\lambda_1 + \lambda_2 + \lambda_3 = 1$ while $f_{\overline{11}}$, $f_{\overline{22}}$, and $f_{\overline{33}}$ are the Cramér–Rao variances for the parameters α, σ, and μ, respectively. Four different cases, in the sense of choices of the λ_k, will be considered. In all cases, as a starting point for iteratively minimizing Q with respect to x, eight random numbers are generated uniformly distributed on $[-3, 3]$. In each case, this is repeated a number of times and the resulting best solution for x in the sense of the criterion is selected.

Because the expectation model is an even function, the designs computed are not unique: if the design x is optimal so is $-x$.

Case 1 *Design uniformly minimizing the Cramér–Rao variances for all parameters*

To minimize the Cramér–Rao variances for all parameters uniformly, the following weights are chosen:

$$\lambda_1 = \lambda_2 = \lambda_3 = \frac{1}{3}\,. \tag{4.256}$$

Tables 4.1. and 4.2. show the results of minimizing Q numerically for this choice. For comparison, the results for eight equidistant measurement points between -3 and 3 have been included. Table 4.1. shows that the reduction of the criterion is mainly caused by a substantial reduction of the Cramér–Rao variance for α. Note that the optimal design, shown in Table 4.2., is concentrated in four points: -1.71, -0.12, 0.43, and 1.72. At these points, two, three, one, and two observations are made, respectively.

Table 4.1. Criterion value and Cramér–Rao variances for an equidistant design and for a design minimizing the variances uniformly for all parameters

Measurement Points	Criterion	Cramér–Rao Variances For α	For σ	For μ
Equidistant	1.39×10^{-4}	0.207×10^{-3}	0.071×10^{-3}	0.138×10^{-3}
Optimal	1.03×10^{-4}	0.107×10^{-3}	0.065×10^{-3}	0.137×10^{-3}

Table 4.2. Equidistant design and design minimizing the variances uniformly for all parameters

Measurement Points	Design x_1	x_2	x_3	x_4	x_5	x_6	x_7	x_8
Equidistant	-3	$-2\frac{1}{7}$	$-1\frac{2}{7}$	$-\frac{3}{7}$	$\frac{3}{7}$	$1\frac{2}{7}$	$2\frac{1}{7}$	3
Optimal	-1.71	-1.71	-0.12	-0.12	-0.12	0.43	1.72	1.72

Case 2 *Optimal design if the amplitude is the target parameter*

Suppose that the purpose of the experiment is the measurement of the parameter α. Then, α is the target parameter and σ and μ are nuisance parameters. This objective is expressed by taking $\lambda_1 \gg \lambda_2, \lambda_3$. The purpose of this choice is, of course, reducing the Cramér–Rao variance for α without caring too much about the Cramér–Rao variances for σ and μ. Here, the chosen weights are

$$\lambda_1 = 0.98 \quad \lambda_2 = \lambda_3 = 0.01 \,. \tag{4.257}$$

Tables 4.3. and 4.4. show the results for this choice. Table 4.3. shows that the optimal design reduces the criterion and the Cramér–Rao variance of the target parameter α substantially. The table also shows that the variances for the parameters σ and μ have increased considerably as compared with the same quantities for the equidistant design. The optimal design shown in Table 4.4. is concentrated in three points only: -0.86, 0.09, and 2.18. No less than six observations are taken at $x_n = 0.09$.

Table 4.3. Criterion value and Cramér–Rao variances for equidistant design and for optimal design if amplitude is target parameter

Measurement		Cramér–Rao Variances		
Points	Criterion	For α	For σ	For μ
Equidistant	2.053×10^{-4}	0.207×10^{-3}	0.071×10^{-3}	0.138×10^{-3}
Optimal	0.682×10^{-4}	0.063×10^{-3}	0.193×10^{-3}	0.449×10^{-3}

Table 4.4. Equidistant design and optimal design if amplitude is target parameter

Measurement	Design							
Points	x_1	x_2	x_3	x_4	x_5	x_6	x_7	x_8
Equidistant	-3	$-2\frac{1}{7}$	$-1\frac{2}{7}$	$-\frac{3}{7}$	$\frac{3}{7}$	$1\frac{2}{7}$	$2\frac{1}{7}$	3
Optimal	-0.86	0.09	0.09	0.09	0.09	0.09	0.09	2.18

Case 3 *Optimal design if the width is the target parameter*

The chosen weights are

$$\lambda_1 = 0.01 \quad \lambda_2 = 0.98 \quad \lambda_3 = 0.01 \,. \tag{4.258}$$

Tables 4.5. and 4.6. summarize the results. Table 4.5. shows that the optimal design reduces the criterion and the Cramér–Rao variance for target parameter σ significantly. Table 4.6. shows that the optimal design is concentrated in three points only: -2.21, 0, and 2.21. At each of the points -2.21 and 2.21, three observations are made.

Case 4 *Optimal design if the location is the target parameter*

The chosen weights are

$$\lambda_1 = \lambda_2 = 0.01 \quad \lambda_3 = 0.98 \,. \tag{4.259}$$

Tables 4.7. and 4.8. show the results. These are more less analogous to the results for α and σ as target parameters. ■

With respect to the four optimal designs of Example 4.21, the following remarks may be made.

Table 4.5. Criterion value and Cramér–Rao variances for equidistant design and for optimal design if width is target parameter

Measurement Points	Criterion	Cramér–Rao Variances For α	For σ	For μ
Equidistant	7.334×10^{-5}	0.207×10^{-3}	0.071×10^{-3}	0.138×10^{-3}
Optimal	4.327×10^{-5}	0.200×10^{-3}	0.041×10^{-3}	0.157×10^{-3}

Table 4.6. Equidistant design and optimal design if the width is the target parameter

Measurement Points	Design x_1	x_2	x_3	x_4	x_5	x_6	x_7	x_8
Equidistant	-3	$-2\frac{1}{7}$	$-1\frac{2}{7}$	$\frac{3}{7}$	$\frac{3}{7}$	$1\frac{2}{7}$	$2\frac{1}{7}$	3
Optimal	-2.21	-2.21	-2.21	0	0	2.21	2.21	2.21

Table 4.7. Criterion value and Cramér–Rao variances for equidistant design and for optimal design if location is target parameter

Measurement Points	Criterion	Cramér–Rao Variances For α	For σ	For μ
Equidistant	1.377×10^{-4}	0.207×10^{-3}	0.071×10^{-3}	0.138×10^{-3}
Optimal	0.813×10^{-4}	0.580×10^{-3}	0.180×10^{-3}	0.075×10^{-3}

Table 4.8. Equidistant design and optimal design if location is target parameter

Measurement Points	Design x_1	x_2	x_3	x_4	x_5	x_6	x_7	x_8
Equidistant	-3	$-2\frac{1}{7}$	$-1\frac{2}{7}$	$\frac{3}{7}$	$\frac{3}{7}$	$1\frac{2}{7}$	$2\frac{1}{7}$	3
Optimal	-1.55	-1.55	-1.55	-0.52	1.37	1.37	1.37	1.37

First, the optimal designs are computed to optimize the precision of *estimates* of parameters in a suitable sense. However, for this computation, the *exact values* of these parameters are needed. Theoretically, this circularity is unavoidable. However, the importance of optimal design for practice is that it enables the experimenter to compute optimal designs for *nominal* or *measured* values of the parameters. Thus, the precision obtained with the design used by the experimenter may be compared with that achieved with the optimal design. This comparison enables the experimenter to decide whether it is worthwhile to change the design. Without such a quantitative comparison, the extent to which the designs in use could be improved remains unclear. The optimal design shows what precision is *ultimately* achievable.

Second, all optimal designs computed and shown in Tables 4.2., 4.4., 4.6., and 4.8. are concentrated in less than the available eight measurement points. One may wonder why for estimating the target parameter the eight measurement points have not all simply been taken as in Example 4.20, that is, at $x_n = 0,\ 2$, and $\sqrt{2}$ for α, σ, and μ, respectively. However, this would imply estimating three parameters using only one measurement point. As discussed in Section 4.9, this would make the Fisher information matrix singular and, therefore, the parameters unidentifiable.

The improved precision of the estimates of the parameters of the expectation model also propagates to functions of the parameters. This is illustrated in the following example.

■ **EXAMPLE 4.22**

Influence of optimal design on estimation of area and location of a Gaussian peak from Poisson distributed observations.

Suppose that in Example 4.21 the purpose is estimating the area and the location of the peak. Simple calculations show that the area under peak is equal to

$$(2\pi)^{\frac{1}{2}}\,\kappa\alpha\sigma, \tag{4.260}$$

which, for $\kappa = 2500$, $\alpha = 1$, and $\sigma = 1$, is equal to 6267. The vector function to be estimated is

$$\rho(\theta) = \begin{pmatrix} (2\pi)^{\frac{1}{2}}\,\kappa\alpha\sigma \\ \mu \end{pmatrix} \tag{4.261}$$

and hence,

$$\frac{\partial\rho(\theta)}{\partial\theta^T} = \begin{pmatrix} (2\pi)^{\frac{1}{2}}\,\kappa\sigma & (2\pi)^{\frac{1}{2}}\,\kappa\alpha & 0 \\ 0 & 0 & 1 \end{pmatrix}. \tag{4.262}$$

The Cramér–Rao lower bound matrix for area and location is described by

$$\frac{\partial\rho(\theta)}{\partial\theta^T}F_\theta^{-1}\frac{\partial\rho(\theta)^T}{\partial\theta} \tag{4.263}$$

where F_θ^{-1} is the Cramér–Rao lower bound matrix for the parameters α, σ, and μ. Then, Table 4.9. shows the Cramér–Rao variances for the area and the location for the equidistant and the optimal design of Table 4.2., respectively. The table shows that the variance for the area is clearly reduced by the optimal design. ■

Table 4.9. Cramér–Rao variances for peak area and location for equidistant design and optimal design

Measurement	Cramér–Rao Variances	
Points	For Peak Area	For Peak Location
Equidistant	5.373×10^3	0.138×10^{-3}
Optimal	3.656×10^3	0.137×10^{-3}

4.11 COMMENTS AND REFERENCES

In a section on statistical inference in [13], Goldberger provides a useful short survey on bias, standard deviation, mean squared error, and related notions, comparable to but somewhat more comprehensive than Section 4.2.

Books by Goldberger [13] and by Dhrymes [7] discuss properties of covariance matrices such as described in Section 4.3. A similar discussion in [15] also includes properties of complex covariance matrices addressed in Subsection 4.3.2.

Fisher information, defined as in Section 4.4, originates from [11].

The original versions of the Cramér–Rao inequality and corresponding proofs are found in [5] and [26]. Goodwin and Payne [14] offer a simple proof of the inequality for vector parameters. Dhrymes [7] proves the inequality for vector functions of vector parameters and independent observations. We combined both proofs to prove the inequality for vector functions of vector parameters and not necessarily independent observations. In the literature, sometimes the condition is made that the dimension of the vector function may not exceed the dimension of the vector of parameters. In our proof, we do not make such a condition. However, if the condition is not met, the Cramér–Rao lower bound matrix is singular. This is not serious since the produced Cramér–Rao variances remain correct.

Goodwin and Payne [14] derive conditions for an estimator of vector parameters to be efficient unbiased. In Subsection 4.5.2, we have generalized this result to vector functions of vector parameters.

The Cramér–Rao lower bound for biased parameters discussed in Subsection 4.6.6 is included in Cramér's original result in [5]. The complex Cramér–Rao lower bound discussed in Section 4.7 originates from the article [31] by the author.

Well-known books on statistical experimental design are [9] and [2]. These books, and most of the statistical experimental design literature, are, however, strongly linear model oriented. This has been the main motive to include Section 4.10 directed to numerically computing experimental designs for expectation models nonlinear in the parameters.

4.12 PROBLEMS

4.1 The *correlation matrix* R of a vector of stochastic variables $w = (w_1 \dots w_N)^T$ with positive definite covariance matrix C has as its (m,n)th element:

$$r_{mn} = \frac{c_{mn}}{\sigma_m \sigma_n}$$

with $m, n = 1, \dots, N$, where σ_m and σ_n are the standard deviations of w_m and w_n, respectively. Thus, $r_{mn} = 1$ if $m = n$. Show that R is positive definite.

4.2

(a) Let C be a symmetric $N \times N$ matrix and define the $N \times 1$ vector x as

$$(0 \dots 0 \, x_p \, 0 \dots 0 \, x_q \, 0 \dots 0)^T .$$

Compute the quadratic form $x^T C x$.

(b) From the result under(a), derive that if C is positive semidefinite but not positive definite, then $c_{pq}^2 \leq c_{pp}\, c_{qq}$, and derive that if C is positive definite, then $c_{pq}^2 < c_{pp}\, c_{qq}$ for $p \neq q$.

4.3 Use the result of Problem 4.2(b) to show that if a diagonal element of a symmetric and positive semidefinite matrix is equal to zero, then all elements in the same column and the same row as that diagonal element are equal to zero.

4.4 Use the results of Problems 4.1 and 4.2 to show that elements of the correlation matrix R satisfy: $\mid r_{pq} \mid \leq 1$.

4.5 Which of the following matrices could be a covariance matrix of a vector of real stochastic variables?

(a)

$$\begin{pmatrix} 10 & 10 \\ 10 & 10 \end{pmatrix}.$$

(b)

$$\begin{pmatrix} 10 & 10 \\ 10 & 0 \end{pmatrix}.$$

(c)

$$\begin{pmatrix} 10 & 8 \\ 8 & 5 \end{pmatrix}.$$

(d)

$$\begin{pmatrix} 10 & 4 \\ 5 & 10 \end{pmatrix}.$$

(e)

$$\begin{pmatrix} 10 & -5 \\ -5 & 5 \end{pmatrix}.$$

(f)

$$\begin{pmatrix} 10 & 0 \\ 0 & 0 \end{pmatrix}.$$

4.6 Suppose that u is a $K \times 1$ vector of stochastic variables with $Eu = o$ and v is an $L \times 1$ vector of stochastic variables with $Ev \neq o$. Show that the $K \times L$ covariance matrix $\text{cov}\,(u, v) = E u\, v^T$.

4.7 The correlation matrix R of a vector of complex stochastic variables $z = (z_1 \ldots z_N)^T$ with $z_n = x_n + jy_n$ and positive definite covariance matrix C has as its (m, n)th element

$$r_{mn} = \frac{c_{mn}}{\sigma_m \sigma_n},$$

where $\sigma_m = (\text{var}\, z_m)^{1/2}$ with $\text{var}\, z_m = E\left[(z_m - Ez_m)(z_m - Ez_m)^*\right] = c_{mm}$ and $c_{mn} = \text{cov}(z_m, z_n) = E\left[(z_m - Ez_m)(z_n - Ez_n)^*\right]$. Thus, $r_{mn} = 1$ if $m = n$. Show that R is positive definite.

4.8

(a) Let C be an Hermitian $N \times N$ matrix and define the complex $N \times 1$ vector z as

$$z = (0 \ldots 0 \, z_p \, 0 \ldots 0 \, z_q \, 0 \ldots 0)^T$$

with $z_p = x_p + j y_p$. Compute the quadratic form $z^H C z$.

(b) Derive from the result under (a) that if C is positive semidefinite but not positive definite then: $|c_{pq}|^2 \leq c_{pp} c_{qq}$, and if C is positive definite: $|c_{pq}|^2 < c_{pp} c_{qq}$ for $p \neq q$.

4.9 Use the result of Problem 4.8(b) to show that if a diagonal element of an Hermitian and positive semidefinite matrix is equal to zero, then all elements in the same row and the same column as that diagonal element are equal to zero.

4.10 Use the results of Problems 4.7 and 4.8(b) to show that the elements of the correlation matrix R satisfy: $| \, r_{pq} \, | \leq 1$.

4.11 Which of the following matrices could be a covariance matrix of a vector of complex stochastic variables?

(a)

$$\begin{pmatrix} 16 & 4+2j \\ 4-2j & 2 \end{pmatrix}.$$

(b)

$$\begin{pmatrix} 4 & 2+j \\ 2+j & 10 \end{pmatrix}.$$

(c)

$$\begin{pmatrix} 9 & 2+j & 0 \\ 2-j & 5 & 0 \\ 0 & 0 & 0 \end{pmatrix}.$$

(d)

$$\begin{pmatrix} 5 & 3+2j \\ 3-2j & 2 \end{pmatrix}.$$

4.12 Show that the Fisher information matrix is equal to

$$F_\theta = -E \frac{\partial s_\theta}{\partial \theta^T} = -E \frac{\partial s_\theta^T}{\partial \theta}.$$

4.13 Prove Lemma C.2 and Theorem C.9.

4.14 Derive the alternative form of the Fisher information matrix for real and complex parameters:

$$F_\theta = -E \frac{\partial^2 q\,(\omega; \theta)}{\partial \theta^* \partial (\theta^*)^H}.$$

4.15 Suppose that the observations $w = (w_1 \dots w_N)^T$ are made at the measurement points $x = (x_1 \dots x_N)^T$. Furthermore, suppose that $Ew_n = \theta x_n$ where the scalar parameter θ has to be estimated. Derive expressions for the Fisher information if the observations are:

(a) normally distributed with diagonal covariance matrix,

(b) independent and Poisson distributed,

(c) independent and binomially distributed like in Problem 3.3.

(d) independent and exponentially distributed like in Problem 3.7,

and if the observations $(w_1 \dots w_{N-1})^T$ are:

(e) multinomially distributed like in Subsection 3.5.1.

4.16 Suppose that the independent and Poisson distributed observations w_n have been made at the points $x = (x_1 \dots x_N)^T$ with all $x_n \geq 0$. Furthermore, suppose that $Ew_n = \theta_1 x_n + \theta_2$, where $\theta = (\theta_1\ \theta_2)^T$ is the vector of unknown parameters with $\theta_1, \theta_2 > 0$.

(a) Derive an expression for the Fisher information matrix.

(b) Which design maximizes the Fisher information for θ_1 and which that for θ_2?

4.17 Suppose that the observations $w = (w_1 \dots w_N)^T$ are normally distributed with $Ew_n = \theta_1 \cos(\theta_2 x_n)$ and a covariance matrix $\sigma^2 I$, where I is the identity matrix of order N. The unknown parameters are $\theta = (\theta_1\ \theta_2)^T$.

(a) Derive an expression for the Fisher information matrix for θ.

(b) Which measurement points x_n maximize the Fisher information inflow for θ_1 and which that for θ_2?

4.18 Suppose that the observations $w = (w_1 \dots w_N)^T$ are independent and Poisson distributed with $Ew = g(\theta)$.

(a) Derive an expression for the Fisher information inflow for each parameter.

(b) Which measurement points $x_n \geq 0$ maximize the Fisher information inflow for θ_1 and which measurement points maximize the Fisher information inflow for θ_2 if $g_n(\theta) = \theta_1 \exp(-\theta_2 x_n)$ with $\theta_1, \theta_2 > 0$?

4.19 The observations $w = (w_1 \dots w_{N-1})^T$ are multinomially distributed with $w_N = M - \sum_{n=1}^{N-1} w_n$ and expectations $Ew_n = g_n(\theta)$. Show that their Fisher information matrix F_θ is described by

$$\begin{aligned} F_\theta &= \frac{\partial g^T(\theta)}{\partial \theta} \operatorname{diag}\left[\frac{1}{g_1(\theta)} \cdots \frac{1}{g_{N-1}(\theta)}\right] \frac{\partial g(\theta)}{\partial \theta^T} \\ &+ \frac{1}{g_N(\theta)} \sum_n \frac{\partial g_n(\theta)}{\partial \theta} \sum_n \frac{\partial g_n(\theta)}{\partial \theta^T} \end{aligned}$$

with $g(\theta) = [g_1(\theta) \dots g_{N-1}(\theta)]^T$, $g_N(\theta) = M - \sum_n g_n(\theta)$, and $n = 1, \dots, N-1$.

4.20 Suppose that the observations $w = (w_1 \dots w_N)^T$ are independent and linear exponential family distributed with $Ew_n = g_n(\theta)$. Furthermore, let $C(\theta)$ be the covariance matrix of w.

(a) Derive an expression for the contribution of an observation w_{N+1} to the Fisher information for each parameter.

(b) Show that the derived expression agrees with the expressions (4.84) and (4.85).

4.21

(a) Let $w = (w_1 \dots w_N)^T$ be a set of observations. Suppose that the estimator $t(w)$ of the $K \times 1$ vector θ meets the necessary and sufficient conditions for efficiency and unbiasedness. Consider the linear transformation $R\theta$ where R is an $L \times K$ matrix independent of θ. Show that the estimator $r(w) = Rt(w)$ meets the necessary and sufficient conditions for efficient unbiased estimation of the $L \times 1$ vector $\rho(\theta) = R\theta$.

(b) Verify the result under (a) by showing that $r(w)$ is an unbiased estimator of $\rho(\theta)$ and that the covariance matrix of $r(w)$ is equal to the Cramér–Rao lower bound matrix for unbiased estimation of $\rho(\theta)$.

4.22 The observations $w = (w_1 \dots w_N)^T$ are normally distributed with covariance matrix C and $Ew = g(\theta)$, where θ is a $K \times 1$ vector of unknown parameters. Suppose that $g(\theta) = X\theta$, where the elements of the $N \times K$ matrix X are known constants.

(a) Specify the necessary and sufficient conditions for the existence of an efficient unbiased estimator of linear combinations of the parameters θ described by $\rho(\theta) = R\theta$, where R is a known $L \times K$ matrix independent of θ.

(b) From the conditions derived under (a), derive the efficient unbiased estimator of $\rho(\theta)$.

(c) From the expression derived under (b), derive an expression for the efficient unbiased estimator if $\rho(\theta) = \theta$.

4.23 Suppose that the probability (density) function of the observations $w = (w_1 \dots w_N)^T$ is a linear exponential family and that $Ew_n = \theta$ for all n, where θ is a scalar parameter. Show that for all distributions concerned, the efficient unbiased estimator of θ is the average of the w_n if these are uncorrelated and have an equal variance $\sigma^2(\theta)$.

4.24 Suppose that the observations $w = (w_1 \dots w_N)^T$ have a Poisson distribution with $Ew_n = \kappa[\alpha_1 \exp(-\beta_1 x_n) + \alpha_2 \exp(-\beta_2 x_n)]$, where κ is a known scale factor. The parameters $\theta = (\alpha_1\ \alpha_2\ \beta_1\ \beta_2)^T$ are unknown but the ratio $\lambda = \alpha_2/\alpha_1$ is known.

(a) Suppose that the true values of the parameters are $\theta = (1.6\ \ 0.9\ \ 1\ \ 0.6)^T$ and, consequently, that $\lambda = 0.9/1.6 = 0.5625$. Furthermore, suppose that $\kappa = 4000$ and let the measurement points be $x_n = (n-1) \times 0.1$, $n = 1, \dots, 51$. For these values, compute the Cramér–Rao standard deviations for the parameters θ numerically if the knowledge of λ is not used and if it is used, respectively.

(b) What conclusion may be drawn from the results?

4.25 Suppose that the Cramér–Rao lower bound matrix for unbiased estimation of the parameters $\theta = (\theta_1\ \theta_2)^T$ is described by

$$F_\theta^{-1} = \begin{pmatrix} \operatorname{var} \check{t}_1 & \operatorname{cov}(\check{t}_1, \check{t}_2) \\ \operatorname{cov}(\check{t}_1, \check{t}_2) & \operatorname{var} \check{t}_2 \end{pmatrix},$$

where $\check{t} = (\check{t}_1\ \check{t}_2)^T$ is a hypothetical efficient unbiased estimator.

(a) If, instead of θ_1 and θ_2, the parameters $\rho_1 = (\theta_1 + \theta_2)/\sqrt{2}$ and $\rho_2 = (\theta_1 - \theta_2)/\sqrt{2}$ are estimated, then show that $\operatorname{var} \check{t}_1 + \operatorname{var} \check{t}_2 = \operatorname{var} \check{r}_1 + \operatorname{var} \check{r}_2$, where $\operatorname{var} \check{r}_1$ and $\operatorname{var} \check{r}_2$ are the Cramér–Rao variances for ρ_1 and ρ_2, respectively.

(b) Show that if $\operatorname{var} \check{t}_1$ and $\operatorname{var} \check{t}_2$ are approximately equal and the correlation coefficient $\operatorname{cov}(\check{t}_1, \check{t}_2)/[(\operatorname{var} \check{t}_1)^{\frac{1}{2}} (\operatorname{var} \check{t}_2)^{\frac{1}{2}}]$ is close to one, then $\operatorname{var} \check{r}_1 \approx 2 \operatorname{var} \check{t}_1$ and both $\operatorname{var} \check{r}_2$ and $\operatorname{cov}(\check{r}_1, \check{r}_2)$ are small by comparison. Also show that if the correlation coefficient is close to minus one, then $\operatorname{var} \check{r}_2 \approx 2 \operatorname{var} \check{t}_2$ and $\operatorname{var} \check{r}_1$ and $\operatorname{cov}(\check{r}_1, \check{r}_2)$ are small.

4.26 In an experiment, the expectation of the nth observation is described by $Ew_n = g_n(\varphi)$ with real parameters $\varphi = (\varphi_1\ \varphi_2\ \varphi_3)^T$. The 3×3 Cramér–Rao lower bound matrix for unbiased estimation of φ is Ψ. Express the Cramér–Rao variances for the complex parameter vector $\theta = (\varphi_1\ \ \varphi_2 + j\varphi_3\ \ \varphi_2 - j\varphi_3)^T$ in those for φ.

4.27 The frequency response method measures the complex transfer function $H(j\omega)$ of a linear dynamic system as the ratio of an estimate of the complex Fourier coefficient $\gamma_y = \alpha_y + j\beta_y$ of the steady–state response to the estimate of the corresponding Fourier coefficient $\gamma_u = \alpha_u + j\beta_u$ of the periodic input of the system for the angular frequency ω. This is inspired by the exact relation $H(j\omega) = \gamma_y/\gamma_u$. Suppose that the 4×4 matrix Ψ is the Cramér–Rao lower bound matrix for unbiased estimation of $\varphi = (\alpha_y\ \alpha_u\ \beta_y\ \beta_u)^T$. Express the Cramér–Rao lower bound for unbiased estimation of $H(j\omega)$ in γ_u, $H(j\omega)$, and the elements of Ψ.

4.28 In an experiment, the expectation of the nth observation is described by

$$Ew_n = g_n(\theta) = \alpha \cos[\omega(x_n - \tau)] + \beta \sin[\omega(x_n - \tau)],$$

where $n = 1, \ldots, N$ and $\theta = (\alpha\ \beta\ \omega\ \tau)^T$ is the vector of unknown parameters.

(a) Show that this four-parameter model may be reparameterized by three parameters.

(b) Show that the columns of the Jacobian matrix $\partial g(\theta)/\partial \theta^T$ are linearly dependent as opposed to those of the Jacobian matrix for the three-parameter model.

4.29 Show that the elements of the Fisher score vector of the observations in Example 4.18 are linearly dependent stochastic variables if the expectation model may be reparameterized by a number of parameters smaller than the dimension of θ.

4.30 The observations $w = (w_1 \ldots w_N)^T$ are Poisson distributed with

$$Ew_n = \alpha_1 \exp(-\beta_1 x_n) + \alpha_2 \exp(-\beta_2 x_n),$$

where $\theta = (\alpha_1\ \alpha_2\ \beta_1\ \beta_2)^T$ are the unknown parameters.

Derive expressions for all quantities involved in the numerical optimization of the experimental design criterion (4.247)

$$Q = \sum_{k=1}^{4} \lambda_k \, f_{\bar{k}k}$$

if the Newton method would be used. See Subsection 4.10.2.

CHAPTER 5

PRECISE AND ACCURATE ESTIMATION

5.1 INTRODUCTION

In Chapter 4, the concepts estimator and estimate have been introduced. In this chapter, the properties and use of two specific estimators will be discussed. These are the maximum likelihood estimator and the least squares estimator. They have been chosen because, in the author's opinion, these are the most important estimators for practice.

The maximum likelihood estimator is defined in Section 5.2. Properties described in Section 5.3 show that this estimator is not only practically feasible but also optimal in a number of respects. In Section 5.4, maximum likelihood estimation from normally distributed observations is discussed and its relation to least squares estimation is explained. Maximum likelihood estimation from Poisson distributed observations and from multinomially distributed observations are the subjects of Section 5.5 and Section 5.6, respectively. Maximum likelihood estimation from exponential family distributed observations is discussed in Section 5.7. Maximum likelihood estimation is based on the assumptions that the distribution of the observations is known and that the expectation model is correct. A test if the latter assumption has to be rejected is the likelihood ratio test discussed in Section 5.8.

Most of the remainder of the chapter is devoted to least squares estimation. After a general introduction to least squares estimation in Section 5.9, theoretical results on nonlinear least squares estimation are presented in Section 5.10. This is least squares estimation of parameters of expectation models that are nonlinear in one or more of these parameters. Sections 5.11–5.19 are devoted to linear least squares estimation. This is least squares estimation of parameters of expectation models that are linear in all of these parameters.

After a general introduction and a derivation of the main linear least squares results in Sections 5.11–5.13, optimal linear least squares estimation is presented in Sections 5.14 and 5.15. Linear least squares estimation of complex parameters from complex observations is discussed in Section 5.16. Section 5.17 is an intermediate summary of the most important ingredients of the linear least squares theory presented in the Sections 5.11–5.16. Estimators that update the estimate with every additional observation are called recursive estimators. Two examples of recursive linear least squares estimators are presented in Sections 5.18 and 5.19, respectively.

5.2 MAXIMUM LIKELIHOOD ESTIMATION

Central in maximum likelihood estimation is the concept *likelihood function*. Suppose that a set of N observations is available described by

$$w = (w_1 \ldots w_N)^T . \tag{5.1}$$

Furthermore, suppose that the joint probability (density) function of these observations is

$$p(\omega; \theta) , \tag{5.2}$$

where $\theta = (\theta_1 \ldots \theta_K)^T$ is the vector of unknown parameters to be estimated from the observations while the elements of the vector of independent variables

$$\omega = (\omega_1 \ldots \omega_N)^T \tag{5.3}$$

correspond to those of the vector of observations (5.1). Then, the likelihood function of the parameters t given the observations w is defined as

$$p(w; t) . \tag{5.4}$$

This expression has been obtained from (5.2) by substituting w for ω and the vector of independent variables $t = (t_1 \ldots t_K)^T$ for the exact parameters θ. Thus, the independent variables ω have been replaced by observations—that is, by *numbers* —and the supposedly fixed, exact parameters θ have been replaced by *independent variables* t. The likelihood function is, therefore, a function of the parameters t considered as independent variables and is parametric in the observations w.

Using this definition of the likelihood function, we define the maximum likelihood estimator of the parameters θ as follows. The maximum likelihood estimator $\tilde{t}$ of the parameters θ from observations w is that value of t that maximizes the likelihood function. Formally,

$$\boxed{\tilde{t} = \arg\max_t p(w; t)} . \tag{5.5}$$

An interpretation of this definition of the maximum likelihood estimator is that the probability (density) function $p(\omega; \tilde{t})$ generates observations around or equal to w with a higher probability than any other $p(\omega; t)$. Furthermore, the definition shows that the maximum likelihood estimator requires the probability (density) function of the observations and its dependence on the unknown parameters to be known.

Often, it is convenient to use the logarithm of the likelihood function instead of the likelihood function itself. Then, (5.5) may be written

$$\boxed{\tilde{t} = \arg\max_t q(w; t)} , \tag{5.6}$$

where

$$\boxed{q\,(w;t) = \ln p\,(w;t)} \tag{5.7}$$

is called *log-likelihood function*. The expressions (5.5) and (5.6) are equivalent because the logarithmic function is monotonic.

Suppose that $q\,(w;t)$ is differentiable with respect to t. Then, (3.17) shows that

$$\boxed{s_t = \frac{\partial q(w;t)}{\partial t}}\,, \tag{5.8}$$

where s_t is the Fisher score vector with θ replaced by t. From here on, s_t will be the standard notation for the gradient of the log-likelihood function $q(w;t)$ with respect to t. Since the point $t = \tilde{t}$ is a maximum of the log-likelihood function, it is a stationary point. Therefore, a *necessary* condition is that at $t = \tilde{t}$:

$$\boxed{s_t = \frac{\partial q(w;t)}{\partial t} = o}\,, \tag{5.9}$$

where o is the $K \times 1$ null vector. The equations (5.9) are called the *likelihood equations.* The maximum likelihood estimate $\tilde{t}$ is the solution or one of the solutions of the likelihood equations.

If the observations $w = (w_1 \ldots w_N)^T$ are realizations of *independent* stochastic variables, their probability (density) function is described by

$$p\,(\omega;\theta) = p_1\,(\omega_1;\theta)\,p_2\,(\omega_2;\theta)\ldots p_N\,(\omega_N;\theta)\,, \tag{5.10}$$

where $p_n(\omega_n;\theta)$ is the marginal probability (density) function of w_n. Then, the likelihood function is

$$p\,(w;t) = \prod_n p_n\,(w_n;t) \tag{5.11}$$

and the log-likelihood function

$$q\,(w;t) = \sum_n \ln p_n\,(w_n;t) = \sum_n q_n\,(w_n;t)\,. \tag{5.12}$$

This is the form of log-likelihood function most often met in the literature. However, it is a special case since the observations are considered to be independent.

We will now present three examples of this type of log-likelihood function and its use for maximum likelihood estimation.

■ **EXAMPLE 5.1**

Maximum likelihood estimation of straight-line parameters from independent normally distributed observations

Let the observations $w = (w_1 \ldots w_N)^T$ have expectations

$$Ew_n = g_n(\theta) = \theta_1 x_n + \theta_2\,, \tag{5.13}$$

where $\theta = (\theta_1 \; \theta_2)^T$ is the vector of unknown parameters. Furthermore, suppose that the observations are *iind* around the expectation values with variance σ^2. Then, the joint log-probability density function is described by (3.32):

$$q(\omega;\theta) = -\frac{N}{2}\ln 2\pi - N\ln\sigma - \frac{1}{2\sigma^2}\sum_n [\omega_n - g_n(\theta)]^2 \,. \tag{5.14}$$

Hence, the log-likelihood function of $t = (t_1 \; t_2)^T$ is described by

$$q(w;t) = -\frac{N}{2}\ln 2\pi - N\ln\sigma - \frac{1}{2\sigma^2}\sum_n (w_n - t_1 x_n - t_2)^2 \,. \tag{5.15}$$

Then, the likelihood equations are

$$s_{t_1} = \frac{1}{\sigma^2}\sum_n (w_n - t_1 x_n - t_2)x_n = 0 \tag{5.16}$$

and

$$s_{t_2} = \frac{1}{\sigma^2}\sum_n (w_n - t_1 x_n - t_2) = 0. \tag{5.17}$$

Therefore, $\tilde{t}_1$ and $\tilde{t}_2$ are the solutions of the system of linear equations

$$\begin{pmatrix} \sum_n x_n^2 & \sum_n x_n \\ \sum_n x_n & N \end{pmatrix} \begin{pmatrix} \tilde{t}_1 \\ \tilde{t}_2 \end{pmatrix} = \begin{pmatrix} \sum_n w_n x_n \\ \sum_n w_n \end{pmatrix}. \tag{5.18}$$

Furthermore, the Hessian matrix of $q(w;t)$ is equal to

$$-\frac{1}{\sigma^2}\begin{pmatrix} \sum_n x_n^2 & \sum_n x_n \\ \sum_n x_n & N \end{pmatrix} = -\frac{1}{\sigma^2}P^T P \tag{5.19}$$

with

$$P^T = \begin{pmatrix} x_1 & \dots & x_N \\ 1 & \dots & 1 \end{pmatrix}. \tag{5.20}$$

Then, by Theorem C.4, the matrix $P^T P$ in the right-hand member of (5.19) is positive definite if P is nonsingular—that is, if its columns are linearly independent. This condition is met if the x_n are not all equal. Therefore, under this assumption, the Hessian matrix of $q(w;t)$ is negative definite and $\tilde{t}$ is a unique maximum. ■

■ **EXAMPLE 5.2**

Maximum likelihood estimation of straight-line parameters from independent Poisson distributed observations.

Let, again, the observations $w = (w_1 \dots w_N)^T$ have expectations

$$Ew_n = g_n(\theta) = \theta_1 x_n + \theta_2 \,, \tag{5.21}$$

where $\theta = (\theta_1 \; \theta_2)^T$ is the vector of unknown parameters. However, suppose now that the observations are independent and Poisson distributed. Then, the joint log-probability function of the observations w is described by (3.43):

$$\sum_n q(\omega_n;\theta) = \sum_n -g_n(\theta) + \omega_n \ln g_n(\theta) - \ln \omega_n! \,. \tag{5.22}$$

Hence, the log-likelihood function of $t = (t_1\ t_2)^T$ is described by

$$q(w;t) = \sum_n -t_1 x_n - t_2 + w_n \ln(t_1 x_n + t_2) - \ln w_n! \,. \tag{5.23}$$

Then, the likelihood equations are

$$s_{t_1} = \sum_n x_n \left(-1 + \frac{w_n}{t_1 x_n + t_2} \right) = 0 \tag{5.24}$$

and

$$s_{t_2} = \sum_n \left(-1 + \frac{w_n}{t_1 x_n + t_2} \right) = 0. \tag{5.25}$$

Equations (5.24) and (5.25) are nonlinear in t_1 and t_2. As a result, they cannot be transformed into closed-form expressions for $\tilde{t}_1$ and $\tilde{t}_2$ and can be solved iteratively only. ■

■ **EXAMPLE 5.3**

Maximum likelihood estimation of the parameters of a multisinusoidal function from uniformly distributed observations

Suppose that the observations $w = (w_1 \dots w_N)^T$ have multisinusoidal expectations

$$Ew_n = g_n(\theta) = \sum_k \alpha_k \cos(\gamma_k x_n) + \beta_k \sin(\gamma_k x_n) \tag{5.26}$$

with θ the vector of unknown parameters

$$\theta = (\alpha_1 \beta_1 \gamma_1 \dots \alpha_K \beta_K \gamma_K)^T, \tag{5.27}$$

where the α_k and the β_k are amplitudes, and the γ_k are the, not necessarily harmonically related, angular frequencies. The expectations $g_n(\theta)$ are nonlinear in the parameters γ_k. Furthermore, suppose that the observations are independent and identically uniformly distributed around these expectations. Then, the probability density function of the observation w_n is described by

$$\frac{1}{\rho} I_{[g_n(\theta) - \frac{\varrho}{2},\, g_n(\theta) + \frac{\varrho}{2}]}(\omega_n), \tag{5.28}$$

where ρ is the width of the uniform distribution and $I_{\mathcal{A}}(\omega_n)$ is the *indicator function*

$$I_{\mathcal{A}}(\omega_n) = \begin{cases} 1 \text{ if } & \omega_n \in \mathcal{A} = \left[g_n(\theta) - \dfrac{\rho}{2},\, g_n(\theta) + \dfrac{\rho}{2} \right], \\ 0 \text{ if } & \omega_n \notin \mathcal{A} = \left[g_n(\theta) - \dfrac{\rho}{2},\, g_n(\theta) + \dfrac{\rho}{2} \right]. \end{cases} \tag{5.29}$$

In words: $I_{\mathcal{A}}(\omega_n)$ is equal to one if ω_n lies on the closed interval $[g_n(\theta) - \frac{\varrho}{2},\, g_n(\theta) + \frac{\varrho}{2}]$ and vanishes elsewhere. Then, the joint probability density function of the N observations is described by

$$p(\omega;\theta,\rho) = \frac{1}{\rho^N} I_{[g_1(\theta)-\frac{\varrho}{2},\, g_1(\theta)+\frac{\varrho}{2}]}(\omega_1) \dots I_{[g_N(\theta)-\frac{\varrho}{2},\, g_N(\theta)+\frac{\varrho}{2}]}(\omega_N). \tag{5.30}$$

The likelihood function of the parameters t and r, corresponding to θ and ρ, is obtained from (5.30) by substituting the available observations w for the independent variables ω, t for θ, and r for ρ:

$$p(w;t,r) = \frac{1}{r^N} I_{[g_1(t)-\frac{r}{2},\, g_1(t)+\frac{r}{2}]}(w_1) \ldots I_{[g_N(t)-\frac{r}{2},\, g_N(t)+\frac{r}{2}]}(w_N). \tag{5.31}$$

If t and r are such that one or more of the w_n are *not* located on the corresponding interval $[g_n(t) - \frac{r}{2},\, g_n(t) + \frac{r}{2}]$, then the likelihood function $p(w;t,r)$ vanishes. Only if *all* w_n are located on the corresponding intervals, the likelihood function is different from zero and equal to $1/r^N$. The likelihood function thus defined is not differentiable with respect to the parameters t and r. As a result, the procedure to find the maximum by inspecting the solutions of the likelihood equations cannot be applied. Therefore, a different approach is followed.

To make the quantity $1/r^N$ as large as possible, r should be chosen as small as possible while at the same time for all n:

$$g_n(t) - \frac{r}{2} \leq w_n \leq g_n(t) + \frac{r}{2} \tag{5.32}$$

or

$$-\frac{r}{2} \leq w_n - g_n(t) \leq \frac{r}{2}. \tag{5.33}$$

Then, the smallest allowable r is equal to twice the absolutely largest deviation $d_n(t) = w_n - g_n(t)$. Therefore, the maximum likelihood estimator of θ is that value $\tilde{t}$ of t that minimizes the absolutely largest deviation or

$$\tilde{t} = \arg \min_t \max_n |d_n(t)| \tag{5.34}$$

and

$$\tilde{r} = 2 \max_n \left|d_n(\tilde{t})\right|. \tag{5.35}$$

A solution like (5.34) is called a *minimax* estimate. If ρ is known, the maximum likelihood estimate of θ is any $\tilde{t}$ such that

$$-\frac{\rho}{2} \leq w_n - g_n(\tilde{t}) \leq \frac{\rho}{2} \tag{5.36}$$

for all n. Therefore, this estimate is not unique. ∎

Examples 5.1–5.3 illustrate a number of aspects of maximum likelihood estimation. In particular, they show how the model parameters enter the probability (density) function of the observations.

In Example 5.1, the log-likelihood function is differentiable and quadratic in the parameters. As a result, the likelihood equations are a system of linear equations. The maximum likelihood estimate is computed by solving this system. No iterations are required.

In Example 5.2, the log-likelihood function is also differentiable but the likelihood equations are no longer linear in the parameters because the log-likelihood function is not quadratic. Therefore, the maximum likelihood estimates have to be computed iteratively. The example also shows that an expectation model that is linear in the parameters does not necessarily imply that the maximum likelihood estimator of the parameters is closed-form.

Finally, in Example 5.3, the log-likelihood function is not differentiable. However, the optimization of the likelihood function may be reformulated as a nonlinear minimax problem. For the solution of this type of problem specialized iterative numerical methods exist.

5.3 PROPERTIES OF MAXIMUM LIKELIHOOD ESTIMATORS

In this section, properties of maximum likelihood estimators that are important for the purposes of this book will be summarized. Some of the proofs will be omitted since they are mathematically and statistically very demanding and, consequently, a serious treatment is outside the scope of this book. Moreover, in most statistics books, even at the advanced level, these proofs are absent or are treated heuristically.

5.3.1 The invariance property of maximum likelihood estimators

Theorem 5.1 *Suppose that $\hat{t} = (\hat{t}_1 \ldots \hat{t}_K)^T$ is the maximum likelihood estimator of the vector of parameters $\theta = (\theta_1 \ldots \theta_K)^T$. Furthermore, suppose that $\rho(\theta)$ is a, not necessarily one-to-one, scalar function of the parameters. Then, $\rho(\hat{t})$ is the maximum likelihood estimator of $\rho(\theta)$.*

Proof. Define

$$\ell(w; \rho) = \max_{\{t:\ \rho(t)=\rho\}} q(w; t)\,, \tag{5.37}$$

where $\ell(w; \rho)$ is the *log-likelihood function induced by the function* $\rho(t)$. The scalar ρ is equal to the value of $\rho(t)$ for an arbitrary, allowable value of t. Equation (5.37) implies that the largest of the values of $q(w; t)$ in all points t satisfying $\rho(t) = \rho$ is selected. Thus, functions $\rho(\theta)$ not being one-to-one are covered. Furthermore,

$$\max_{\{t:\ \rho(t)=\rho\}} q(w; t) \leq \max_t q(w; t) \tag{5.38}$$

since $\{t : \rho(t) = \rho\}$ is a subset of all allowable values of t. The right-hand member of (5.38) is, by definition, equal to $q(w; \hat{t})$, where $\hat{t}$ is the maximum likelihood estimate of θ. Then,

$$q(w; \hat{t}) = \max_{\{t:\ \rho(t)=\rho(\hat{t})\}} q(w; t) = \ell\left(w; \rho(\hat{t})\right). \tag{5.39}$$

Hence, from (5.37), (5.38), and (5.39):

$$\ell(w; \rho(\hat{t})) \geq \ell(w; \rho). \tag{5.40}$$

This shows that the log-likelihood function induced by $\rho(t)$ is maximized by $\rho = \rho(\hat{t})$. In this sense, $\rho(\hat{t})$ is the maximum likelihood estimator of $\rho(\theta)$. ■

Corollary 5.1 *Suppose that $\rho(\theta)$ is a vector function defined as*

$$\rho(\theta) = [\rho_1(\theta) \ldots \rho_M(\theta)]^T\,. \tag{5.41}$$

Then,

$$\rho(\hat{t}) = \left[\rho_1(\hat{t}) \ldots \rho_M(\hat{t})\right]^T \tag{5.42}$$

is the maximum likelihood estimator of $\rho(\theta)$.

Proof. The result follows directly from Theorem 5.1. ■

The property that $\rho(\hat{t})$ is the maximum likelihood estimator of $\rho(\theta)$ if $\hat{t}$ is the maximum likelihood estimator of θ is called the *invariance property*.

5.3.2 Connection of efficient unbiased estimators and maximum likelihood estimators

Theorem 5.2 *An efficient unbiased estimator is also the maximum likelihood estimator.*

Proof. Suppose that there exists an efficient unbiased estimator $r(w)$ for the vector $\rho(\theta)$. Then, by Theorem 4.5,

$$\frac{\partial \rho(\theta)}{\partial \theta^T} F_\theta^{-1} s_\theta = r(w) - \rho(\theta), \tag{5.43}$$

where F_θ is the Fisher information matrix, which is a function of θ but not of w, and s_θ is the Fisher score vector. Then, for any allowable value of t,

$$\frac{\partial \rho(t)}{\partial t^T} F_t^{-1} s_t = r(w) - \rho(t). \tag{5.44}$$

If t is taken equal to the maximum likelihood estimator $\tilde{t}$, the left-hand members of these equations vanish since

$$s_{\tilde{t}} = o \tag{5.45}$$

where o is the appropriate null vector. Hence,

$$r(w) = \rho(\tilde{t}). \tag{5.46}$$

Since $\tilde{t}$ is the maximum likelihood estimator of θ, $\rho(\tilde{t})$ is, by the Invariance Property, the maximum likelihood estimator of $\rho(\theta)$. This completes the proof. ■

5.3.3 Consistency of maximum likelihood estimators

Under very general conditions, the maximum likelihood estimator is *consistent.*

In Section 4.2, an estimator has been defined as consistent if the number of observations can be chosen such that, if exceeded, the probability that an estimate deviates absolutely more than specified from the true parameter value becomes arbitrarily small. The most general and rigorous proof of consistency of maximum likelihood available in the literature does not require the likelihood function to be differentiable. However, this proof and other, more heuristic or specialized ones assume the observations to be independent and identically distributed around their expectations. This does not mean that a maximum likelihood estimator is necessarily not consistent if the observations are not *iid* around their expectations. Also, in the rigorous proof, there are a number of further conditions to be met, some of which are, unfortunately, quite demanding for the nonmathematician, are difficult to verify in practice, or both. Nevertheless, it is generally agreed in the literature that although not *all* maximum likelihood estimators are consistent, the conditions under which they *are* consistent are very general. Furthermore, an estimator that is not consistent may nevertheless suit the experimenter's ends. This is demonstrated by the following example.

■ **EXAMPLE 5.4**

Maximum likelihood estimation of the amplitude and decay constant of a monoexponential decay model from Poisson distributed observations

Suppose that the expectations of the independent and Poisson distributed observations $w = (w_0 \dots w_{N-1})^T$ are described by

$$Ew_n = g_n(\theta) = \alpha \exp(-\beta x_n), \tag{5.47}$$

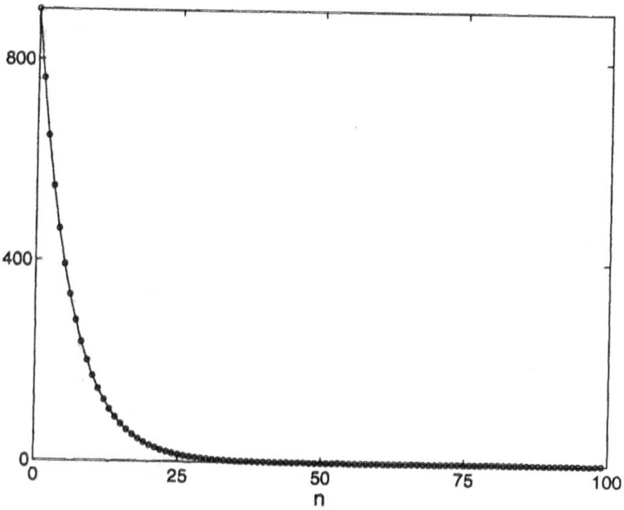

Figure 5.1. Monoexponential expectation model (solid line) and its values at the measurement points (circles). (Example 5.4)

where $\theta = (\alpha\ \beta)^T$. In this numerical example, $\alpha = 900$, $\beta = 1$, $x_n = n/6$, $n = 0, \ldots, N-1$. Three cases are considered. Case 1: $N = 25$, Case 2: $N = 50$, and Case 3: $N = 100$. The observations in Case 1 are taken as the first 25 of the observations used in Case 3 and the observations used in Case 2 as the first 50 of the observations used in Case 3. As a result, the sets of observations used in the three cases are not independent. Figure 5.1. shows the expectation model of the observations and its values at the measurement points. In a simulation experiment, 2500 sets of 100 observations are generated and from each set the maximum likelihood estimates of the parameters α and β are computed for the three cases. Figure 5.2. shows such a set of 100 observations. As a further example, Fig. 5.3. (a) shows a set of 25 observations. From the maximum likelihood estimates $\tilde{t} = (\tilde{a}\ \tilde{b})^T$ computed from these observations, the value of the expectation model at the measurement points is estimated by

$$g_n(\tilde{t}) = g(x_n; \tilde{t}) = \tilde{a}\exp(-\tilde{b}x_n). \tag{5.48}$$

These values are shown in Fig. 5.3.(b). The *residuals* shown in Fig. 5.3.(c) are the deviations of the estimated model from the observations:

$$d_n(\tilde{t}) = w_n - g_n(\tilde{t}). \tag{5.49}$$

Therefore, the residuals are estimated fluctuations of the observations. The numerical method used for computing the maximum likelihood estimates $\tilde{a}$ and $\tilde{b}$ is the Fisher scoring method described in Section 6.5.

Results of the simulation experiment are summarized in Table 5.1. For the three cases, this table shows the mean value and estimated variance and corresponding standard deviation of the maximum likelihood estimates $\tilde{a}$ and $\tilde{b}$. The precision of the mean and the variance follows from their estimated standard deviations that are also shown. The table shows that increasing the number of observations from 25 to 50 reduces the variance of $\tilde{a}$ and $\tilde{b}$ visibly and statistically significantly. However, increasing the number of observations from 50 to

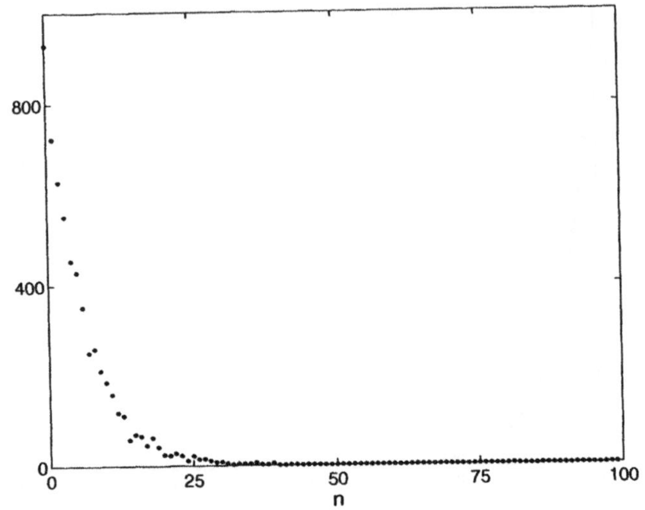

Figure 5.2. Poisson distributed monoexponential decay observations. (Example 5.4)

100 has no such effect: The variances remain the same. They would also remain the same if the number of observations would be further increased. So, in spite of an increasing number of observations, the width of the probability density function of the estimates remains the same. Therefore, the probability that an estimate deviates absolutely from the true value by

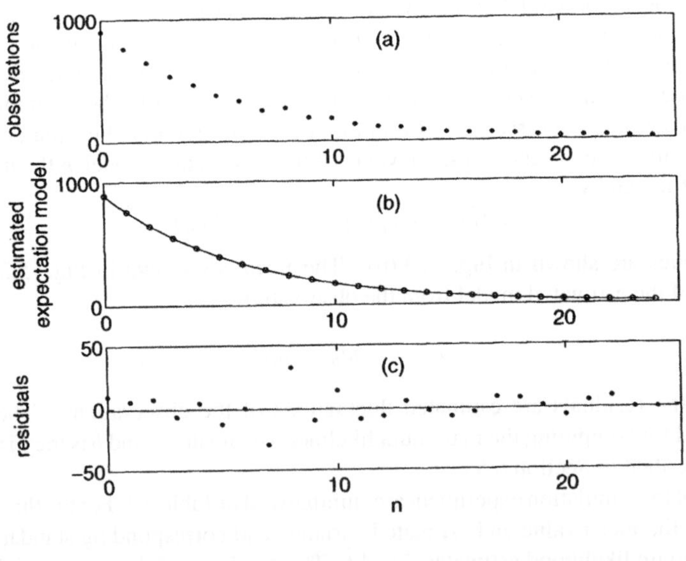

Figure 5.3. (a) Poisson distributed monoexponential decay observations. (b) Maximum likelihood estimate of the expectation model (solid line) and its values at the measurement points (circles). (c) Residuals. (Example 5.4)

more than a specified amount cannot be made arbitrarily small by increasing the number of observations the way described. This shows that the estimator is not consistent. Finally, the mean values of the parameter estimates and their estimated standard deviations show that the bias of the parameter estimates is negligible compared with their standard deviation. ■

Table 5.1. Mean, standard deviation of the mean, variance, standard deviation of the variance, and standard deviation of maximum likelihood estimates of the amplitude and decay constant of a monoexponential obtained from 2500 independent sets of Poisson distributed observations. The true values of the amplitude and the decay constant are 900 and 1, respectively. (Example 5.4)

	Maximum Likelihood Estimator		
	$\tilde{a}$	$\tilde{b}$	Number of Observations
Mean	900.6	1.0004	
Standard Deviation of the Mean	0.3	0.0003	
Variance	286.8	2.47×10^{-4}	25
Standard Deviation of the Variance	8.1	0.07×10^{-4}	
Standard Deviation	16.9	0.016	
Mean	900.4	1.0002	
Standard Deviation of the Mean	0.3	0.0003	
Variance	255.9	1.76×10^{-4}	50
Standard Deviation of the Variance	7.2	0.05×10^{-4}	
Standard Deviation	16.0	0.013	
Mean	900.4	1.0001	
Standard Deviation of the Mean	0.3	0.0003	
Variance	256.4	1.76×10^{-4}	100
Standard Deviation of the Variance	7.2	0.05×10^{-4}	
Standard Deviation	16.0	0.013	

Example 5.4 demonstrates that an estimator that is not consistent may, nevertheless, be adequate for the purposes of the experimenter. Table 5.1. shows that for a number of 25 observations the standard deviations of the maximum likelihood estimates $\tilde{a}$ and $\tilde{b}$ are approximately equal to 2% of the true values of the parameters while the bias is negligible. Depending on the application, such a precision may be more than sufficient. If a higher precision is required, this could be achieved by increasing the number of observations at and between the measurement points. The precision could also be improved by increasing the number of counts in all points—that is, by increasing the amplitude α in (5.47). This improves the relative precision of all observations since, by (3.39), the ratio of the standard deviation of a Poisson distributed observation w_n to its expectation $g_n(\theta)$ is equal to

$$\frac{1}{[g_n(\theta)]^{\frac{1}{2}}} = \frac{1}{[\alpha \exp(-\beta x_n)]^{\frac{1}{2}}}. \tag{5.50}$$

5.3.4 Asymptotic normality of maximum likelihood estimators

Under general conditions, the probability density function of a maximum likelihood estimator tends asymptotically to a normal probability density function with the true parameter values as expectations and the Cramér–Rao lower bound matrix as covariance matrix. That is, for N sufficiently large, the elements of the vector $\tilde{t} - \theta$ are distributed as

$$\mathcal{N}(o;\ F_\theta^{-1}), \tag{5.51}$$

where o is the $K \times 1$ null vector and F_θ^{-1} is the inverse of the Fisher information matrix, that is, the Cramér–Rao lower bound matrix.

Like the proofs of consistency of maximum likelihood estimators, the proofs of asymptotic normality found in the literature apply to observations that are *iid* around their expectations. In addition, a number of other assumptions is made including the assumption that the first and second order derivatives of the log-likelihood function with respect to the parameters exist. Furthermore, the likelihood function has to meet regularity conditions comparable to those made in the derivation of the Cramér–Rao lower bound in Section 4.5.

EXAMPLE 5.5

Estimated probability density function of the maximum likelihood estimates of the amplitude and decay constant of a monoexponential model from Poisson distributed observations

In this example, maximum likelihood estimates $\tilde{a}$ and $\tilde{b}$ of the parameters α and β for Case 1 of Example 5.4 are presented computed from 25 000 simulated independent sets of observations. The estimates are processed as follows. First, they are arranged into a bivariate histogram consisting of 25 by 25 cells and covering the ranges of the estimates found. Next, the probability is computed that an estimate occurs within the boundaries of each cell under the assumption that the estimates have a bivariate normal distribution with the Cramér–Rao lower bound matrix F_θ^{-1} as covariance matrix and the true values α and β as expectations. Expression (4.44) is used for computing the elements of F_θ. The bivariate normal probability density function concerned is described by

$$\frac{1}{2\pi \left(\det F_\theta^{-1}\right)^{1/2}} \exp\left(-\frac{1}{2} \left(\tilde{t} - \theta\right)^T F_\theta \left(\tilde{t} - \theta\right)\right) \tag{5.52}$$

with $\tilde{t} = (\tilde{a}\ \tilde{b})^T$ and $\theta = (\alpha\ \beta)^T$. This is the asymptotic probability density function of the estimates. Then, multiplying the values of this probability density function at the midpoints of the cells by the cell area produces probabilities associated with each cell that are sufficiently accurate for our purposes. These probabilities are next multiplied by 25,000 to find the expected number of estimates in each cell. These numbers are compared with the numbers found in the corresponding cells of the histogram produced by the simulation experiment. The purpose is to find out if the measured histogram and the expected histogram agree. This is done by plotting the expected histogram versus the measured histogram and fitting a straight line to the points thus obtained. Figure 5.4. shows the results. The coordinate pair of each point plotted in this figure consists of the expected and the measured number of estimates in a particular cell. The distribution of the 25,000 estimates $\tilde{t}$ over the 625 cells is the multinomial distribution that has been introduced in Section 3.5. The total number of counts M is here 25,000 while the 625 cells correspond to the N different counters referred to in Section 3.5. Then, by (3.59), the variance of the number of counts in

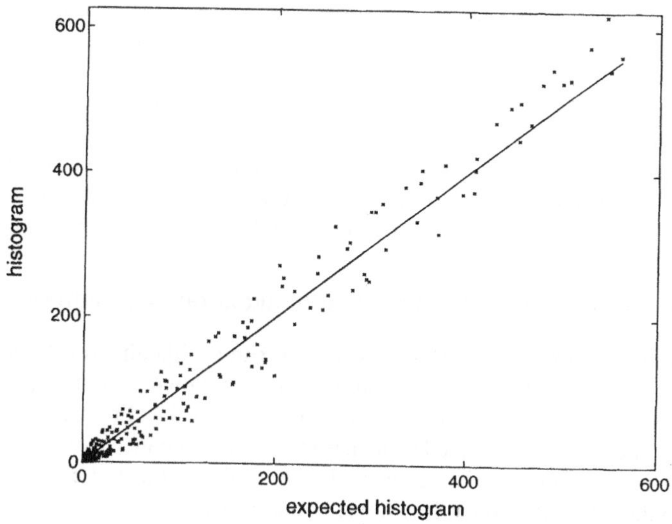

Figure 5.4. Plot of measured histogram of maximum likelihood estimates versus expected histogram. The best fitting straight line through the origin is also shown. (Example 5.5)

the nth cell is equal to $M\varphi_n(1-\varphi_n)$. Since in the experiment all φ_n are less than 0.024, the variance $M\varphi_n(1-\varphi_n)$ is approximately equal to $M\varphi_n$. The quantity $M\varphi_n$ is also the expected value of the multinomial variate, which, in this application, is the expected value of the histogram. Since, by (3.60), the covariance of the contents of the pth cell and those of the qth cell is equal to $-M\varphi_p\varphi_q$, it is small compared with the variances. Therefore, the covariance matrix of the histogram values may be and is taken as the diagonal matrix with diagonal elements $M\varphi_n$. This covariance matrix will be used in the linear least squares procedure used for fitting a straight line to the plot. First, the number of cells taken into consideration is reduced as follows. The ratio of the standard deviation to the expected value in the nth cell is equal to $\sqrt{M\varphi_n}/M\varphi_n = 1/\sqrt{M\varphi_n}$. We now leave out the relatively imprecise cells where this ratio is larger than one or, equivalently, $M\varphi_n$ is smaller than one. The purpose of this reduction is to facilitate the plotting and processing.

Next, a straight line is fitted to the 280 plotted points thus selected. This is done using the best linear unbiased estimator to be introduced in Section 5.14. It produces the straight line shown in Fig. 5.4. The estimated slope of the line is 1.000 with an estimated standard deviation of 0.015. The final conclusion is, therefore, that the agreement of the expected histogram with the measured histogram is striking. ■

In Example 5.5, 25 observations are used for the estimation of the parameters α and β. In spite of this relatively small number, the bivariate probability density function of the maximum likelihood estimates $\tilde{a}$ and $\tilde{b}$ of the parameters α and β is already strikingly similar to a bivariate normal probability density function with the true values α and β as expectations and the relevant Cramér–Rao lower bound F_θ^{-1} as covariance matrix. Thus, it is clear that in the case considered, the asymptotic probability density function may be confidently used.

The following question then arises: Under what conditions, in general, is the use of the asymptotic distribution justified? To the author's knowledge, no such conditions are available in the literature. Therefore, the safest way to investigate the behavior of a maximum likelihood estimator for relatively small numbers of observations is to carry out careful computer simulations of the observations for nominal values of the parameters. From a sufficiently large number of independent sets of simulated observations, the maximum likelihood estimates of the parameters and characteristic statistical properties of these estimates may be computed. In any case, these properties include the mean value and the covariance matrix of the parameter estimates. The dominant role of computer simulation in parameter estimation problems like the ones treated in this book will be returned to in Chapter 6.

5.3.5 Asymptotic efficiency of maximum likelihood estimators

The asymptotic distribution (5.51) shows that maximum likelihood estimators thus distributed are asymptotically efficient unbiased. In the following example, 25,000 sets of simulated observations are used similar to those of Example 5.4. From the maximum likelihood parameter estimates computed from these sets, the covariance matrix of the maximum likelihood estimator concerned is estimated. Next, this estimated covariance matrix is compared to the relevant Cramér–Rao lower bound matrix.

EXAMPLE 5.6

Comparison of the covariance of maximum likelihood estimates for a relatively small number of observations to the Cramér–Rao lower bound matrix.

For Case 1, the estimated covariance matrix of the parameter estimates computed from the 25,000 simulated sets of observations is

$$\begin{pmatrix} 280.4 & 0.18 \\ 0.18 & 0.000241 \end{pmatrix}. \tag{5.53}$$

The estimated standard deviations of $\tilde{a}$ and $\tilde{b}$ are the square roots of the diagonal elements of this matrix, equal to 16.8 and 0.0155, respectively. The corresponding estimated biases may be neglected since they were found to be 0.3 and 0.0001. The relevant Cramér–Rao lower bound matrix is

$$\begin{pmatrix} 282.3 & 0.19 \\ 0.19 & 0.000241 \end{pmatrix}. \tag{5.54}$$

The estimated variance of $\tilde{a}$ is equal to 280.4 and slightly smaller than the corresponding Cramér–Rao variance, which is equal to 282.3. This is caused by the statistical fluctuations in the estimated variance as repeated experiments show. The estimated variance of $\tilde{b}$ and the corresponding Cramér–Rao variance agree.

For Case 2, the corresponding matrices are

$$\begin{pmatrix} 254.3 & 0.13 \\ 0.13 & 0.000175 \end{pmatrix} \tag{5.55}$$

and

$$\begin{pmatrix} 256.7 & 0.14 \\ 0.14 & 0.000174 \end{pmatrix}, \tag{5.56}$$

respectively. The estimated biases of $\tilde{a}$ and $\tilde{b}$ were found to be equal to 0.4 and 0.0002, respectively. They may be neglected since the corresponding standard deviations here are 16.0 and 0.0132. The behavior of the variances is similar to that in Case 1.

Finally, the corresponding results for Case 3 are

$$\begin{pmatrix} 252.8 & 0.14 \\ 0.14 & 0.000172 \end{pmatrix} \tag{5.57}$$

and

$$\begin{pmatrix} 255.1 & 0.14 \\ 0.14 & 0.000171 \end{pmatrix}, \tag{5.58}$$

respectively. The biases were found to be equal to those in Case 2. The conclusions are the same as in Case 2.

The final conclusion is that in these three cases, that is, for 25, 50, and 100 observations, the difference of the properties of the maximum likelihood estimators $\tilde{a}$ and $\tilde{b}$ from those of the hypothetical asymptotically efficient unbiased estimator is minor. Furthermore, the Cramér–Rao lower bound matrix for Case 2 hardly differs from that for Case 3. This shows that the inflow of Fisher information as a result of increasing the number of observations from 50 to 100 is negligible. ■

5.4 MAXIMUM LIKELIHOOD FOR NORMALLY DISTRIBUTED OBSERVATIONS

In Example 5.1, straight-line parameters were estimated from independent, normally distributed observations with equal variance. The linearity of the expectation model in the parameters as well as the particularity of the distribution of the observations, contributed to the simplicity of the resulting estimator. In this section, the general problem of maximum likelihood estimation of parameters of nonlinear expectation models from normally distributed observations with any covariance matrix is discussed. It will, however, be assumed throughout that this covariance matrix does not depend on the unknown parameters of the expectation model.

5.4.1 The likelihood function for normally distributed observations

For normally distributed observations, the log-likelihood function follows directly from (3.31):

$$\boxed{q(w;t) = -\frac{N}{2}\ln 2\pi - \frac{1}{2}\ln\det C - \frac{1}{2}d^T(t)\,C^{-1}\,d(t)}\,, \tag{5.59}$$

where $d(t) = [d_1(t)\dots d_N(t)]^T$ with $d_n(t) = w_n - g_n(t)$. The gradient of this log-likelihood function with respect to the parameters follows directly from the Fisher score vector defined by (3.35) and is described by

$$\boxed{s_t = \frac{\partial g^T(t)}{\partial t}C^{-1}d(t)}\,. \tag{5.60}$$

The likelihood equations for normally distributed observations are obtained by equating (5.60) to the corresponding null vector. The result is

$$s_t = \frac{\partial g^T(t)}{\partial t} C^{-1} d(t) = o\,, \tag{5.61}$$

where o is the $K \times 1$ null vector. For uncorrelated observations, this system of equations becomes

$$\sum_n \frac{1}{\sigma_n^2} \frac{\partial g_n(t)}{\partial t} d_n(t) = o \tag{5.62}$$

and for uncorrelated observations with equal variances we obtain

$$\sum_n \frac{\partial g_n(t)}{\partial t}\, d_n(t) = o. \tag{5.63}$$

A direct consequence of the particular form of the log-likelihood function (5.59) is described by the following theorem:

Theorem 5.3 *Suppose that the observations are jointly normally distributed. Then, the maximum likelihood estimator of the unknown parameters of the expectation model is the weighted least squares estimator with the inverse of the covariance matrix of the observations as weighting matrix.*

Proof. Since under the conditions made both first terms of the log-likelihood function (5.59) do not depend on t, the particular value $\tilde{t}$ of t that maximizes the log-likelihood function is identical to the value $\hat{t}$ of t that *minimizes*

$$J(t) = d^T(t) C^{-1} d(t)\,. \tag{5.64}$$

Since this is the weighted least squares criterion with C^{-1} as weighting matrix, this completes the proof. ■

This theorem does not necessarily imply the truth of statements like: "For normally distributed observations, the least squares estimator is equivalent to the maximum likelihood estimator." For this to be true, the additional condition has to be made that the covariance matrix of the observations does not depend on the unknown parameters. In addition, the presence of the weighting matrix in the least squares criterion has to be mentioned.

If the observations are uncorrelated, the covariance matrix C is described by

$$C = \text{diag}(\sigma_1^2 \ldots \sigma_N^2). \tag{5.65}$$

Then, the least squares criterion (5.64) becomes

$$J(t) = \sum_{n=1}^{N} \frac{1}{\sigma_n^2} d_n^2(t). \tag{5.66}$$

If the observations are uncorrelated and their variances are all equal to σ^2, the least squares criterion (5.66) becomes

$$J(t) = \frac{1}{\sigma^2} \sum_{n=1}^{N} d_n^2(t). \tag{5.67}$$

Below, this criterion and the corresponding estimator will be called ordinary least squares criterion and estimator, respectively

We conclude this section by the following summary:

- If the observations are jointly normally distributed and correlated, the maximum likelihood estimator of the parameters of the expectation model is the weighted least squares estimator with the inverse of the covariance matrix of the observations as weighting matrix.
- If the observations are jointly normally distributed and uncorrelated, the maximum likelihood estimator of the parameters of the expectation model is the least squares estimator with the reciprocals of the variances of the observations as weights.
- If the observations are jointly normally distributed, are uncorrelated, and have equal variances, the maximum likelihood estimator of the parameters of the expectation model is the ordinary least squares estimator.

5.4.2 Properties of maximum likelihood estimators for normally distributed observations

5.4.2.1 Linearity and nonlinearity of the likelihood equations

All that needs to be done to find the maximum likelihood estimate $\hat{t}$ is finding the solution of the likelihood equations that maximizes the log-likelihood function absolutely. However, the problem is that the likelihood equations are nonlinear whenever the expectation model is nonlinear in one or more of the unknown parameters. This is demonstrated in the following example.

■ **EXAMPLE 5.7**

Maximum likelihood estimation of the parameters of a sinusoidal function from normally distributed observations

Suppose that the expectations of the observations $w = (w_1 \ldots w_N)^T$ are described by

$$Ew_n = g_n(\theta) = \alpha \cos(2\pi\gamma x_n) + \beta \sin(2\pi\gamma x_n), \tag{5.68}$$

where $\theta = (\alpha\ \beta\ \gamma)^T$ is the vector of unknown parameters. Also suppose that the w_n are jointly normally distributed with covariance matrix C. Then, in (5.60), the $3 \times N$ matrix $\partial g^T(t)/\partial t$ is described by

$$\frac{\partial g^T(t)}{\partial t} = \begin{pmatrix} \cos(2\pi c x_1) & \cdots & \cos(2\pi c x_N) \\ \sin(2\pi c x_1) & \cdots & \sin(2\pi c x_N) \\ -2\pi x_1[a\sin(2\pi c x_1) - b\cos(2\pi c x_1)] & \cdots & -2\pi x_N[a\sin(2\pi c x_N) - b\cos(2\pi c x_N)] \end{pmatrix} \tag{5.69}$$

and the $N \times 1$ vector $d(t) = w - g(t)$ by

$$d(t) = \begin{pmatrix} w_1 - a\cos(2\pi c x_1) - b\sin(2\pi c x_1) \\ \vdots \\ w_N - a\cos(2\pi c x_N) - b\sin(2\pi c x_N) \end{pmatrix}. \tag{5.70}$$

Substituting these expressions in (5.61) yields a system of three equations in a, b, and c. Closer examination reveals that both first equations of this system are linear in a and b and nonlinear in c. The third equation is nonlinear in a, b, and c. ■

Example 5.7 shows that maximum likelihood estimation of α, β, and γ using the likelihood equations implies finding all solutions of three nonlinear equations in three unknowns and investigating which solution represents the absolute maximum of the likelihood function. Since the equations to be solved are nonlinear, generally no closed-form solution is available. This is the reason why this kind of estimation problem has to be solved by iterative numerical methods. Usually, the numerical method chosen consists of directly maximizing the log-likelihood function instead of solving the likelihood equations for all possible stationary points and selecting the absolute maximum. Suitable numerical optimization methods are described in Chapter 6. On the other hand, the alternative problem of estimating the parameters α and β for a known parameter γ requires the solution of both first equations only. Since these are two linear equations in two unknowns, the solution is closed-form and, generally, unique. Then, the maximum likelihood estimate of the parameters is computed in one step and no iterations are needed. This attractive closed-form and, typically, unique maximum likelihood estimator occurs whenever the observations are normally distributed and, in addition, the expectation model is linear in the parameters. This important special case will be the subject of Subsection 5.4.3. Finally, we mention that parameters like α and β are usually called *linear* parameters and parameters like γ *nonlinear* parameters.

5.4.2.2 *The asymptotic covariance matrix for maximum likelihood estimation from normally distributed observations* An important characteristic of the maximum likelihood estimator of the parameters of nonlinear models is its asymptotic covariance matrix or, equivalently, the pertinent Cramér–Rao lower bound matrix. The Fisher information matrix for normally distributed observations is described by (4.40):

$$F_\theta = \frac{\partial g^T(\theta)}{\partial \theta} C^{-1} \frac{\partial g(\theta)}{\partial \theta^T} . \tag{5.71}$$

As has been shown in Example 4.18, F_θ thus defined is nonsingular if and only if $\partial g(\theta)/\partial \theta^T$ is. Under this condition, the asymptotic covariance matrix of the maximum likelihood estimator of θ from normally distributed observations is described by

$$F_\theta^{-1} = \left(\frac{\partial g^T(\theta)}{\partial \theta} C^{-1} \frac{\partial g(\theta)}{\partial \theta^T} \right)^{-1} . \tag{5.72}$$

5.4.2.3 *Use of the covariance matrix of the observations* The log-likelihood function for normal observations (5.59) depends on the covariance matrix C. Up to now this covariance matrix has been assumed known. If it is not known, the experimenter may decide to use the ordinary least squares estimator since this does not require knowledge of the covariance matrix. Thus, the least squares criterion (5.67) is used instead of the weighted least squares criterion (5.64). The following example shows the differences between both estimators if the observations are correlated.

■ **EXAMPLE 5.8**

Comparison of the efficiency of a maximum likelihood estimator and an ordinary least squares estimator for correlated, normally distributed observations.

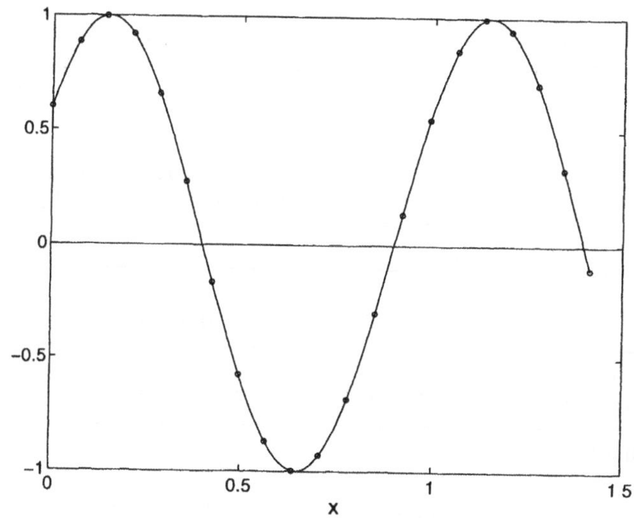

Figure 5.5. Sinusoidal expectation model (solid line) and its values at the measurement points (circles). (Example 5.8)

Suppose that the expectations of the sinusoidal observations $w = (w_1 \ \dots \ w_N)^T$ are described by (5.68) with parameters $\theta = (\alpha \ \beta \ \gamma)^T = (0.6 \ 0.8 \ 1)^T$. Furthermore, let the equidistant measurement points be $x_n = (n-1)\sqrt{2}/20$ with $n = 1, \dots, 21$. Figure 5.5. shows the expectation model and its values at the measurement points. Also suppose that the observations are jointly normally distributed with a covariance matrix C defined by its (p,q)th element

$$c_{pq} = \sigma^2 \rho^{|p-q|} \tag{5.73}$$

with $p, q = 1, \dots, 21$. This means that (5.73) is the covariance of the observations w_p and w_q and that the variance of all observations is equal to σ^2. The quantity ρ is the correlation coefficient of two adjacent observations. The correlation coefficient of any two observations is seen to decrease exponentially with their distance. In this example, $\rho = 0.75$ and $\sigma = 0.05$.

In a simulation experiment, 10 000 sets of observations $w = (w_1 \dots w_{21})^T$ are generated. From each set, the maximum likelihood estimates of the parameters α, β and γ are estimated by numerically minimizing the weighted least squares criterion (5.64) with the matrix C defined by (5.73). The numerical method used in this example is the Gauss–Newton method described in Section 6.7. From the 10 000 estimates of the parameters thus obtained, the variances of the maximum likelihood estimator are estimated. From the same sets of observations, the parameters are estimated by numerically minimizing the ordinary least squares criterion (5.67). The variances of the ordinary least squares estimator are estimated from the parameter estimates so obtained. Table 5.2. shows the Cramér–Rao variances and the estimated variances of both estimators.

The table shows that the maximum likelihood estimator closely approximates the Cramér–Rao lower bound. The ordinary least squares estimator, on the other hand, is less precise. These results show how, with correlated, normally distributed observations, a priori knowl-

Table 5.2. The Cramér–Rao variances for the parameters of a sinusoidal function. The observations are normally distributed and correlated. The estimated variances of the maximum likelihood estimator and those of the ordinary least squares estimator of the same parameters are also shown. (Example 5.8)

	Cramér–Rao Variances and Estimated Variances of Estimators of Sinusoid Parameters		
	α	β	γ
Cramér–Rao Variances	1.18×10^{-3}	7.51×10^{-4}	4.80×10^{-5}
Variance Maximum Likelihood Estimator	1.19×10^{-3}	7.55×10^{-4}	4.75×10^{-5}
Variance Ordinary Least Squares Estimator	1.43×10^{-3}	9.95×10^{-4}	5.87×10^{-5}

Table 5.3. Estimated bias, standard deviation, and efficiency of maximum likelihood estimates and ordinary least squares estimates of the parameters of a sinusoidal function from correlated, normally distributed observations. The efficiencies have been computed from the estimated mean squared errors and the Cramér–Rao variances. (Example 5.8)

		Estimated Bias, Standard Deviation, and Efficiency of Estimators of Sinusoid Parameters		
		α	β	γ
Maximum Likelihood Estimator	Bias	−0.0007	−0.0007	−0.00007
	Standard Deviation	0.0345	0.0274	0.00689
	Efficiency	0.99	0.99	1.01
Ordinary Least Squares Estimator	Bias	0.0005	−0.0014	−0.00022
	Standard Deviation	0.0378	0.0316	0.00766
	Efficiency	0.83	0.75	0.82

edge of the covariance matrix of the observations may be used to construct the maximum likelihood estimator and to attain the Cramér–Rao lower bound.

We next consider the bias of both estimators. Table 5.3. shows their estimated bias and, in addition, their estimated standard deviation and efficiency. The bias has been estimated by subtracting the true value of the parameter concerned from the average of the 10 000 estimates. The standard deviations are the square roots of the estimated variances shown in Table 5.2. As estimates of the mean squared errors needed for the computation of the efficiencies, the sums of the estimated variance and the square of the estimated bias have been taken. Table 5.3. shows that in all cases the maximum likelihood estimator and the ordinary least squares estimator are very accurate in the sense that their bias is much smaller than their standard deviation. ■

If the covariance matrix is unknown, one might wonder if it can be estimated from the observations along with the parameters of the expectation model. Unfortunately, this is, typically, impossible. To see this consider the general form of the covariance matrix of a number of N correlated observations:

$$C = \begin{pmatrix} c_{11} & c_{12} & \cdots & c_{1N} \\ c_{12} & c_{22} & & \\ \vdots & & \ddots & \\ c_{1N} & & & c_{NN} \end{pmatrix}. \tag{5.74}$$

This matrix contains $N \times (N+1)/2$ different unknown elements. Therefore, if the expectation model has K parameters, a total of $K + N \times (N+1)/2$ parameters would have to be estimated uniquely from N observations, which is impossible. In Example 5.8, however, the covariance matrix C is characterized by only two parameters: σ and ρ. Then, the total number of unknown parameters appearing in the normal log-likelihood function (5.59) is equal to $K + 2$. Therefore, if C has a known structure defined by a fixed number of parameters, such as (5.73), these may be estimated along with the target parameters if N is sufficiently large.

5.4.2.4 Efficient unbiased estimators The Fisher score vector of normally distributed observations is described by (3.35):

$$s_\theta = \frac{\partial g^T(\theta)}{\partial \theta} C^{-1}[w - g(\theta)]. \tag{5.75}$$

Their Fisher information matrix is described by (4.40):

$$F_\theta = \frac{\partial g^T(\theta)}{\partial \theta} C^{-1} \frac{\partial g(\theta)}{\partial \theta^T}. \tag{5.76}$$

The necessary and sufficient condition for nonsingularity of F_θ thus defined has been derived in Example 4.18. This condition is that $\partial g(\theta)/\partial \theta^T$ is nonsingular. It is assumed to be met in the following discussion of efficient unbiased estimation from normally distributed observations. By Theorem 4.5, an estimator $r(w)$ is an efficient unbiased estimator for $\rho(\theta)$ if and only if

$$\frac{\partial \rho(\theta)}{\partial \theta^T} F_\theta^{-1} s_\theta = r(w) - \rho(\theta). \tag{5.77}$$

Combination of (5.75), (5.76), and (5.77) yields the necessary and sufficient conditions for $t(w)$ to be efficient unbiased if the observations are normally distributed.

An expectation model is called *linear (expectation) model* if

$$Ew = g(\theta) = X\theta, \tag{5.78}$$

where $g_n(\theta) = g(x_n; \theta)$ and X is a known $N \times K$ matrix with elements independent of θ. The linear model implies that the expectation of the nth observation is described by

$$Ew_n = g(x_n; \theta) = x_{n1}\theta_1 + \cdots + x_{nK}\theta_K = x_n^T\theta \tag{5.79}$$

with

$$x_n = (x_{n1} \ldots x_{nK})^T. \tag{5.80}$$

First, consider estimating $\rho(\theta) = \theta$. Then, $\partial\rho(\theta)/\partial\theta^T = I$, and $r(w) = t(w)$. Furthermore, for $g(\theta) = X\theta$,

$$\frac{\partial g(\theta)}{\partial \theta^T} = X. \tag{5.81}$$

Since, by assumption, X is nonsingular, $N \geq K$. Substituting (5.81) in (5.75) and in (5.76) yields

$$s_\theta = X^T C^{-1}(w - X\theta) \tag{5.82}$$

and

$$F_\theta = X^T C^{-1} X. \tag{5.83}$$

As a result, the necessary and sufficient conditions (5.77) become

$$(X^T C^{-1} X)^{-1} X^T C^{-1}(w - X\theta) = t(w) - \theta. \tag{5.84}$$

Then,

$$t(w) = (X^T C^{-1} X)^{-1} X^T C^{-1} w. \tag{5.85}$$

This is the efficient unbiased estimator of the parameters of the linear model from normally distributed observations.

Next, consider estimating a linear combination $\rho(\theta) = v^T\theta = v_1\theta_1 + \cdots + v_K\theta_K$ where the v_k are known scalars. Then, the necessary and sufficient condition for an estimator $r(w)$ to be efficient unbiased is

$$v^T(X^T C^{-1} X)^{-1} X^T C^{-1}(w - X\theta) = r(w) - v^T\theta\,, \tag{5.86}$$

where $v = (v_1 \ldots v_K)^T$. Then, after some rearrangements, we obtain

$$r(w) = v^T t(w)\,, \tag{5.87}$$

where $t(w)$ is the efficient unbiased estimator (5.85) of θ. This means that the efficient unbiased estimator of a linear combination of the parameters of the linear model from normally distributed observations is equal to the linear combination of the efficient unbiased estimators of the individual parameters.

Typically, no efficient unbiased estimator exists for nonlinear parameters from normally distributed observations. Fortunately, as we have seen, maximum likelihood estimators may already exhibit their asymptotic properties for a limited number of observations. Then, they are, in fact, efficient unbiased.

■ **EXAMPLE 5.9**

A linear and a nonlinear expectation model

In Example 5.8, the expectations of the observations are described by

$$Ew_n = g_n(\theta) = \alpha\cos(2\pi\gamma x_n) + \beta\sin(2\pi\gamma x_n). \tag{5.88}$$

This implies that the model is nonlinear in the parameters if the vector of unknown parameters is $\theta = (\alpha\ \beta\ \gamma)^T$ since the model is nonlinear in the parameter γ. Then, an efficient unbiased estimator of θ may be shown not to exist, but the results of Example 5.8 show how close the maximum likelihood estimator may approximate the efficient unbiased estimator. The expectation model is linear if γ is known and $\theta = (\alpha\ \beta)^T$ is the vector of unknown parameters. Then, (5.85) is the efficient unbiased estimator of θ. ■

By Theorem 5.2, an efficient unbiased estimator is also the maximum likelihood estimator. This will be verified in the next subsection for estimation of the parameters of the linear model from normally distributed observations.

5.4.3 Maximum likelihood estimation of the parameters of linear models from normally distributed observations

5.4.3.1 The general solution and its properties Suppose that the observations $w = (w_1 \dots w_N)^T$ are jointly normally distributed with known, positive definite covariance matrix C. Also suppose that $Ew = g(\theta) = X\theta$. Then, the maximum likelihood estimate $\tilde{t}$ of θ from observations w is a solution of the likelihood equations (5.61):

$$\frac{\partial g^T(t)}{\partial t} C^{-1}[w - g(t)] = X^T C^{-1}(w - Xt) = o. \tag{5.89}$$

Hence,

$$\tilde{t} = (X^T C^{-1} X)^{-1} X^T C^{-1} w, \tag{5.90}$$

where $X^T C^{-1} X$ is assumed nonsingular. Since the covariance matrix C and, therefore, C^{-1} are positive definite, nonsingularity implies that X is nonsingular. See Theorem C.4 and Corollary C.1. The matrix X is singular if its columns are linearly dependent. For example, this occurs if the number of parameters K exceeds the number of observations N.

The expectation of $\tilde{t}$ is described by

$$\begin{aligned} E\tilde{t} &= E\left[(X^T C^{-1} X)^{-1} X^T C^{-1} w\right] \\ &= (X^T C^{-1} X)^{-1} X^T C^{-1} E w \\ &= (X^T C^{-1} X)^{-1} X^T C^{-1} X\theta = \theta, \end{aligned} \tag{5.91}$$

which shows that $\tilde{t}$ is unbiased.

Furthermore,

$$\tilde{t} - E\tilde{t} = (X^T C^{-1} X)^{-1} X^T C^{-1}(w - Ew). \tag{5.92}$$

Therefore,

$$\begin{aligned} \operatorname{cov}(\tilde{t}, \tilde{t}) &= E[(\tilde{t} - E\,\tilde{t})(\tilde{t} - E\tilde{t})^T] \\ &= (X^T C^{-1} X)^{-1} X^T C^{-1} \\ &\quad \times E[(w - Ew)(w - Ew)^T] C^{-1} X (X^T C^{-1} X)^{-1} \\ &= (X^T C^{-1} X)^{-1} X^T C^{-1} C C^{-1} X (X^T C^{-1} X)^{-1} \\ &= (X^T C^{-1} X)^{-1}. \end{aligned} \tag{5.93}$$

On the other hand, if the observations are normally distributed, the Fisher information matrix for θ is described by (4.40):

$$F_\theta = \frac{\partial g^T(\theta)}{\partial \theta} C^{-1} \frac{\partial g(\theta)}{\partial \theta^T}. \tag{5.94}$$

For the linear model, this is equal to $F_\theta = X^T C^{-1} X$. Therefore, the Cramér–Rao lower bound matrix is described by

$$F_\theta^{-1} = (X^T C^{-1} X)^{-1}. \tag{5.95}$$

Comparison of (5.93) and (5.95) shows that $\tilde{t}$ attains the Cramér–Rao lower bound matrix and does so for *any* number of observations $N \geq K$. These results agree with those of Subsection 5.4.2.4, which were obtained in a different way.

Finally, (5.90) shows that $\tilde{t}$ is a linear combination of the jointly normally distributed observations $w = (w_1 \dots w_N)^T$. Since any linear combination of normally distributed stochastic variables is normally distributed, the elements of $\tilde{t}$ are jointly normally distributed with expectation θ and covariance matrix $(X^T C^{-1} X)^{-1}$.

5.4.3.2 Uncorrelated observations with equal variance Suppose next that the normally distributed observations are uncorrelated and have an equal variance σ^2. Then, their covariance matrix is

$$C = \sigma^2 I_N , \tag{5.96}$$

where I_N is the identity matrix of order N. Then, the estimator (5.90) of the parameters θ of the expectations $Ew = X\theta$ simplifies to

$$\tilde{t} = (X^T X)^{-1} X^T w \tag{5.97}$$

with covariance matrix

$$\operatorname{cov}(\tilde{t}, \tilde{t}) = \sigma^2 (X^T X)^{-1}. \tag{5.98}$$

Next, consider maximum likelihood estimation of both the parameters θ and the standard deviation σ. For these parameters and the covariance matrix (5.96), the normal log-likelihood function (5.59) becomes

$$q(w; t, s) = -\frac{N}{2} \ln 2\pi - N \ln s - \frac{1}{2s^2} (w - Xt)^T (w - Xt) \tag{5.99}$$

where the variable s corresponds to σ. This expression shows that the maximum likelihood estimator $\tilde{t}$ is not affected by the simultaneous estimation of σ and is described by

$$\tilde{t} = (X^T X)^{-1} X^T w. \tag{5.100}$$

The likelihood equation with respect to s is

$$-\frac{N}{s} + \frac{1}{s^3} (w - Xt)^T (w - Xt) = 0 \tag{5.101}$$

and, therefore,

$$\tilde{s}^2 = \tilde{s^2} = \frac{1}{N} (w - X\tilde{t})^T (w - X\tilde{t}). \tag{5.102}$$

We will now compute the expectation of the latter estimator. Here, the fluctuations of the observations are $d(\theta) = w - Ew = w - X\theta$. Furthermore, under the assumptions made, $\operatorname{cov}[d(\theta), d(\theta)] = \operatorname{cov}(w, w) = \sigma^2 I_N$. See Subsection 2.4.1. Then, substituting (5.100) for $\tilde{t}$ and rearranging shows that

$$w - X\tilde{t} = (I_N - P)\, d(\theta), \tag{5.103}$$

where the $N \times N$ matrix $P = X(X^T X)^{-1} X^T$. Therefore, P is symmetric, $PP = P$, and $PX\theta = X\theta$. Substituting (5.103) in (5.102) and using the properties of P yields

$$E\tilde{s^2} = \frac{1}{N} E[d^T(\theta)\, (I_N - P)\, d(\theta)]. \tag{5.104}$$

In this expression,

$$E[d^T(\theta)\, I_N\, d(\theta)] = N\sigma^2 \tag{5.105}$$

since $Ed(\theta) = o$ and $\text{var}\, d_n(\theta) = \sigma^2$ for all n. Furthermore,

$$E[d^T(\theta)\, P\, d(\theta)] = \sum_{n=1}^{N} p_{nn}\sigma^2 = \sigma^2 \text{tr} P = K\sigma^2 \tag{5.106}$$

since the elements of $d(\theta)$ are uncorrelated and, by Theorem B.3,

$$\text{tr}\, P = \text{tr}\, X(X^T X)^{-1} X^T = \text{tr}(X^T X)^{-1} X^T X = \text{tr}\, I_K = K\,. \tag{5.107}$$

Finally, substituting (5.105) and (5.106) in (5.104) yields

$$E\tilde{s^2} = (1 - \frac{K}{N})\sigma^2. \tag{5.108}$$

This shows that the estimator $\tilde{s^2}$ is a biased estimator of σ^2 and that its bias is equal to

$$-\frac{K}{N}\sigma^2. \tag{5.109}$$

Also, (5.102) and (5.108) show that

$$\frac{1}{N-K}(w - X\tilde{t})^T(w - X\tilde{t}) \tag{5.110}$$

is an unbiased estimator of σ^2. It is, however, not a maximum likelihood estimator since it does not maximize the likelihood function.

5.5 MAXIMUM LIKELIHOOD FOR POISSON DISTRIBUTED OBSERVATIONS

The subject of this section is maximum likelihood estimation of parameters of expectation models from independent Poisson distributed observations. Suppose that $w = (w_1 \ldots w_N)^T$ is the vector of observations and $\theta = (\theta_1 \ldots \theta_K)^T$ the vector of unknown parameters. Then, the log-probability function of the observations is defined by (3.43):

$$q(\omega; \theta) = \sum_n -g_n(\theta) + \omega_n \ln g_n(\theta) - \ln \omega_n!\,. \tag{5.111}$$

It shows that the log-likelihood function of the parameters is

$$\boxed{q(w;t) = \textstyle\sum_n -g_n(t) + w_n \ln g_n(t) - \ln w_n!}\,, \tag{5.112}$$

where $t = (t_1 \ldots t_K)^T$. The gradient of this expression is

$$s_t = \sum_n \frac{\partial g_n(t)}{\partial t}\left(\frac{w_n}{g_n(t)} - 1\right). \tag{5.113}$$

Alternatively, s_t may be obtained directly from the relevant Fisher score vector s_θ defined by (3.47):

$$\boxed{s_t = \frac{\partial g^T(t)}{\partial t} C^{-1} d(t)} \tag{5.114}$$

combined with expression (3.44) for the covariance matrix of the observations:

$$C = \operatorname{diag} g(t). \tag{5.115}$$

Simple calculations show that (5.113) and (5.114) are identical. The corresponding system of K likelihood equations is

$$s_t = \frac{\partial g^T(t)}{\partial t} C^{-1} d(t) = o\,. \tag{5.116}$$

The following example shows that these likelihood equations are nonlinear in the parameters t even if the expectation model is linear.

■ **EXAMPLE 5.10**

The likelihood equations for estimating straight-line parameters from Poisson distributed observations

In Example 5.2, the likelihood equations have been derived for estimating the parameters of the straight-line expectation model from Poisson distributed observations. With this model, the expectations of the observations are described by

$$Ew_n = g_n(\theta) = \theta_1 x_n + \theta_2. \tag{5.117}$$

The likelihood equations are (5.24) and (5.25). Clearly, these equations are nonlinear although the expectation model (5.117) is linear. ■

Numerical methods for maximizing nonlinear, nonquadratic log-likelihood functions are described in Chapter 6.

5.6 MAXIMUM LIKELIHOOD FOR MULTINOMIALLY DISTRIBUTED OBSERVATIONS

In this section, maximum likelihood estimation of the parameters of expectation models from multinomially distributed observations is discussed. The log-probability function is defined by (3.67):

$$q(\omega;\theta) = \ln M! - M \ln M - \sum_{n=1}^{N} \ln \omega_n! + \sum_{n=1}^{N} \omega_n \ln g_n(\theta) \tag{5.118}$$

with $\omega_N = M - \sum_{n=1}^{N-1} \omega_n$ and $g_N(\theta) = M - \sum_{n=1}^{N-1} g_n(\theta)$. It shows that the log-likelihood function of the parameters t is

$$\boxed{q(w;t) = \ln M! - M \ln M - \textstyle\sum_{n=1}^{N} \ln w_n! + \sum_{n=1}^{N} w_n \ln g_n(t)} \tag{5.119}$$

with $w_N = M - \sum_{n=1}^{N-1} w_n$ and $g_N(t) = M - \sum_{n=1}^{N-1} g_n(t)$. The gradient of this expression follows directly from (3.68):

$$s_t = \sum_{n=1}^{N-1} \frac{\partial g_n(t)}{\partial t} \left(\frac{w_n}{g_n(t)} - \frac{w_N}{g_N(t)} \right). \tag{5.120}$$

Alternatively, s_t may be obtained from (3.73):

$$\boxed{s_t = \frac{\partial g^T(t)}{\partial t} C^{-1} d(t)} \tag{5.121}$$

combined with expression (3.64) for the covariance matrix of the observations

$$C = \frac{1}{M} \begin{pmatrix} g_1(M - g_1) & -g_1 g_2 & \cdots & -g_1 g_{N-1} \\ -g_1 g_2 & g_2(M - g_2) & \cdots & -g_2 g_{N-1} \\ \vdots & & \ddots & \vdots \\ -g_1 g_{N-1} & \cdots & \cdots & g_{N-1}(M - g_{N-1}) \end{pmatrix} \tag{5.122}$$

with closed-form inverse (3.71):

$$C^{-1} = \frac{1}{g_N} \begin{pmatrix} \frac{g_N}{g_1} + 1 & 1 & 1 & \cdots & 1 \\ 1 & \frac{g_N}{g_2} + 1 & 1 & \cdots & 1 \\ \vdots & \vdots & \ddots & & \vdots \\ 1 & 1 & 1 & & \frac{g_N}{g_{N-1}} + 1 \end{pmatrix} \tag{5.123}$$

where, both in (5.122) and (5.123), $g_n = g_n(t)$. Straightforward calculations show that (5.120) and (5.121) are identical. The corresponding system of K likelihood equations is

$$s_t = \frac{\partial g^T(t)}{\partial t} C^{-1} d(t) = o\,. \tag{5.124}$$

5.7 MAXIMUM LIKELIHOOD FOR EXPONENTIAL FAMILY DISTRIBUTED OBSERVATIONS

In Section 3.6, exponential families of distributions have been introduced as all distributions that can be described as

$$p(\omega; \theta) = \alpha(\theta)\beta(\omega)\exp\left(\gamma^T(\theta)\delta(\omega)\right), \tag{5.125}$$

where $\alpha(\theta)$ is a scalar function of the elements of θ, $\beta(\omega)$ is a scalar function of the elements of the vector ω, $\gamma(\theta)$ is an $L \times 1$ vector of functions of the elements of θ, and $\delta(\omega)$ is an $L \times 1$ vector of functions of the elements of ω. The log-likelihood function corresponding to (5.125) is

$$q(w; t) = \ln\alpha(t) + \ln\beta(w) + \gamma^T(t)\delta(w) \tag{5.126}$$

while that for *linear* exponential families is

$$\boxed{q(w; t) = \ln\alpha(t) + \ln\beta(w) + \gamma^T(t)w}\,. \tag{5.127}$$

For linear exponential families, the Fisher score vector is described by expression (3.99):

$$s_\theta = \frac{\partial g^T(\theta)}{\partial \theta} C^{-1} d(\theta). \tag{5.128}$$

Then, by (5.8), the general expression for the gradient of $q(w;t)$ for linear exponential families of distributions is

$$\boxed{s_t = \frac{\partial g^T(t)}{\partial t} C^{-1} d(t)} \tag{5.129}$$

where, generally, the covariance matrix of the observations C depends on t. The corresponding likelihood equations are described by

$$s_t = \frac{\partial g^T(t)}{\partial t} C^{-1} d(t) = o\,. \tag{5.130}$$

This expression is identical to the likelihood equations (5.61) for normal, (5.116) for Poisson, and (5.124) for multinomial observations. This is, of course, no surprise since these three distributions are linear exponential families. It illustrates the usefulness of the concept linear exponential family. The Fisher score vector, the likelihood equations, and the Fisher information matrix for a particular family are easily derived from the generic expressions. The fact that many distributions frequently occurring in practice *are* linear exponential families contributes further to the importance of this concept. An alternative and simpler expression for the gradient of $q(w;t)$ for linear exponential families follows directly from (3.108):

$$\boxed{s_t = \frac{\partial \gamma^T}{\partial t} d(t)}\,. \tag{5.131}$$

We have seen earlier that, under general conditions, the asymptotic distribution of maximum likelihood estimators is the normal distribution with the Cramér–Rao lower bound matrix as covariance matrix. For linear exponential families, the general expression for the Cramér–Rao lower bound matrix is (4.216):

$$F_\theta^{-1} = \left(\frac{\partial g^T(\theta)}{\partial \theta} C^{-1} \frac{\partial g(\theta)}{\partial \theta^T} \right)^{-1}, \tag{5.132}$$

where, generally, the elements of the covariance matrix C of the observations are functions of the elements of θ. Thus, the relatively simple expression (5.132) constitutes the asymptotic covariance matrix of the maximum likelihood estimator for all linear exponential family distributed observations.

5.8 TESTING THE EXPECTATION MODEL: THE LIKELIHOOD RATIO TEST

5.8.1 Model testing for arbitrary distributions

Maximum likelihood estimation of expectation model parameters is based on two hypotheses. The first is that the distribution of the observations is known. The second is that the expectation model is correct. The subject of this section is the statistical testing of the latter hypothesis.

Concluding from the available observations that the chosen expectation model is correct is not possible since models can always be found that perfectly fit, for example, polynomial models of a sufficiently high degree. However, what *can* be tested is whether there is reason to reject the chosen model. For that purpose, in this section a test is described that is closely connected with the maximum likelihood estimator: the *likelihood ratio test*.

Suppose that the probability (density) function of the observations is parametric in the parameters $\theta = (\theta_1 \dots \theta_K)^T$. Generally, the purpose of the likelihood ratio test is to test *constraints* on these parameters. A well-known example is the *equality* constraint. Then, the parameters θ_k, $k = 1, \dots, K'$ with $1 \leq K' \leq K$, are assumed to be known—that is, equal to specified values.

Suppose that from the set of observations w two sets of maximum likelihood estimates of the parameters are computed: one set *restricted* by constraints and one set *unrestricted*, that is, without the constraints. Furthermore, let $\tilde{t}$ and $\tilde{\tilde{t}}$ be the restricted and the unrestricted maximum likelihood estimates, respectively. Then, the *log-likelihood ratio* ℓ is defined as

$$\boxed{\ell = \ln \frac{p(w;\tilde{t})}{p(w;\tilde{\tilde{t}})} = q(w;\tilde{t}) - q(w;\tilde{\tilde{t}})}\,, \tag{5.133}$$

where $q(w;t)$ is the log-likelihood function of t. Thus, if the constraints are equality constraints, $\tilde{t} = (\theta_1 \dots \theta_{K'}\, \tilde{t}_{K'+1} \dots \tilde{t}_K)^T$ and $\tilde{\tilde{t}} = (\tilde{\tilde{t}}_1 \dots \tilde{\tilde{t}}_K)^T$. The log-likelihood ratio is negative since the constrained parameter values are a subset of the unconstrained parameter values. The quantity ℓ is a stochastic variable since it is a function of the stochastic variables w. If we would know the distribution of ℓ, we would know which values of ℓ were improbable. Thus, we could use ℓ as a test statistic to test the *null hypothesis* that the parameters are restricted as specified as opposed to being arbitrary. In this respect, the following theorem is helpful.

Theorem 5.4 (Wilks) *Let in a parameter estimation problem ℓ be the log-likelihood ratio and suppose that the number of parameters is K and the number of constraints is K' with $1 \leq K' \leq K$. Then, under the null hypothesis that the parameters are constrained as specified, the quantity -2ℓ is asymptotically chi-square distributed with K' degrees of freedom.*

Proof. A reference to a proof of this theorem is given in Section 5.20. ∎

With respect to the theorem, the following remarks may be made. First, the log-likelihood functions involved belong to the same parametric family. The restricted log-likelihood function is a special case of the unrestricted log-likelihood function. Second, the result is asymptotic, that is, the maximum likelihood estimators involved must possess their asymptotic properties.

■ EXAMPLE 5.11

Testing the absence of a background function

In many practical applications, the observations are made chronologically and the measurement points x_n are time instants. Then, as a result of nonstationarities, a background function such as $\alpha x_n + \beta$ may be present in the observations. It is worthwhile to test if it may be considered absent because estimating the nuisance background parameters $(\alpha\ \beta)^T$ influences the precision of the estimates of the target parameters adversely. Suppose that Poisson distributed observations $w = (w_1 \dots w_N)^T$ have been made with expectations

$$Ew_n = \gamma\theta_1 \exp\left(-\frac{1}{2}\left(\frac{x_n - \theta_2}{\theta_3}\right)^2\right), \tag{5.134}$$

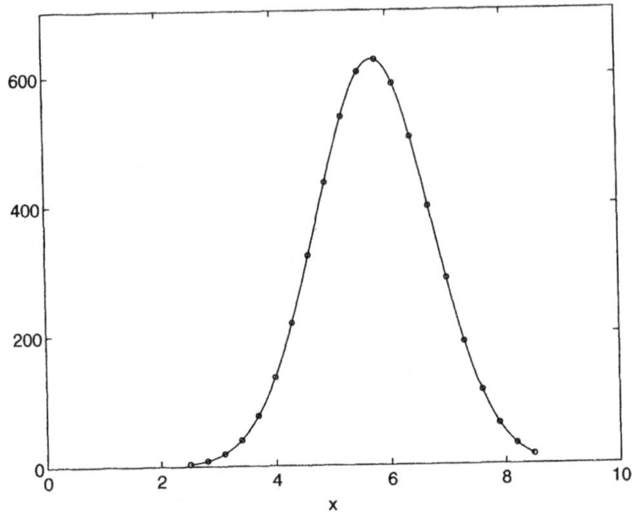

Figure 5.6. Gaussian peak expectation model (solid line) and its values at the measurement points (circles). (Example 5.11)

where $\gamma = 625$ is a known scale factor, $\theta = (1 \quad 5.75 \quad 1)^T$ are the parameters, and the measurement points are described by $x_n = 2.5 + 0.3 \times (n-1)$, with $n = 1, \ldots, 21$. Figure 5.6. shows the expectation model and its values at the measurement points. Suppose that the following observations have been made:

$$w = (1 \ 7 \ 24 \ 46 \ 84 \ 146 \ 218 \ 335 \ 396 \ 526 \ 545 \ 647 \ 610 \ 480 \ 417 \ 285 \ 188 \ 130 \ 52 \ 39 \ 17)^T. \tag{5.135}$$

In the test of the absence of a linear background contribution, the expectations

$$Ew_n = g_n(\theta) = \gamma\theta_1 \exp\left(-\frac{1}{2}\left(\frac{x_n - \theta_2}{\theta_3}\right)^2\right) + \theta_4 x_n + \theta_5 \tag{5.136}$$

with parameter vector $\theta = (\theta_1 \ldots \theta_5)^T$ correspond to the *unrestricted* model. The expectations (5.134), on the other hand, correspond to the *restricted* model since the parameter vector is $(\theta_1 \; \theta_2 \; \theta_3 \; 0 \; 0)^T$. Next, the Fisher scoring method to be described in Section 6.5 is used to compute the maximum likelihood estimates of the parameters of both models from the available observations. The results are as follows. The maximum likelihood estimates of the parameters of the restricted model are: $\tilde{t} = (0.977 \quad 5.758 \quad 1.021 \quad 0 \quad 0)^T$ and those of the unrestricted model: $\tilde{\tilde{t}} = (0.974 \quad 5.752 \quad 1.033 \quad -3.685 \quad 0.269)^T$. The expression (5.112) for the Poisson log-likelihood function shows that

$$\begin{aligned} \ell &= q(w;\tilde{t}) - q(w;\tilde{\tilde{t}}) \\ &= \sum_n -g_n(\tilde{t}) + g_n(\tilde{\tilde{t}}) + w_n \ln g_n(\tilde{t}) - w_n \ln g_n(\tilde{\tilde{t}}) \end{aligned} \tag{5.137}$$

with $g_n(t)$ described by (5.136). Then, substituting $\tilde{t}$ and $\tilde{\tilde{t}}$ in this expression shows that $-2\ell = 2.499$ and, if (5.134) is the true model, this is supposed to be a sample value of a chi-square distributed stochastic variable with two degrees of freedom. Generally, the *size* of a hypothesis test is the probability that a true hypothesis is rejected. The size is chosen by the experimenter and, as is often done, we take the size as 0.05. Tables of the chi-square distribution with two degrees of freedom show that the probability that the chi-square variable is smaller than 5.991 is 0.95. Consequently, this is the value exceeded with probability 0.05 if the restricted model is true. It is said to be the 0.05 *quantile*. Since $-2\ell = 2.499 < 5.991$, there is no reason to reject the hypothesis that there is no background function. ■

We now apply Wilks's theorem to the testing of the chosen expectation model itself. Suppose that N observations $w = (w_1 \dots w_N)^T$ are available with expectations $Ew = \mu = (\mu_1 \dots \mu_N)^T$. Then, these expectations may be unrestrictedly estimated from w. On the other hand, if $Ew = g(\theta)$ is assumed to be true, then we obtain

$$\begin{aligned} g_1(\theta) &= \mu_1 \\ g_2(\theta) &= \mu_2 \\ &\vdots \\ g_N(\theta) &= \mu_N. \end{aligned} \tag{5.138}$$

Suppose that K of these equations are selected and, without loss of generality, assume that these are the first K equations:

$$\begin{aligned} g_1(\theta) &= \mu_1 \\ g_2(\theta) &= \mu_2 \\ &\vdots \\ g_K(\theta) &= \mu_K. \end{aligned} \tag{5.139}$$

Typically, these equations establish a one-to-one relation of the K elements of θ and the K elements $\mu' = (\mu_1 \dots \mu_K)^T$. Subsequently, denote this relation as $\theta' = \theta(\mu')$. Then, substituting θ' for θ in the last $N - K$ equations of (5.138) yields

$$\begin{aligned} g_{K+1}(\theta') &= \mu_{K+1} \\ g_{K+2}(\theta') &= \mu_{K+2} \\ &\vdots \\ g_N(\theta') &= \mu_N. \end{aligned} \tag{5.140}$$

This is a system of $N - K$ equations relating the first K elements of the vector μ to the last $N - K$ elements of the same vector. Therefore, it is a set of $N - K$ constraints on the elements of the parameter vector μ. These considerations are the basis of the following theorem:

Theorem 5.5 (Den Dekker) *Suppose that the observations $w = (w_1 \dots w_N)^T$ are assumed to possess expectations $Ew = \mu = g(\theta)$, where θ is a $K \times 1$ vector of parameters with $K < N$. Let $q(w; m)$ be the log-likelihood function of the expectation parameters m with elements corresponding to those of μ. Define $\tilde{m}$ as the maximum likelihood estimate of μ and $\tilde{t}$ as the maximum likelihood estimate of θ, respectively. Then, the log-likelihood ratio for the assumed expectation model is described by*

$$\ell = q\left(w; g(\tilde{t})\right) - q(w; \tilde{m}) \tag{5.141}$$

and the quantity -2ℓ is asymptotically chi-square distributed with $N-K$ degrees of freedom.

Proof. The proof follows directly from Wilks's theorem and the number of constraints imposed by $g(\theta)$. ■

Theorem 5.5 is true under the null hypothesis, that is, if the model is correct. The following examples illustrate the form of ℓ and its use for normal, Poisson, and multinomial observations.

■ **EXAMPLE 5.12**

The log-likelihood ratio for normally distributed observations

The log-likelihood function $q(w;m)$ for normally distributed observations follows directly from (5.59):

$$q(w;m) = -\frac{N}{2}\ln 2\pi - \frac{1}{2}\ln\det C - \frac{1}{2}(w-m)^T C^{-1}(w-m) \tag{5.142}$$

where $m = (m_1 \ldots m_N)^T$. Thus, the unrestricted maximum likelihood estimate $\tilde{m}_n$ of μ_n is equal to the nth observation w_n. The corresponding value of the log-likelihood function is equal to

$$q(w;\tilde{m}) = q(w;w) = -\frac{N}{2}\ln 2\pi - \frac{1}{2}\ln\det C. \tag{5.143}$$

For the restricted maximum likelihood estimate of μ the log-likelihood function is equal to

$$q(w;g(\tilde{t})) = -\frac{N}{2}\ln 2\pi - \frac{1}{2}\ln\det C - \frac{1}{2}\left[w-g(\tilde{t})\right]^T C^{-1}\left[w-g(\tilde{t})\right]. \tag{5.144}$$

Then,

$$\ell = q(w;g(\tilde{t})) - q(w;\tilde{m}) = -\frac{1}{2}\left[w-g(\tilde{t})\right]^T C^{-1}\left[w-g(\tilde{t})\right]. \tag{5.145}$$

Therefore, the quantity

$$-2\ell = \left[w-g(\tilde{t})\right]^T C^{-1}\left[w-g(\tilde{t})\right] \tag{5.146}$$

is asymptotically chi-square distributed with $N-K$ degrees of freedom. ■

■ **EXAMPLE 5.13**

The log-likelihood ratio for Poisson distributed observations

For Poisson distributed observations, the log-likelihood function $q(w;m)$ follows directly from (5.112):

$$q(w;m) = \sum_n -m_n + w_n \ln m_n - \ln w_n!. \tag{5.147}$$

Differentiating this expression with respect to m_n shows that the unrestricted maximum likelihood estimate $\tilde{m}_n$ of μ_n is equal to the nth observation w_n. The corresponding value of the log-likelihood function is

$$q(w;\tilde{m}) = q(w;w) = \sum_n -w_n + w_n \ln w_n - \ln w_n!\,. \tag{5.148}$$

For the restricted estimate of μ the log-likelihood function is equal to

$$q\left(w;g(\tilde{t})\right) = \sum_n -g_n\left(\tilde{t}\right) + w_n \ln g_n(\tilde{t}) - \ln w_n!\,. \tag{5.149}$$

Then,

$$\begin{aligned} \ell &= q(w;g(\tilde{t})) - q(w;\tilde{m}) \\ &= \sum_n w_n - g_n(\tilde{t}) + w_n \ln g_n(\tilde{t}) - w_n \ln w_n. \end{aligned} \tag{5.150}$$

If $w_n = 0$, the term $w_n \ln w_n$ in this expression vanishes since $\lim_{x\to 0} x \ln x = 0$. The quantity

$$-2\ell = 2\sum_n g_n(\tilde{t}) - w_n + w_n \ln w_n - w_n \ln g_n(\tilde{t}) \tag{5.151}$$

is asymptotically chi-square distributed with $N - K$ degrees of freedom. ■

■ **EXAMPLE 5.14**

Testing a Gaussian peak model

Suppose that the Poisson distributed observations

$$\begin{aligned} w = {} & (7\ 9\ 18\ 45\ 80\ 156\ 227\ 319\ 425\ 559\ 609\ 620\ 586 \\ & \quad 506\ 382\ 275\ 176\ 124\ 53\ 29\ 13)^T \end{aligned} \tag{5.152}$$

correspond to the expectations (5.134) in Example 5.11. The measurement points are also those in Example 5.11. The maximum likelihood estimates $\tilde{t}$ of $\theta = (\theta_1\ \theta_2\ \theta_3)^T$ produced by the Fisher scoring method are $\tilde{t} = (0.995\ \ 5.723\ \ 1.007)^T$. The corresponding quantity -2ℓ described by (5.151) is equal to 9.428. This is supposed to be a sample value of a chi-square distributed stochastic variable with 18 degrees of freedom. The 0.05 quantile of such a distribution is 28.869. Therefore, there is no reason to reject the model. ■

■ **EXAMPLE 5.15**

The log-likelihood ratio for multinomially distributed observations

For multinomially distributed observations, the log-likelihood function $q(w;m)$ follows directly from (5.119):

$$q(w;m) = \ln M! - M \ln M - \sum_{n=1}^{N} \ln w_n! + \sum_{n=1}^{N} w_n \ln m_n \tag{5.153}$$

with $w_N = M - \sum_{n=1}^{N-1} w_n$ and $m_N = M - \sum_{n=1}^{N-1} m_n$. Differentiating this expression with respect to m_n and equating the result to zero produces

$$\frac{\partial q(w;t)}{\partial m_n} = \frac{w_n}{m_n} - \frac{w_N}{m_N} = 0 \tag{5.154}$$

with $m_N = M - \sum_{n=1}^{N-1} m_n$ and $n = 1, \ldots, N-1$. Equation (5.154) is solved by $\tilde{m}_n = w_n$, $n = 1, \ldots, N-1$. Therefore,

$$q(w;\tilde{m}) = q(w;w) = \ln M! - M \ln M - \sum_{n=1}^{N} \ln w_n! + \sum_{n=1}^{N} w_n \ln w_n . \tag{5.155}$$

For the restricted estimate of μ the log-likelihood function is equal to

$$q\left(w; g(\tilde{t})\right) = \ln M! - M \ln M - \sum_{n=1}^{N} \ln w_n! + \sum_{n=1}^{N} w_n \ln g_n(\tilde{t}), \tag{5.156}$$

where $g_N(\tilde{t}) = M - \sum_{n=1}^{N-1} g_n(\tilde{t})$. Then,

$$\ell = q\left(w; g(\tilde{t})\right) - q(w;\tilde{m}) = \sum_{n=1}^{N} w_n \ln \frac{g_n(\tilde{t})}{w_n} . \tag{5.157}$$

The quantity -2ℓ is asymptotically chi-square distributed with $N - K - 1$ degrees of freedom. ■

5.8.2 Model testing for exponential families of distributions

For linear exponential family distributed observations, the log-likelihood function $q(w;m)$ follows directly from (5.127):

$$q(w;m) = \ln \alpha(m) + \ln \beta(w) + \gamma^T(m) w . \tag{5.158}$$

The likelihood equations for unrestricted estimation of μ follow from (5.130):

$$\frac{\partial m^T}{\partial m} C^{-1} (w - m) = C^{-1} (w - m) = o. \tag{5.159}$$

Therefore, since C^{-1} is nonsingular, the maximum likelihood estimate $\tilde{m}$ of μ is the vector of observations w. Substituting this estimate in the log-likelihood function (5.127) for linear exponential families yields

$$q(w;w) = \ln \alpha(w) + \ln \beta(w) + \gamma^T(w)\, w. \tag{5.160}$$

On the other hand, the log-likelihood function for the restricted estimation of the expectations is

$$q\left(w; g(\tilde{t})\right) = \ln \alpha\left(g(\tilde{t})\right) + \ln \beta(w) + \gamma^T\left(g(\tilde{t})\right)\, w. \tag{5.161}$$

Equations (5.160) and (5.161) show that the log-likelihood ratio for testing expectation models of linear exponential family distributed observations is described by

$$\ell = \ln \frac{\alpha\left(g(\tilde{t})\right)}{\alpha(w)} + \left[\gamma\left(g(\tilde{t})\right) - \gamma(w)\right]^T w. \tag{5.162}$$

The quantity -2ℓ is asymptotically chi-square distributed with $N - K$ degrees of freedom.

■ EXAMPLE 5.16

The log-likelihood ratio for normally distributed observations

Example 3.1 shows that the normal distribution is a linear exponential family of distributions and that

$$\alpha(m) = \frac{1}{(2\pi)^{\frac{N}{2}}(\det C)^{\frac{1}{2}}} \exp -\frac{1}{2}m^T C^{-1} m \tag{5.163}$$

and

$$\gamma(m) = C^{-1}m. \tag{5.164}$$

Then,

$$\ln \frac{\alpha\left(g(\tilde{t})\right)}{\alpha(w)} = -\frac{1}{2}g^T(\tilde{t})C^{-1}g(\tilde{t}) + \frac{1}{2}w^T C^{-1} w \tag{5.165}$$

and

$$\left[\gamma\left(g(\tilde{t})\right) - \gamma(w)\right]^T w = g^T(\tilde{t})\,C^{-1}w - w^T C^{-1} w. \tag{5.166}$$

Substituting this in (5.162) shows that

$$\ell = -\frac{1}{2}\left[w - g(\tilde{t})\right]^T C^{-1}\left[w - g(\tilde{t})\right]. \tag{5.167}$$

This agrees with (5.145). ■

■ EXAMPLE 5.17

The log-likelihood ratio for Poisson distributed observations

Example 3.2 shows that the Poisson distribution is a linear exponential family of distributions and that

$$\alpha(m) = \exp\left(-\sum\nolimits_n m_n\right) \tag{5.168}$$

and

$$\gamma(m) = (\ln m_1 \ldots \ln m_N)^T. \tag{5.169}$$

Then,

$$\ln \frac{\alpha\left(g(\tilde{t})\right)}{\alpha(w)} = \sum_n w_n - g_n(\tilde{t}) \tag{5.170}$$

and

$$\left[\gamma(g(\tilde{t})) - \gamma(w)\right]^T w = \sum_n \left[\ln g_n(\tilde{t}) - \ln w_n\right] w_n. \tag{5.171}$$

Substituting this in (5.162) shows that

$$\ell = \sum_n w_n - g_n(\tilde{t}) + w_n \ln g_n(\tilde{t}) - w_n \ln w_n\,. \tag{5.172}$$

This agrees with (5.150). ■

5.9 LEAST SQUARES ESTIMATION

The maximum likelihood estimator has attractive properties but it requires a priori knowledge about the probability (density) function of the observations. If such knowledge is absent, the least squares estimator may be an alternative.

Various forms of the least squares criterion have already been encountered in Subsection 5.4.1 dealing with maximum likelihood estimation from normally distributed observations. The general form is the *weighted least squares criterion*

$$\boxed{J(t) = d^T(t) R\, d(t)}, \tag{5.173}$$

where $d(t) = [d_1(t) \ldots d_N(t)]^T$ with $d_n(t) = w_n - g_n(t)$. The symmetric and positive definite $N \times N$ matrix R is called the *weighting matrix*. Expression (5.173) may alternatively be written

$$J(t) = \sum_p \sum_q r_{pq}\, d_p(t)\, d_q(t) \tag{5.174}$$

with $p, q = 1, \ldots, N$. Expression (5.174) is a positive definite quadratic form in the deviations $d(t)$ with the elements of R as coefficients. The least squares method consists in minimizing the criterion $J(t)$ with respect to t and taking the value $\hat{t}$ of t that minimizes $J(t)$ as estimate of the parameters θ. There are no particular connections with statistical properties of the observations in this formulation. This is a difference with the least squares problems described in Subsection 5.4.1 where maximum likelihood estimators from normally distributed observations were found to be various forms of least squares estimators.

If the weighting matrix is diagonal, (5.173) becomes

$$J(t) = d^T(t)\, R\, d(t) \tag{5.175}$$

with

$$R = \text{diag}\,(r_{11} \ldots r_{NN}). \tag{5.176}$$

An alternative expression for this least squares criterion is

$$J(t) = \sum_n r_{nn} d_n^2(t). \tag{5.177}$$

Finally, if all weights r_{nn} in this expression are equal to one, the corresponding least squares criterion is the *ordinary* least squares criterion

$$J(t) = d^T(t)\, d(t) = \sum_n d_n^2(t). \tag{5.178}$$

Since the least squares solution $\hat{t}$ for t is the absolute minimum of the least squares criterion, it is a stationary point. Therefore, a necessary but not a sufficient condition for $\hat{t}$ to be the least squares solution is that at $t = \hat{t}$ the gradient vector of $J(t)$ vanishes. The gradient vector of the weighted least squares criterion (5.173) is equal to

$$\boxed{\frac{\partial J(t)}{\partial t} = -2\frac{\partial g^T(t)}{\partial t} R\, d(t)}, \tag{5.179}$$

where $\partial g(t)/\partial t^T$ is the $N \times K$ Jacobian matrix of $g(t)$ with respect to t. This expression may also be written

$$\frac{\partial J(t)}{\partial t} = -2 \sum_p \sum_q r_{pq}\, d_p(t) \frac{\partial g_q(t)}{\partial t} \tag{5.180}$$

with $p, q = 1, \ldots, N$. Equation (5.179) shows that at $t = \hat{t}$

$$\frac{\partial g^T(t)}{\partial t} R\, d(t) = o\,, \tag{5.181}$$

where o is the $K \times 1$ null vector. The K equations in the elements of t defined by (5.181) are called the *normal equations* associated with the least squares criterion. Equation (5.180) shows that the normal equations may also be written

$$\sum_p \sum_q r_{pq}\, d_p(t) \frac{\partial g_q(t)}{\partial t} = o\,. \tag{5.182}$$

If R is diagonal, the gradient vector is described by

$$\frac{\partial J(t)}{\partial t} = -2 \frac{\partial g^T(t)}{\partial t} R\, d(t) \tag{5.183}$$

with R defined by (5.176) or, equivalently, by

$$\frac{\partial J(t)}{\partial t} = -2 \sum_n r_{nn}\, d_n(t) \frac{\partial g_n(t)}{\partial t}. \tag{5.184}$$

Then, the normal equations are

$$\frac{\partial g^T(t)}{\partial t} R\, d(t) = o \tag{5.185}$$

with R defined by (5.176) or, equivalently,

$$\sum_n r_{nn} d_n(t) \frac{\partial g_n(t)}{\partial t} = o. \tag{5.186}$$

Finally, the gradient vector for the ordinary least squares criterion (5.178) is described by

$$\frac{\partial J(t)}{\partial t} = -2 \frac{\partial g^T(t)}{\partial t}\, d(t) \tag{5.187}$$

or, equivalently, by

$$\frac{\partial J(t)}{\partial t} = -2 \sum_n d_n(t) \frac{\partial g_n(t)}{\partial t}. \tag{5.188}$$

Then, the normal equations are described by

$$\frac{\partial g^T(t)}{\partial t}\, d(t) = o \tag{5.189}$$

or, equivalently, by

$$\sum_n d_n(t) \frac{\partial g_n(t)}{\partial t} = o\,. \tag{5.190}$$

The main subdivision of least squares estimators is in *linear* and *nonlinear* least squares estimators. A least squares estimator is called *linear least squares estimator* when applied to the linear expectation model.

■ **EXAMPLE 5.18**

The straight-line expectation model

A simple example of a linear expectation model is the straight-line model. With this model, the expectations of the observations are

$$Ew_n = g(x_n; \theta) = \theta_1 s_n + \theta_2 , \tag{5.191}$$

where $\theta = (\theta_1 \; \theta_2)^T$ with θ_1 the slope and θ_2 the intercept of the straight line. Parameters such as θ_1 and θ_2 are called *linear* parameters. The $x_n^T = (s_n \; 1)$ are the vector measurement points. ■

On the other hand, a least squares estimator is called *nonlinear least squares estimator* if it is applied to an expectation model that is nonlinear in one or more of the elements of θ.

■ **EXAMPLE 5.19**

The multiexponential expectation model

A well-known nonlinear expectation model is the multiexponential model. With this model, the expectations of the observations are

$$Ew_n = g(x_n; \theta) = \alpha_1 \exp(-\beta_1 s_n) + \cdots + \alpha_K \exp(-\beta_K s_n) \tag{5.192}$$

with $\theta = (\alpha_1 \ldots \alpha_K \; \beta_1 \ldots \beta_K)^T$, where the amplitudes $\alpha_k > 0$ are the *linear* parameters and the decay constants $\beta_k > 0$ are the *nonlinear* parameters of the expectation model. The $x_n = s_n$ are the scalar measurement points. ■

Nonlinear least squares estimation will be discussed in Section 5.10 and linear least squares estimation in Sections 5.11–5.19.

5.10 NONLINEAR LEAST SQUARES ESTIMATION

At the measurement points, the general nonlinear expectation model is described by

$$Ew_n = g_n(\theta) = g(x_n; \theta), \tag{5.193}$$

where $g(x_n; \theta)$ is nonlinear in at least one of the elements of the parameter vector θ and x_n is the known scalar or vector measurement point described by

$$x_n = (x_{n1} \ldots x_{nL})^T . \tag{5.194}$$

■ **EXAMPLE 5.20**

Estimating parameters of Gaussian peaks

Suppose that the observations $w = (w_1 \dots w_N)^T$ are available with the expectations

$$\begin{aligned} Ew_n &= g(x_n;\theta) \\ &= \alpha_1 \exp\left(-\frac{(\xi_n - \mu_{11})^2 + (\eta_n - \mu_{21})^2}{2\rho^2} \right) \\ &\quad + \dots + \alpha_K \exp\left(-\frac{(\xi_n - \mu_{1K})^2 + (\eta_n - \mu_{2K})^2}{2\rho^2} \right), \end{aligned} \tag{5.195}$$

which are K Gaussian peaks. In (5.195), α_k is the height of the kth peak while μ_{1k} and μ_{2k} are the location parameters of its maximum. Furthermore, ρ is the common half-width of the peaks. Suppose that the heights, locations, and the half-width are unknown. Then, the $(3K + 1) \times 1$ vector of unknown parameters is

$$\theta = (\alpha_1 \; \mu_{11} \; \mu_{21} \; \dots \; \alpha_K \; \mu_{1K} \; \mu_{2K} \; \rho)^T. \tag{5.196}$$

The vector measurement points are

$$x_n = (x_{n1} \; x_{n2})^T = (\xi_n \; \eta_n)^T \tag{5.197}$$

with $n = 1, \dots, N$. ■

A useful theoretical result concerning nonlinear least squares estimation is the following.

Theorem 5.6 (Jennrich) *Suppose that the observations $w = (w_1 \; \dots \; w_N)^T$ are independent and identically distributed with unknown equal variance σ^2 around the expectations $E\, w_n = g_n(\theta)$. Then, under general conditions, the asymptotic distribution of the ordinary least squares estimator $\hat{t}$ of θ is the normal distribution with expectation θ and asymptotic covariance matrix*

$$\sigma^2 \left(\frac{\partial g^T(\theta)}{\partial \theta} \frac{\partial g(\theta)}{\partial \theta^T} \right)^{-1}, \tag{5.198}$$

where $g(\theta) = [g_1(\theta) \dots g_N(\theta)]^T$.

Proof. A reference to a proof of this theorem is given in Section 5.20. ■

Expression (5.198) for the asymptotic covariance matrix of the ordinary least squares estimator coincides with expression (5.72) for the asymptotic covariance matrix of the maximum likelihood estimator from uncorrelated, normally distributed observations with equal variance σ^2, that is, for $C = \sigma^2 I$. The reason is that, in that particular case, the maximum likelihood estimator *is* the ordinary least squares estimator. However, for other distributions, the expression for the asymptotic covariance matrix of the maximum likelihood estimator will typically differ from (5.198) since then the maximum likelihood estimator is not the ordinary least squares estimator.

Expression (5.198) is also useful for experimental design. However, such a design relates exclusively to the use of the ordinary least squares estimator.

Theorem 5.6 may be generalized in the following way:

Theorem 5.7 (Jennrich) *Suppose that the observations* $(w_1 \ \dots \ w_N)^T$ *are independent and have expectations* $E\, w_n = g_n(\theta)$. *Also suppose that the variance of the observation* w_n *is described by*

$$\sigma_n^2 = \frac{\sigma^2}{\rho_n}, \tag{5.199}$$

where the ρ_n *are known positive scalars and* σ^2 *may be unknown. Then, the asymptotic distribution of the weighted least squares estimator of* θ *with weighting matrix* $R = \text{diag}\,(\rho_1 \dots \rho_N)$ *is the normal distribution with expectation* θ *and asymptotic covariance matrix*

$$\sigma^2 \left(\frac{\partial g^T(\theta)}{\partial \theta} R \frac{\partial g(\theta)}{\partial \theta^T} \right)^{-1}, \tag{5.200}$$

where $g = [g_1(\theta) \dots g_N(\theta)]^T$.

Proof. Transform the observations w_n into $w_n' = \sqrt{\rho_n}\, w_n$ and the expectations $g_n(\theta)$ into $g_n'(\theta) = \sqrt{\rho_n}\, g_n(\theta)$ with $n = 1, \dots N$. Then, the variances of these transformed observations are all equal to σ^2. If the least squares criterion

$$J(t) = \sum_n \rho_n \left[w_n - g_n(t)\right]^2 = \sum_n \left[w_n' - g_n'(t)\right]^2 \tag{5.201}$$

is subsequently minimized with respect to t, this is ordinary least squares estimation of the parameters θ from independent observations $w' = (w_1' \dots w_N')^T$ with equal variance σ^2 and expectations $g'(\theta)$. Then, Theorem 5.6 applies. Therefore, the least squares estimator of θ is asymptotically normally distributed with covariance matrix

$$\sigma^2 \left(\frac{\partial g'^T(\theta)}{\partial \theta} \frac{\partial g'(\theta)}{\partial \theta^T} \right)^{-1} = \sigma^2 \left(\frac{\partial g^T(\theta)}{\partial \theta} R \frac{\partial g(\theta)}{\partial \theta^T} \right)^{-1}. \tag{5.202}$$

This completes the proof. ■

The asymptotic covariance matrix (5.202) may be estimated by substituting $\hat{t}$ for θ and

$$\widehat{s^2} = \frac{1}{N-K} \sum_n \left[w_n' - g_n'(\hat{t})\right]^2 \tag{5.203}$$

for σ^2. The sum of the squares of the differences is divided by $N - K$, instead of by N, to avoid bias. See Subsection 5.4.3.2. Then, the estimated asymptotic covariance matrix of the estimates is equal to

$$\widehat{s^2} \left(\frac{\partial g^T(t)}{\partial t} R \frac{\partial g(t)}{\partial t^T} \right)^{-1}_{t=\hat{t}}. \tag{5.204}$$

Most of the literature on nonlinear least squares problems concerns numerical methods for their solution rather than analytical results like Jennrich's theorems described in this section. Numerical methods for the solution of nonlinear estimation problems, including nonlinear least squares problems, are addressed in Chapter 6.

5.11 LINEAR LEAST SQUARES ESTIMATION

Suppose that the expectation of the vector observations $w = (w_1 \ldots w_N)^T$ is described by the linear model (5.78):

$$Ew = g(\theta) = X\theta \,. \tag{5.205}$$

Then,

$$Ew_n = g_n(\theta) = \theta_1 x_{n1} + \cdots + \theta_K x_{nK} = x_n^T \theta. \tag{5.206}$$

The vector

$$x_n = (x_{n1} \ldots x_{nK})^T \tag{5.207}$$

is the nth measurement point. Its transpose x_n^T is the nth row of the $N \times K$ matrix

$$X = \begin{pmatrix} x_{11} & x_{12} & \ldots & x_{1K} \\ x_{21} & x_{22} & \ldots & x_{2K} \\ \vdots & \vdots & & \vdots \\ x_{N1} & x_{N2} & \ldots & x_{NK} \end{pmatrix} = \begin{pmatrix} x_1^T \\ x_2^T \\ \vdots \\ x_N^T \end{pmatrix} . \tag{5.208}$$

We present three examples to show the importance of the linear model.

■ **EXAMPLE 5.21**

Polynomial

Suppose that

$$Ew_n = g_n(\theta) = \theta_1 + \theta_2 s_n + \theta_3 s_n^2 + \cdots + \theta_K s_n^{K-1} \,. \tag{5.209}$$

These are points of a polynomial expectation model linear in the parameters. The measurement points are described by

$$x_n = (1 \; s_n \; s_n^2 \ldots s_n^{K-1})^T \tag{5.210}$$

and

$$X = \begin{pmatrix} 1 & s_1 & s_1^2 & \ldots & s_1^{K-1} \\ 1 & s_2 & s_2^2 & \ldots & s_2^{K-1} \\ \vdots & \vdots & \vdots & \vdots & \vdots \\ 1 & s_N & s_N^2 & \ldots & s_N^{K-1} \end{pmatrix} . \tag{5.211}$$

■

■ **EXAMPLE 5.22**

Moving average

The expectation of the observation w_n of the response of a moving average dynamic system is described by

$$Ew_n = g_n(\theta) = \theta_1 u_n + \theta_2 u_{n-1} + \cdots + \theta_K u_{n-K+1}. \tag{5.212}$$

The index n denotes discrete equidistant time instants. At time instant n, $g_n(\theta)$ is the response of the dynamic system to the known input $u_n, \ldots, u_{n-K+1}$. If the unit impulse

$u_n = \delta_{1,n}$ is chosen as input, $u_n = 1$ for $n = 1$ and $u_n = 0$ for $n \neq 1$. Then, the response for $n = 1, 2, \ldots$ is the *impulse response* $\theta_1, \ldots, \theta_K$, 0, 0, ... of the system. The expectations $g(\theta) = [g_1(\theta) \ldots g_N(\theta)]^T$ of the observations $w = (w_1 \ldots w_N)^T$ of the response to arbitrary inputs are described by the linear model $X\theta$ with

$$x_n = \begin{pmatrix} u_n & u_{n-1} & u_{n-2} & \ldots & u_{n-K+1} \end{pmatrix}^T \tag{5.213}$$

and

$$X = \begin{pmatrix} u_1 & u_0 & \ldots & u_{2-K} \\ u_2 & u_1 & \ldots & u_{3-K} \\ & & \vdots & \\ u_N & u_{N-1} & \ldots & u_{N-K+1} \end{pmatrix}. \tag{5.214}$$

■

■ **EXAMPLE 5.23**

Nonstandard Fourier analysis

Suppose that the expectation of the observation w_n is the real Fourier series described by

$$Ew_n = g_n(\theta) = \sum_k \alpha_k \cos(k\omega s_n) + \beta_k \sin(k\omega s_n), \tag{5.215}$$

where $\theta = (\alpha_1\, \beta_1 \ldots \alpha_K\, \beta_K)^T$ with α_k and β_k the Fourier cosine and sine coefficient of the kth harmonic, respectively. The constant ω is the known angular fundamental frequency. Unlike in standard discrete Fourier analysis, the known sampling points s_n may occur anywhere. Here,

$$x_n^T = [\cos(\omega s_n)\ \sin(\omega s_n)\ \cos(2\omega s_n)\ \sin(2\omega s_n) \ldots \cos(K\omega s_n)\ \sin(K\omega s_n)] \tag{5.216}$$

and

$$X = \begin{pmatrix} \cos(\omega s_1) & \sin(\omega s_1) & \ldots & \cos(K\omega s_1) & \sin(K\omega s_1) \\ \cos(\omega s_2) & \sin(\omega s_2) & \ldots & \cos(K\omega s_2) & \sin(K\omega s_2) \\ & & \vdots & & \\ \cos(\omega s_N) & \sin(\omega s_N) & \ldots & \cos(K\omega s_N) & \sin(K\omega s_N) \end{pmatrix}. \tag{5.217}$$

Since the α_k and β_k are the only unknown parameters, the model is linear. ■

5.12 WEIGHTED LINEAR LEAST SQUARES ESTIMATION

In Section 5.11, examples of linear expectation models were presented. In this section, a general expression is derived for the least squares estimator of the parameters of such models. The least squares criterion chosen is the general weighted least squares criterion described by (5.173) and (5.174):

$$\begin{aligned} J(t) &= [w - g(t)]^T R\, [w - g(t)] \\ &= \sum_p \sum_q r_{pq} [w_p - g_p(t)]\, [w_q - g_q(t)] . \end{aligned} \tag{5.218}$$

Suppose that the expectations $g(\theta)$ are described by (5.78). Then,

$$g(t) = Xt. \tag{5.219}$$

Therefore, the least squares criterion for this model is

$$\begin{aligned} J(t) &= (w - Xt)^T R (w - Xt) \\ &= \sum_p \sum_q r_{pq} \left(w_p - x_p^T t\right) \left(w_q - x_q^T t\right), \end{aligned} \tag{5.220}$$

where $x_n^T = (x_{n1} \ldots x_{nK})$ and $t = (t_1 \ldots t_K)^T$.

EXAMPLE 5.24

Least squares estimation of the parameters of a straight line

Suppose that the observations $w = (w_1 \ldots w_N)^T$ have straight-line expectations

$$Ew_n = \theta_1 s_n + \theta_2 \,. \tag{5.221}$$

Then, the nth measurement point is $x_n = (s_n \; 1)^T$. The expectations are, therefore, described by $Ew = X\theta$ with $\theta = (\theta_1 \; \theta_2)^T$ and

$$X = \begin{pmatrix} s_1 & 1 \\ s_2 & 1 \\ \ldots & \ldots \\ s_N & 1 \end{pmatrix}. \tag{5.222}$$

Suppose that for estimating θ the ordinary least squares estimator is chosen. Then, R is equal to the identity matrix and the least squares criterion is described by

$$J(t) = \sum (w_n - t_1 s_n - t_2)^2, \tag{5.223}$$

where the summation is over n. A necessary condition for a minimum $\hat{t}$ of $J(t)$ is

$$\left.\frac{\partial J(t)}{\partial t_1}\right|_{t=\hat{t}} = -2 \sum (w_n - \hat{t}_1 s_n - \hat{t}_2) s_n = 0 \tag{5.224}$$

and

$$\left.\frac{\partial J(t)}{\partial t_2}\right|_{t=\hat{t}} = -2 \sum (w_n - \hat{t}_1 s_n - \hat{t}_2) = 0. \tag{5.225}$$

This is a system of two linear equations

$$\begin{pmatrix} \sum s_n^2 & \sum s_n \\ \sum s_n & N \end{pmatrix} \begin{pmatrix} \hat{t}_1 \\ \hat{t}_2 \end{pmatrix} = \begin{pmatrix} \sum w_n s_n \\ \sum w_n \end{pmatrix} \tag{5.226}$$

equivalent to

$$X^T X \hat{t} = X^T w \,, \tag{5.227}$$

or

$$\hat{t} = \left(X^T X\right)^{-1} X^T w \,, \tag{5.228}$$

where it has been assumed that the matrix X is nonsingular, that is, the s_n are not all equal. ■

In the derivation of a general expression for the linear least squares estimator, three simple lemmas will be used that will be presented first.

Lemma 5.1 *Consider a pair of* $N \times 1$ *vectors* a *and* x. *Then,*

$$\frac{\partial\left(a^T x\right)}{\partial x} = \frac{\partial\left(x^T a\right)}{\partial x} = a\,. \tag{5.229}$$

Proof. By definition,

$$a^T x = x^T a = \sum_n a_n x_n. \tag{5.230}$$

Then,

$$\frac{\partial\left(a^T x\right)}{\partial x_n} = \frac{\partial\left(x^T a\right)}{\partial x_n} = a_n \tag{5.231}$$

and, therefore,

$$\frac{\partial\left(a^T x\right)}{\partial x} = \frac{\partial\left(x^T a\right)}{\partial x} = a. \tag{5.232}$$

This completes the proof. ■

Lemma 5.2 *Let* x *be an* $N \times 1$ *vector and let* A *be a symmetric* $N \times N$ *matrix. Then,*

$$\frac{\partial(x^T A x)}{\partial x} = 2Ax. \tag{5.233}$$

Proof. The scalar quadratic form $x^T Ax$ is described by

$$x^T Ax = \sum_p \sum_q a_{pq} x_p x_q\,. \tag{5.234}$$

Then,

$$\frac{\partial(x^T Ax)}{\partial x_n} = 2\sum_p a_{np} x_p. \tag{5.235}$$

Therefore,

$$\frac{\partial(x^T Ax)}{\partial x} = 2Ax. \tag{5.236}$$

This completes the proof. ■

Lemma 5.3 *Let* x *be an* $N \times 1$ *vector and let* A *be a symmetric* $N \times N$ *matrix. Then,*

$$\frac{\partial(x^T Ax)}{\partial x \partial x^T} = 2A. \tag{5.237}$$

Proof. The scalar quadratic form $x^T Ax$ is described by

$$x^T Ax = \sum_p \sum_q a_{pq} x_p x_q\,. \tag{5.238}$$

Then,

$$\frac{\partial(x^T Ax)}{\partial x_m \partial x_n} = 2a_{mn}. \tag{5.239}$$

Therefore,

$$\frac{\partial(x^T Ax)}{\partial x \partial x^T} = 2A. \tag{5.240}$$

This completes the proof. ■

We now return to the main subject of this section. This is deriving the least squares estimator of the parameters of the linear model (5.219) defined as that value $\hat{t}$ of t that minimizes the least squares criterion (5.220).

Theorem 5.8 *Let the least squares estimator of the parameters of the expectations $X\theta$ of observations $w = (w_1 \ldots w_N)^T$ be that value $\hat{t}$ of t that minimizes the least squares criterion*

$$J(t) = (w - Xt)^T R(w - Xt), \tag{5.241}$$

where the $N \times N$ matrix R is a symmetric and positive definite weighting matrix. Then,

$$\boxed{\hat{t} = \left(X^T RX\right)^{-1} X^T Rw} \tag{5.242}$$

if X is nonsingular.

Proof. The least squares criterion $J(t)$ is the sum of four scalar terms:

$$J(t) = w^T Rw - t^T X^T Rw - w^T RXt + t^T X^T RXt. \tag{5.243}$$

Since a scalar is equal to its transpose, the second and the third term of this expression are equal. Then,

$$J(t) = w^T Rw - 2t^T X^T Rw + t^T X^T RXt. \tag{5.244}$$

The gradient of $J(t)$ is

$$\frac{\partial J(t)}{\partial t} = -2X^T Rw + 2X^T RXt, \tag{5.245}$$

where Lemma 5.1 and Lemma 5.2 have been used for differentiating the second and the third term of (5.244), respectively. A necessary condition for a point $t = \hat{t}$ to be stationary is that at it the gradient vanishes. Thus,

$$-2X^T Rw + 2X^T RX\hat{t} = o, \tag{5.246}$$

where o is the $K \times 1$ null vector. This implies that $\hat{t}$ is the solution of the system of K linear equations in the K elements of $\hat{t}$:

$$X^T RX\hat{t} = X^T Rw. \tag{5.247}$$

This system has a unique solution only if X is nonsingular—that is, if the columns of X are linearly independent. Then, the matrix $X^T RX$ is positive definite and, therefore, nonsingular since R is positive definite. See Theorems C.2 and C.4. This solution is

$$\hat{t} = \left(X^T RX\right)^{-1} X^T Rw. \tag{5.248}$$

Furthermore, applying Lemma 5.3 to (5.244) shows that

$$\frac{\partial^2 J(t)}{\partial t \partial t^T} = 2X^T RX \tag{5.249}$$

for all t. Then, $\partial^2 J(t)/\partial t \partial t^T$ is positive definite. As will be explained in Chapter 6, this implies that $J(t)$ is *minimum* for $t = \hat{t}$. This completes the proof. ■

In the linear least squares literature, the parameters θ are called *estimable* from w if X is nonsingular.

5.13 PROPERTIES OF THE LINEAR LEAST SQUARES ESTIMATOR

In this section, properties of the weighted linear least squares estimator (5.242) are presented in the form of theorems.

Theorem 5.9 *The weighted linear least squares estimator is linear in the observations.*

Proof. The estimator is of the form $\hat{t} = Aw$ with $A = \left(X^T RX\right)^{-1} X^T R$. ■

This theorem shows that each of the elements of $\hat{t}$ is a linear combination of the observations. Because linear combinations of normally distributed stochastic variables are also normally distributed, the elements of $\hat{t}$ are jointly normally distributed if the observations are.

Theorem 5.10 *The weighted linear least squares estimator is unbiased.*

Proof. The expectation of $\hat{t}$ is equal to

$$E\hat{t} = E[(X^T RX)^{-1} X^T Rw] = (X^T RX)^{-1} X^T R\, Ew. \tag{5.250}$$

By (5.219),

$$Ew = X\theta. \tag{5.251}$$

Substituting this in (5.250) yields

$$E\hat{t} = \left(X^T RX\right)^{-1} X^T RX\theta = \theta. \tag{5.252}$$

This completes the proof. ■

The proof of this theorem depends on the assumption that (5.251) is correct. The theorem, therefore, says that then the weighted least squares estimator is accurate in the sense that it contains no systematic error. The theorem also shows that the estimator is unbiased for any weighting matrix R.

Theorem 5.11 *The covariance matrix of the weighted linear least squares estimator is equal to*

$$\text{cov}(\hat{t}, \hat{t}) = \left(X^T RX\right)^{-1} X^T RCRX \left(X^T RX\right)^{-1}, \tag{5.253}$$

where C is the covariance matrix of the observations w.

Proof. Subtracting (5.250) from (5.242) yields

$$\hat{t} - E\hat{t} = \left(X^T R X\right)^{-1} X^T R\,(w - Ew). \tag{5.254}$$

Then, by definition,

$$\begin{aligned}
\operatorname{cov}(\hat{t},\hat{t}) &= E[(\hat{t} - E\hat{t})\;(\hat{t} - E\hat{t})^T] \\
&= E[\left(X^T R X\right)^{-1} X^T R\,(w - Ew)\,(w - Ew)^T R X \left(X^T R X\right)^{-1}] \\
&= \left(X^T R X\right)^{-1} X^T R\, E\left[(w - Ew)\,(w - Ew)^T\right] R X \left(X^T R X\right)^{-1} \\
&= \left(X^T R X\right)^{-1} X^T R C R X \left(X^T R X\right)^{-1},
\end{aligned} \tag{5.255}$$

where use has been made of the symmetry of the matrices $\left(X^T R X\right)^{-1}$ and R. This completes the proof. ■

The theorem shows how the covariance matrix of the weighted least squares estimator depends on the measurement points x_n, the covariance matrix of the observations C, and the weighting matrix R. The dependence on R is of particular importance since this matrix is chosen by the experimenter. Different R will typically lead to different diagonal elements of $\operatorname{cov}(\hat{t},\hat{t})$, that is, to different variances. Then, the following question arises: Which choice of R is best, in the sense of producing the smallest variances? This question will be addressed below, but first the following theorem is presented.

Theorem 5.12 *If, in the weighted linear least squares estimator, the weighting matrix R is chosen as the inverse of the covariance matrix of the observations C, the covariance matrix of the estimator is equal to*

$$\left(X^T C^{-1} X\right)^{-1}. \tag{5.256}$$

Proof. Substituting C^{-1} for R in (5.255) yields

$$\left(X^T C^{-1} X\right)^{-1} X^T C^{-1} C C^{-1} X \left(X^T C^{-1} X\right)^{-1} = \left(X^T C^{-1} X\right)^{-1}. \tag{5.257}$$

■

In the analysis which weighting matrix R yields the most precise results, the *best linear unbiased estimator* is central. An estimator is best linear unbiased if it has minimum variance within the class of estimators that are linear in the observations and are unbiased. It has minimum variance within a class if the difference of the covariance matrix of any estimator of that class and its covariance matrix is positive semidefinite.

5.14 THE BEST LINEAR UNBIASED ESTIMATOR

Theorem 5.13 (Aitken) *The weighted linear least squares estimator is best linear unbiased if the weighting matrix is the inverse of the covariance matrix of the observations.*

Proof. Consider first the estimator $\hat{t}$ for $R = C^{-1}$. Then,

$$\hat{t} = Aw \tag{5.258}$$

with

$$A = \left(X^T C^{-1} X\right)^{-1} X^T C^{-1}. \tag{5.259}$$

This shows that

$$AX = I, \tag{5.260}$$

where I is the identity matrix of order K, and

$$\operatorname{cov}(\hat{t}, \hat{t}) = ACA^T. \tag{5.261}$$

Next, let

$$t' = A'w \tag{5.262}$$

be any unbiased estimator of θ linear in w. Then, the $K \times N$ matrix B may be chosen such that

$$A' = A + B. \tag{5.263}$$

The estimator t' is unbiased if

$$Et' = A'Ew = (A + B)X\theta = (I + BX)\theta = \theta, \tag{5.264}$$

that is, if

$$BX = O, \tag{5.265}$$

where I and O are the identity matrix of order K and the $K \times K$ null matrix, respectively. Equation (5.262) shows that the covariance matrix of t' is equal to

$$\begin{aligned} \operatorname{cov}(t', t') &= A'CA'^T = (A + B)C(A + B)^T \\ &= ACA^T + ACB^T + BCA^T + BCB^T. \end{aligned} \tag{5.266}$$

However, by (5.259),

$$BCA^T = BCC^{-1}X\left(X^T C^{-1} X\right)^{-1} = BX\left(X^T C^{-1} X\right)^{-1} = O \tag{5.267}$$

since C^{-1} and $\left(X^T C^{-1} X\right)^{-1}$ are symmetric, and $BX = O$. Then, also $ACB^T = O$ and hence,

$$\operatorname{cov}(t', t') = ACA^T + BCB^T. \tag{5.268}$$

Then, by (5.261),

$$\operatorname{cov}(t', t') - \operatorname{cov}(\hat{t}, \hat{t}) = BCB^T \succeq O \tag{5.269}$$

since C is positive definite and, therefore, BCB^T is positive semidefinite. See Theorem C.4. Thus, the variances of the elements of arbitrary unbiased estimators t' linear in the observations are always larger than or equal to the corresponding variances of the weighted least squares estimator $\hat{t}$ with the inverse of the covariance matrix of the observations as weighting matrix. See Theorem C.1. Since the latter estimator is also unbiased and linear in the observations, it is the best linear unbiased estimator. ■

Best linear unbiasedness is preserved under linear transformation. This is expressed by the following theorem.

Theorem 5.14 *A linear combination of best linear unbiased estimators of a number of parameters is a best linear unbiased estimator of that linear combination of the parameters.*

Proof. Let the $K \times 1$ vector θ be the parameter vector of the expectations $X\theta$. Furthermore, suppose that A is an $L \times K$ matrix where L is arbitrary. Then, $A\theta$ is an $L \times 1$ vector of linear combinations of elements of θ. Let $\hat{t}$ be the best linear unbiased estimator of θ and let t' be any linear unbiased estimator. Then,

$$E[A\hat{t}] = A\,E\hat{t} = A\theta\,. \tag{5.270}$$

Therefore, $A\hat{t}$ is unbiased for $A\theta$. Similarly, At' is unbiased for $A\theta$. Also, both estimators are linear in the observations since $\hat{t}$ and t' are. The covariance matrices of $A\hat{t}$ and At' are described by

$$A\,\mathrm{cov}(\hat{t},\hat{t})\,A^T \tag{5.271}$$

and

$$A\,\mathrm{cov}(t',t')\,A^T, \tag{5.272}$$

respectively. Then,

$$A\,\mathrm{cov}(t',t')\,A^T - A\,\mathrm{cov}(\hat{t},\hat{t})\,A^T = A\left[\mathrm{cov}(t',t') - \mathrm{cov}(\hat{t},\hat{t})\right]A^T \succeq O \tag{5.273}$$

since, by definition, the matrix $\mathrm{cov}(t',t') - \mathrm{cov}(\hat{t},\hat{t})$ is positive semidefinite. See Theorem C.5. The conclusion is that $A\hat{t}$ is best linear unbiased for $A\theta$. ■

5.15 SPECIAL CASES OF THE BEST LINEAR UNBIASED ESTIMATOR AND A RELATED RESULT

In this section, first two special cases of the best linear unbiased estimator will be studied. Then, the best linear unbiased estimator is compared with the maximum likelihood estimator of the parameters of the linear model from linear exponential family distributed observations.

5.15.1 The Gauss–Markov theorem

Suppose that the observations are uncorrelated and have an equal variance σ^2. Then, the following theorem applies.

Theorem 5.15 (Gauss–Markov) *For uncorrelated observations with equal variance, the ordinary least squares estimator of the parameters of the linear model is best linear unbiased.*

Proof. The covariance matrix of the observations is described by

$$C = \sigma^2 I\,, \tag{5.274}$$

where I is the identity matrix of order N. Substituting this expression in that for the best linear unbiased estimator (5.258) yields

$$\hat{t} = (X^T X)^{-1} X^T w\,, \tag{5.275}$$

which is the ordinary least squares estimator of the parameters θ of the expectations $X\theta$. ■

Of course, if the covariance matrix of the observations is not described by (5.274), the ordinary least squares estimator is linear in the observations and unbiased but it is not best.

5.15.2 Normally distributed observations

Up to now, in the discussion of the linear least squares estimator presented in this chapter, the distribution of the observations has been left out of consideration. However, if the observations are normally distributed, all linear unbiased estimators have the following property.

Theorem 5.16 *For normally distributed observations with covariance matrix C, any linear unbiased estimator Aw of the parameters of the linear model is normally distributed with expectation θ and covariance ACA^T.*

Proof. By definition, the estimator is a linear combination of the observations, which, in this particular case, are normally distributed. Since any linear combination of normally distributed stochastic variables is normally distributed, the estimator is normally distributed. Since the estimator is unbiased, its expectation is equal to θ. Furthermore, its covariance matrix is equal to

$$\begin{aligned} & E[\,(Aw - A\,Ew)(Aw - A\,Ew)^T] \\ & = A\,E[\,(w - Ew)(w - Ew)^T]\,A^T = ACA^T\,. \end{aligned} \tag{5.276}$$

This completes the proof. ■

The best linear unbiased estimator possesses some further useful properties which are summarized by the following theorem.

Theorem 5.17 *For normally distributed observations, the best linear unbiased estimator of the parameters of the linear model is normally distributed, identical to the maximum likelihood estimator, and efficient unbiased.*

Proof. See Subsections 5.4.3 and 5.4.2.4, and Theorem 5.16. ■

5.15.3 Exponential family distributed observations

Since the normal distribution is a linear exponential family of distributions, the following question arises: Do the best linear unbiased estimator and the maximum likelihood estimator coincide for all distributions that are linear exponential families? The answer follows from the general expression for the likelihood equations for linear exponential families (5.130):

$$\frac{\partial g^T(t)}{\partial t} C^{-1}\,[w - g(t)] = o. \tag{5.277}$$

For the linear model,

$$g(t) = Xt \tag{5.278}$$

and, therefore, the likelihood equations are

$$X^T C^{-1}(w - Xt) = o. \tag{5.279}$$

If X is nonsingular, this produces the maximum likelihood estimator

$$\tilde{t} = (X^T C^{-1} X)^{-1} X^T C^{-1} w. \tag{5.280}$$

At the first sight, this is the best linear unbiased estimator. However, since the elements of C are generally functions of θ, the elements of C in (5.280) depend on $\tilde{t}$, making this expression a system of nonlinear equations in the elements of $\tilde{t}$ that cannot be transformed in closed-form expressions for these elements and is different from the best linear unbiased estimator. This is illustrated by the following simple example.

■ EXAMPLE 5.25

Estimation of straight-line parameters from Poisson distributed observations.

Let the estimation problem be that of Example 5.2. In that example, the observations have a Poisson distribution, which is a linear exponential family of distributions. The expectation model is the straight line. Therefore, the expectations of the observations are described by

$$Ew_n = x_n^T \theta = \theta_1 \xi_n + \theta_2 \tag{5.281}$$

where $x_n = (\xi_n \ 1)^T$ is the nth measurement point. The observations are independent and, since they are Poisson distributed, their variance is equal to their expectation. Therefore,

$$C = \text{diag}(\theta_1 \xi_1 + \theta_2 \ldots \theta_1 \xi_N + \theta_2). \tag{5.282}$$

Then, substituting

$$X = \begin{pmatrix} \xi_1 & 1 \\ \vdots & \vdots \\ \xi_N & 1 \end{pmatrix} \tag{5.283}$$

and

$$C = \text{diag}(t_1 \xi_1 + t_2 \ \ldots \ t_1 \xi_N + t_2) \tag{5.284}$$

in the likelihood equations (5.279) shows that the resulting equations are nonlinear in t_1 and t_2 and cannot be transformed in closed-form expressions for these estimators. This conclusion agrees with the one already drawn in Example 5.2. However, the resemblance of (5.280) to the expression for the closed-form best linear unbiased estimator will be exploited in Chapter 6 to design a simple numerical procedure to solve (5.280) for $\hat{t}$. ■

5.16 COMPLEX LINEAR LEAST SQUARES ESTIMATION

In this section, the linear least squares estimator for a vector of real and complex parameters from a vector of real and complex observations will be derived. For that derivation, the description of complex stochastic variables developed in Section 3.8 will be used.

Suppose that the real vector s of a number of real observations and the real and imaginary parts of a number of complex observations is described by

$$s = \begin{pmatrix} r \\ u \\ v \end{pmatrix}, \tag{5.285}$$

where r is a $P \times 1$ vector of real observations and u and v are $Q \times 1$ vectors representing the real parts and the imaginary parts of the complex observations z, respectively:

$$z = \begin{pmatrix} u_1 + jv_1 \\ \vdots \\ u_Q + jv_Q \end{pmatrix}. \tag{5.286}$$

Then, in the notation of Section 3.8,

$$w = B_w s \tag{5.287}$$

with

$$B_w = \text{diag}\,(I \;\; A_w), \tag{5.288}$$

where I is the identity matrix of order P and

$$A_w = \begin{pmatrix} I & jI \\ I & -jI \end{pmatrix}, \tag{5.289}$$

where I is the identity matrix of order Q and the $(P + 2Q) \times 1$ vector of real and complex observations w is defined as

$$w = \begin{pmatrix} r \\ z \\ z^* \end{pmatrix}. \tag{5.290}$$

Analogously, the $(M + 2L) \times 1$ vector of real and complex parameters is described by

$$\theta = B_\theta \varphi, \tag{5.291}$$

where φ is the $(M + 2L) \times 1$ real vector composed of M real parameters and L real parts of complex parameters followed by the L imaginary parts of the same complex parameters.

Let $N = P + 2Q$ and $K = M + 2L$ and suppose that the expectation of the vector s is described by

$$Es = Y\varphi, \tag{5.292}$$

where the $N \times K$ matrix Y is known and real. Then, the least squares criterion for estimating φ from s is defined as

$$(s - Yf)^T\, U\, (s - Yf), \tag{5.293}$$

where the elements of the real $K \times 1$ vector of variables f correspond to those of φ and U is the chosen real $N \times N$ weighting matrix. This criterion will now be rewritten in terms of the vector of real and complex observations w and the vector t corresponding to the vector of real and complex parameters θ. The matrix B_w is, by definition, nonsingular. Therefore, by (5.287), the observations s are related to the observations w by

$$s = B_w^{-1} w \tag{5.294}$$

while, by (5.291), the vector of variable parameters t corresponding to θ and the vector f corresponding to φ are related by

$$f = B_\theta^{-1} t. \tag{5.295}$$

Substituting these expressions for s and f in the least squares criterion (5.293) yields

$$(w - Xt)^H\, R\, (w - Xt) \tag{5.296}$$

with

$$X = B_w Y B_\theta^{-1} \tag{5.297}$$

and

$$R = B_w^{-H} U B_w^{-1}. \tag{5.298}$$

In this derivation, use has been made of the fact that $(s - Yf)^T = (s - Yf)^H$ since $(s - Yf)$ is real. Similarly, the least squares estimate $\hat{t}$ of θ from the observations w may be shown to be equal to

$$\hat{t} = (X^H R X)^{-1} X^H R w. \tag{5.299}$$

5.17 SUMMARY OF PROPERTIES OF LINEAR LEAST SQUARES ESTIMATORS

In this section, we summarize the most important ingredients of the linear least squares theory presented in Sections 5.11–5.16. Generally, the properties of a linear least squares estimator depend on the statistical properties of the observations. Throughout, the expression for the expectations

$$Ew = X\theta \tag{5.300}$$

is assumed to be correct and the matrix X is assumed to be nonsingular. Then, the general form of the least squares criterion is

$$\begin{aligned} J(t) &= (w - Xt)^T R (w - Xt) \\ &= \sum_p \sum_q r_{pq} \left(w_p - x_p^T t\right) \left(w_q - x_q^T t\right), \end{aligned} \tag{5.301}$$

where x_n is the nth measurement point and the weighting matrix R is positive definite. The linear least squares estimator minimizes this criterion and is described by

$$\hat{t} = \left(X^T R X\right)^{-1} X^T R w. \tag{5.302}$$

Under these assumptions, the linear least squares estimator has the following properties:

- If the distribution and the covariance matrix of the observations are unknown, then all that may be concluded about the weighted linear least squares estimator is that it is linear in the observations and unbiased. This is true for all weighting matrices including the identity matrix that corresponds to the ordinary least squares estimator. Under these conditions, there are no reasons to assume that the variance of $\hat{t}$ has optimal properties.

- If the covariance matrix C of the observations w is known, it may be used to construct the best linear unbiased estimator:

$$\hat{t} = \left(X^T C^{-1} X\right)^{-1} X^T C^{-1} w. \tag{5.303}$$

  Within the class of estimators that are unbiased and linear in the observations, this has minimum variance. Equation (5.303) shows that, strictly, the construction of the best linear unbiased estimator $\hat{t}$ requires the relative magnitudes of the elements of C only. If the observations are uncorrelated, then $C = \text{diag}\,(\sigma_1^2 \ldots \sigma_N^2)$. If the observations are uncorrelated and have an equal, not necessarily known, variance σ^2, then $C = \sigma^2 I$ and the ordinary least squares estimator is best linear unbiased. If the observations are independent, Theorems 5.6 and 5.7 show that $\hat{t}$ is asymptotically normally distributed.

- Finally, if the observations are normally distributed with known covariance matrix C, then the best linear unbiased estimator (5.303) is efficient unbiased and is identical to the maximum likelihood estimator.

5.18 RECURSIVE LINEAR LEAST SQUARES ESTIMATION

Of all linear least squares estimators, the ordinary linear least squares estimator is simplest and used most frequently. It is described by (5.275):

$$\hat{t} = (X^T X)^{-1} X^T w. \tag{5.304}$$

In this expression,

$$X = \begin{pmatrix} x_{11} & x_{12} & \cdots & x_{1K} \\ x_{21} & x_{22} & \cdots & x_{2K} \\ \vdots & \vdots & & \vdots \\ x_{N1} & x_{N2} & \cdots & x_{NK} \end{pmatrix} = \begin{pmatrix} x_1^T \\ x_2^T \\ \vdots \\ x_N^T \end{pmatrix} \tag{5.305}$$

and

$$w = (w_1 \ldots w_N)^T. \tag{5.306}$$

Below, $\hat{t}$, X, and w thus defined will be denoted by $\hat{t}_{(N)}$, X_N, and $w_{(N)}$ to indicate their dependence on the number of observations N. Furthermore, we define

$$P_N = (X_N^T X_N)^{-1}. \tag{5.307}$$

Suppose that an additional observation w_{N+1} is made and that the corresponding measurement point is

$$x_{N+1} = (x_{N+1,1} \; x_{N+1,2} \; \cdots \; x_{N+1,K})^T. \tag{5.308}$$

Then, in this section, an expression will be derived for the linear least squares estimator $\hat{t}_{(N+1)}$ in terms of $\hat{t}_{(N)}$, w_{N+1} and x_{N+1}. This estimator will be called *recursive linear least squares estimator.* The main mathematical tool for the derivation of an expression for the recursive estimator is Corollary B.5.

Theorem 5.18 *The ordinary linear least squares estimate $\hat{t}_{(N+1)}$ may be computed from $\hat{t}_{(N)}$ using the recursive expression*

$$\hat{t}_{(N+1)} = \hat{t}_{(N)} + m_{N+1}(w_{N+1} - x_{N+1}^T \hat{t}_{(N)}), \tag{5.309}$$

where m_{N+1} is a $K \times 1$ vector of weights defined as

$$m_{N+1} = \frac{P_N \, x_{N+1}}{1 + x_{N+1}^T \, P_N \, x_{N+1}}. \tag{5.310}$$

The recursive expression

$$P_N = P_{N-1} - P_{N-1} \frac{x_N \, x_N^T}{1 + x_N^T \, P_{N-1} \, x_N} P_{N-1} \tag{5.311}$$

may be used for computing P_N.

Proof. The linear least squares estimator $\hat{t}_{(N)}$is defined as

$$\hat{t}_{(N)} = (X_N^T X_N)^{-1} X_N^T w_{(N)} = P_N X_N^T w_{(N)}, \tag{5.312}$$

where

$$X_N = \begin{pmatrix} x_{11} & x_{12} & \dots & x_{1K} \\ x_{21} & x_{22} & \dots & x_{2K} \\ \vdots & \vdots & & \vdots \\ x_{N1} & x_{N2} & \dots & x_{NK} \end{pmatrix} \tag{5.313}$$

and

$$w_{(N)} = (w_1 \dots w_N)^T . \tag{5.314}$$

Then,

$$\hat{t}_{(N+1)} = (X_{N+1}^T X_{N+1})^{-1} X_{N+1}^T w_{(N+1)} = P_{N+1} X_{N+1}^T w_{(N+1)}. \tag{5.315}$$

In this expression,

$$X_{N+1} = \begin{pmatrix} X_N \\ x_{N+1}^T \end{pmatrix} \tag{5.316}$$

and

$$w_{(N+1)} = \begin{pmatrix} w_{(N)} \\ w_{N+1} \end{pmatrix}. \tag{5.317}$$

Then, by (5.307) and (5.316),

$$P_{N+1} = (P_N^{-1} + x_{N+1} x_{N+1}^T)^{-1} \tag{5.318}$$

and by (5.316) and (5.317),

$$X_{N+1}^T w_{(N+1)} = X_N^T w_{(N)} + x_{N+1} w_{N+1} . \tag{5.319}$$

By Corollary B.5,

$$P_{N+1} = P_N - P_N \frac{x_{N+1}\, x_{N+1}^T}{1 + x_{N+1}^T\, P_N\, x_{N+1}} P_N . \tag{5.320}$$

This is equivalent to (5.311). Substituting (5.320) and (5.319) in (5.315) yields

$$\begin{aligned} \hat{t}_{(N+1)} &= (P_N - P_N \frac{x_{N+1}\, x_{N+1}^T}{1 + x_{N+1}^T\, P_N\, x_{N+1}} P_N) \\ &\qquad \times (X_N^T w_{(N)} + x_{N+1} w_{N+1}) \\ &= P_N X_N^T w_{(N)} + P_N x_{N+1} w_{N+1} \\ &\quad - P_N \frac{x_{N+1}\, x_{N+1}^T}{1 + x_{N+1}^T\, P_N\, x_{N+1}} P_N X_N^T w_{(N)} \\ &\quad - P_N \frac{x_{N+1}\, x_{N+1}^T}{1 + x_{N+1}^T\, P_N\, x_{N+1}} P_N x_{N+1} w_{N+1} . \end{aligned} \tag{5.321}$$

The first term of this expression is equal to $\hat{t}_{(N)}$. The second term is reduced to the same denominator as the third and the fourth term:

$$\frac{(1 + x_{N+1}^T\, P_N\, x_{N+1}) P_N x_{N+1} w_{N+1}}{1 + x_{N+1}^T\, P_N\, x_{N+1}}, \tag{5.322}$$

which is allowed since the denominator is nonzero because $x_{N+1}^T\ P_N\ x_{N+1} \geq 0$ since P_N is positive definite. In the third term, $\hat{t}_{(N)}$ is again substituted for $P_N X_N^T w_{(N)}$. Finally, the fourth term may be rewritten as

$$\frac{-(x_{N+1}^T\ P_N\ x_{N+1})P_N x_{N+1} w_{N+1}}{1 + x_{N+1}^T\ P_N\ x_{N+1}} \tag{5.323}$$

since $x_{N+1}^T\ P_N\ x_{N+1}$ is scalar. Summing the terms of (5.321) thus modified yields

$$\hat{t}_{(N+1)} = \hat{t}_{(N)} + m_{N+1}(w_{N+1} - x_{N+1}^T \hat{t}_{(N)}) \tag{5.324}$$

with

$$m_{N+1} = \frac{P_N\ x_{N+1}}{1 + x_{N+1}^T\ P_N\ x_{N+1}}. \tag{5.325}$$

This completes the proof. ■

The result (5.309) may be interpreted as follows. The estimator $\hat{t}_{(N+1)}$ consists of two parts: $\hat{t}_{(N)}$ and a correction term. In (5.309), the quantity

$$(w_{N+1} - x_{N+1}^T \hat{t}_{(N)}) \tag{5.326}$$

is the difference of the newly made observation w_{N+1} and its prediction $x_{N+1}^T \hat{t}_{(N)}$ on the basis of the exactly known new measurement point x_{N+1} and the most recent estimate $\hat{t}_{(N)}$ of the parameters θ. Generally, absolutely larger differences may be expected as the standard deviations of the w_n or those of the elements of $\hat{t}_{(N)}$ are larger. The $K \times 1$ vector m_{N+1} is a vector of *weights*. A heuristic description of the behavior of this vector may be given as follows. The matrix P_N appearing in expression (5.310) is decreasing with N in the sense that $P_N \succeq P_{N+1}$ as may be inferred from (5.320). Therefore, as N increases, the vector m_N tends to the null vector since the numerator of its elements tends to zero while their denominator tends to one. This implies that with increasing N, the correction term gradually decreases: The estimates converge.

The steps in the recursive scheme for computation of the linear least squares estimate once the $(N + 1)$th observation has been made are:

1. Compute m_{N+1} by substituting P_N computed in the previous step and the newly obtained measurement point x_{N+1}^T in (5.310).

2. Compute $\hat{t}_{(N+1)}$ by substituting $\hat{t}_{(N)}$, m_{N+1}, the new observation w_{N+1} and x_{N+1}^T in (5.309).

3. Compute P_{N+1} by substituting P_N and x_{N+1}^T in (5.320).

The recursive scheme requires initial values for P_N and $\hat{t}_{(N)}$. A straightforward way to generate these is to use X_{N_ι} and the first N_ι observations $w_{(N_\iota)}$ to compute P_{N_ι} and $\hat{t}_{(N_\iota)}$ nonrecursively.

Advantages of the use of the recursive least squares estimator are:

- The solution of a system of linear equations with every additional observation, as required by the nonrecursive computation, is avoided. Thus, the number of numerical operations associated with including each new observation is reduced.
- The recursive computation requires a very small and constant amount of memory. In fact, it does not require the $N \times K$ matrix X_N and the $N \times 1$ vector $w_{(N)}$ to be stored. Instead, it only requires the $K \times K$ matrix P_{N+1} and the $K \times 1$ vector $\hat{t}_{(N)}$.
- The recursive estimation and the collection of observations may be stopped once a desired degree of convergence of the parameter estimates has been attained.

In the derivation of the recursive estimator $\hat{t}_{(N+1)}$ defined by (5.309)–(5.311) no approximations have been made. Therefore, nonrecursive calculation of $\hat{t}_{(N+1)}$ using (5.315) or recursive calculation should produce the same result if the initial conditions have been generated as suggested. The difference of both approaches is, therefore, the way of computing not of estimating.

The recursive scheme derived in this section may be extended to include schemes for tracking parameters that change during the collection of the observations. Such a scheme is the subject of the next section.

5.19 RECURSIVE LINEAR LEAST SQUARES ESTIMATION WITH FORGETTING

In Section 5.18, the recursively estimated parameters were considered to be constants. In certain problems, however, the parameters change during the collecting of the observations. Then, to reduce the influence of past observations on the current estimate, weights are introduced in the least squares criterion so that only relatively recent observations influence the estimator. An example of such a weighting scheme is *exponential forgetting*. This scheme is the subject of this section.

The least squares criterion employed by exponential forgetting is described by

$$\sum_n \eta^{N-n}(w_n - x_n^T t)^2 \tag{5.327}$$

where $n = 1, \ldots, N$ and $0 < \eta \leq 1$. This is equivalent to the weighted least squares criterion

$$J(t_{(N)}) = \left(w_{(N)} - X_N t_{(N)}\right)^T \Omega_N \left(w_{(N)} - X_N t_{(N)}\right) \tag{5.328}$$

with weighting matrix

$$\Omega_N = \text{diag}\,(\eta^{N-1}\ \eta^{N-2} \ldots \eta\ 1). \tag{5.329}$$

The solution of this linear least squares problem is

$$\hat{t}_{(N)} = \left(X_N^T \Omega_N X_N\right)^{-1} X_N^T \Omega_N w_{(N)} = P_N X_N^T \Omega_N w_{(N)} \tag{5.330}$$

and, therefore,

$$\hat{t}_{(N+1)} = P_{N+1} X_{N+1}^T \Omega_{N+1} w_{(N+1)}. \tag{5.331}$$

The identity

$$\Omega_{N+1} = \text{diag}\,(\eta \Omega_N\ 1) \tag{5.332}$$

shows that

$$\begin{aligned} P_{N+1} &= \left(\eta X_N^T \Omega_N X_N + x_{N+1} x_{N+1}^T\right)^{-1} \\ &= \left(\eta P_N^{-1} + x_{N+1} x_{N+1}^T\right)^{-1} \end{aligned} \tag{5.333}$$

and

$$X_{N+1}^T \Omega_{N+1} w_{(N+1)} = \eta X_N^T \Omega_N w_{(N)} + x_{N+1} w_{N+1}. \tag{5.334}$$

The rest of the derivation of the recursive linear least squares estimator with exponential forgetting is analogous to that of the recursive linear least squares estimator without exponential forgetting, discussed in Section 5.18. Applying Corollary B.5 to (5.333) yields

$$P_{N+1} = \frac{1}{\eta}\left(P_N - P_N \frac{x_{N+1} x_{N+1}^T}{\eta + x_{N+1}^T P_N x_{N+1}} P_N\right). \tag{5.335}$$

Substituting this result and (5.334) in (5.331) yields after some rearrangements

$$\hat{t}_{(N+1)} = \hat{t}_{(N)} + m_{N+1}\left(w_{N+1} - x_{N+1}^T \hat{t}_{(N)}\right) \tag{5.336}$$

with

$$m_{N+1} = \frac{P_N \, x_{N+1}}{\eta + x_{N+1}^T \, P_N \, x_{N+1}}. \tag{5.337}$$

These results are consistent with those of Section 5.18 for $\eta = 1$. To illustrate the properties of the recursive least squares estimator with exponential forgetting we conclude this section by a numerical example.

■ EXAMPLE 5.26

Estimation of the slope of a straight line through the origin using recursive least squares estimation with exponential forgetting

In this example, the slope θ is estimated from observations with an expectation $g_n(\theta) = \theta x_n$. The known measurement points x_n have been generated uniformly over the interval $[0, 1]$ using a random number generator. The simulated observations have been generated by adding $\mathcal{N}(0; 0.0004)$ distributed numbers to the $g_n(\theta)$. The number of observations generated is 125. For the first 30 observations we have $\theta = 1$, and for the next 95 observations we have $\theta = 0.7$. Using the observations thus simulated, two numerical experiments are carried out. In the first experiment, θ is recursively estimated with $\eta = 0.75$ while in the second experiment $\eta = 0.95$. Initial estimates of θ are generated by applying the weighted least squares estimator (5.330) directly to the first ten observations. The initial values thus obtained are equal to 1.008 and 1.003, respectively. The results are shown in Fig. 5.7. The estimator is seen to track the value of the parameter properly for both values of η. However, there are two important differences in behavior for both values. For $\eta = 0.75$, the estimator reacts much more quickly to the jump in the true value of θ than for $\eta = 0.95$. This is so because the effective memory of the estimator is shorter as the value of η is smaller. On the other hand, the fluctuations of the estimates around the true value of the parameter are much larger for $\eta = 0.75$ than for $\eta = 0.95$. This is so because the effective number of observations used by the estimator is smaller in the former case. ■

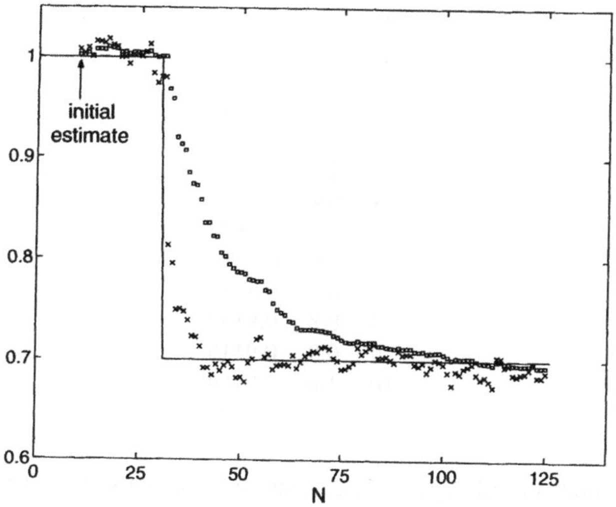

Figure 5.7. Varying value of the parameter (solid line) and recursive least squares estimates with exponential forgetting for $\eta = 0.75$ (crosses) and $\eta = 0.95$ (squares). (Example 5.26)

5.20 COMMENTS AND REFERENCES

Advanced general books on estimation, including maximum likelihood estimation, are, for example, Stuart, Ord, and Arnold [30], Lehmann and Casella [22], Zacks [34, 35], and Cramér [5]. Mood, Graybill, and Boes [24] is an excellent textbook and as such relatively easily accessible. However, it covers vector parameter estimation such as addressed in this book only partly. The book by Jennrich [17] is user-oriented and very practical without neglecting theoretical aspects .

The invariance property of the maximum likelihood estimator of a not necessarily one-to-one scalar function of a scalar parameter has been proved by Zehna [36]. Mood, Graybill, and Boes [24] generalize this result to vector functions of vector parameters. In Subsection 5.3.1, we have simplified their proof by specializing it to a not necessarily one-to-one scalar function of a vector parameter. Then, the extension to a vector function of a vector parameter is self-evident while the condition in [24] that the dimension of the vector function may not exceed the dimension of the parameter vector may be dropped.

A proof of Wilks's Theorem on the asymptotic distribution of the log-likelihood ratio in Subsection 5.8.1 may be found in [30]. Den Dekker's Theorem in the same subsection is presented in [6].

Useful references for Sections 5.9–5.14, dealing with least squares estimation, are Bates and Watts [4] and Jennrich [17]. Bates and Watts is a classical book on linear and nonlinear least squares estimation. It contains many practical examples. The book is expectation model rather than statistics of observations oriented. This is a difference with Jennrich's book that also addresses maximum likelihood estimation and exponential families of distributions. For a proof of the first Jennrich theorem in Section 5.10, see [16]. The second theorem is described in Jennrich's book [17]. The classical reference to recursive linear least squares estimation, discussed in Sections 5.18 and 5.19, is the paper by Fagin [8].

Table 5.4. Problem 5.3

n	x_n	w_n
1	0.1	0.6186
2	0.2	0.6694
3	0.3	0.7203
4	0.4	0.7822
5	0.5	0.7806
6	0.6	0.8012
7	0.7	0.8743
8	0.8	0.9167
9	0.9	0.9115
10	1.0	1.0164

The book by Goodwin and Payne [14] contains an excellent survey of various recursive forgetting schemes.

5.21 PROBLEMS

5.1 The observations $w = (w_1 \dots w_N)^T$ are independent and binomially distributed with expectations $Ew_n = g_n(\theta)$. The number of independent trials is the same for every w_n and equal to M. See Problem 3.3. Derive an expression for the log-likelihood function of the parameters.

5.2 The observations $w = (w_1 \dots w_N)^T$ are independent and exponentially distributed with expectations $Ew_n = g_n(\theta)$. See Problem 3.7. Derive an expression for the log-likelihood function of the parameters.

5.3 The observations w_n are independent and symmetrically uniformly distributed around the values of straight-line expectations $g_n(\theta) = \alpha x_n + \beta$, where $\theta = (\alpha\ \beta)^T$ are the unknown parameters. The width of the uniform distribution is known and is equal to 0.1. Suppose that in a particular experiment the observations are those presented in Table 5.4. Plot these observations and numerically compute and plot the boundary of the collection of all points in the (a, b)-plane that qualify as maximum likelihood estimates $(\tilde{a}, \tilde{b})$ of (α, β), where the variables a and b correspond to the parameters α and β, respectively.

5.4 The observations $w = (w_1 \dots w_N)^T$ are Poisson distributed and have expectations $Ew_n = \gamma\alpha\exp(-\beta x_n)$, where $\theta = (\alpha\ \beta)^T$ with $\alpha,\ \beta > 0$ are the unknown parameters and $\gamma > 0$ is a known scale factor.

(a) Derive the likelihood equations for the parameters θ.

(b) For root finding of scalar functions of one variable, effective and simple numerical methods are available. Show how the maximum likelihood estimates of α and β can be computed with such a numerical method.

(c) In an experiment, the observations are those presented in Table 5.5. Furthermore, $\gamma =$ 900. Plot the observations, compute the maximum likelihood estimates of α and

β using a numerical root finding method, and, for comparison, plot the expectation model with the estimates as parameters in the same figure as the observations.

5.5 The observations $w = (w_1 \dots w_N)^T$ are Poisson distributed and have expectations $Ew_n = \alpha x_n + \beta$ with $\theta = (\alpha\ \beta)^T$. Show that the maximum likelihood estimates of the parameters θ can be numerically computed by root finding like in Problem 5.4.

5.6 The observations $w = (w_1 \dots w_N)^T$, with N even, are independent and normally distributed with equal variance σ^2. Their expectations are described by $Ew_n = \alpha x_n + \beta$ with $\theta = (\alpha\ \beta)^T$.

(a) Compute the covariance matrix of the maximum likelihood estimator $\tilde{t} = (\tilde{a}\ \tilde{b})^T$ of θ.

(b) If the measurement points x_n may be freely chosen on the interval $[-A, A]$, which choice minimizes the variances of both $\tilde{a}$ and $\tilde{b}$?

(c) If, different from the observations in (b) and as a result of constraints, the measurement points $x_n, n = 1, \dots, N$ are located on a sphere with radius R and the origin as center, where should they be chosen to minimize the variances of both $\tilde{a}$ and $\tilde{b}$?

5.7 Suppose that the observations $w = (w_1 \dots w_N)^T$ are independent and exponentially distributed. See Problem 3.7. Also suppose that $Ew_n = \theta x_n$ where θ is an unknown scalar parameter.

(a) Derive an expression for the maximum likelihood estimator $\tilde{t}$ of θ.

(b) Show that $\tilde{t}$ is unbiased.

(c) Show that $\tilde{t}$ meets the necessary and sufficient condition for efficiency and unbiasedness.

5.8 The observations $w = (w_1 \dots w_N)^T$ are independent and normally distributed. Their expectations are described by $Ew_n = \alpha x_n + \beta$ where $(x_n\ 1)^T$ is the nth measurement point. The standard deviation of w_n is equal to γx_n^2 with $\gamma > 0$. Derive closed-form expressions for the maximum likelihood estimators of α, β, and γ.

Table 5.5. Problem 5.4

n	x_n	w_n
1	0	938
2	0.3	660
3	0.6	498
4	0.9	343
5	1.2	264
6	1.5	205
7	1.8	131
8	2.1	117
9	2.4	72
10	2.7	64
11	3.0	42

5.9 An experimenter has reason to suppose that his observations $w = (w_1 \dots w_N)^T$ are uncorrelated and that their unknown variances are equal. Furthermore, $Ew = X\theta$, where the $N \times K$ matrix X is known and θ is the $K \times 1$ vector of unknown parameters. He decides to use the ordinary least squares estimator $\hat{t} = (X^T X)^{-1} X^T w$ for estimating θ.

(a) Under what conditions is $\hat{t}$ the maximum likelihood estimator and under what conditions the best linear unbiased estimator?

(b) What is the particular form of the matrix X if $Ew_n = \alpha x_n + \beta$ and what if $Ew_n = \beta$?

(c) Derive an unbiased estimator for the variance of the w_n if $Ew_n = \alpha x_n + \beta$ and if $Ew_n = \beta$, respectively.

(d) Next assume that the observations are normally distributed. Show that the estimators derived under (c) are *not* the maximum likelihood estimators of the variance.

5.10 Like in Problem 5.4, the observations $w = (w_1 \dots w_N)^T$ are Poisson distributed with $Ew_n = \gamma\alpha \exp(-\beta x_n)$, where $\theta = (\alpha\ \beta)^T$ with $\alpha,\ \beta > 0$ are the unknown parameters, and $\gamma > 0$ is a known scale factor. Different from the maximum likelihood solution chosen in Problem 5.4, the parameters are sometimes estimated by fitting the straight-line model $\ln \gamma + \ln a - bx_n$, in the least squares sense and with respect to $\ln a$ and b, to $(\ln w_1 \dots \ln w_N)^T$.

(a) Compute the Cramér–Rao lower bound for unbiased estimation of α and β for parameter values $\alpha = \beta = 1$, scale factor $\gamma = 225$, and $x_n = (n-1) \times 0.3$ with $n = 1, \dots, 11$.

(b) For the same expectations, parameter values, scale factor, and measurement points numerically generate Poisson distributed observations and use root finding to compute the maximum likelihood estimates of the parameters α and β from these observations. Subsequently, estimate these parameters from the same observations by the straight-line method. Repeat this experiment sufficiently often to be able to compare the bias, variance, and efficiency of both estimators. Comment briefly on the results of this comparison.

5.11 Suppose that the observations $w = (w_1 \dots w_N)^T$ have a binomial distribution such as described in Problem 3.3 (b) and have expectations $g_n(\theta)$. Derive the log-likelihood ratio for this model.

5.12 Suppose that the observations $w = (w_1 \dots w_N)^T$ have an exponential distribution such as described in Problem 3.7(b) and have expectations $g_n(\theta)$. Derive the log-likelihood ratio for this model.

5.13 Suppose that the observations $w = (w_1 \dots w_N)^T$ have a binomial distribution such as described in Problem 3.3(b) and have expectations $g_n(\theta)$. Use the results of Problem 3.4(a), where it is shown that this distribution is a linear exponential family to derive the log-likelihood ratio for this model. Verify that the result is identical to Solution 5.11.

5.14 Suppose that the observations $w = (w_1 \dots w_N)^T$ have an exponential distribution such as described in Problem 3.7(b) and have expectations $g_n(\theta)$. Use the results of Problem 3.8(a), where it is shown that this distribution is a linear exponential family to derive the log-likelihood ratio for this model. Verify that the result is identical to Solution 5.12.

5.15 The expectations of the observations $w = (w_1 \dots w_N)^T$ are described by

$$Ew_n = g_n(\theta) = \alpha_1 h(x_n; \beta_1) + \dots + \alpha_K h(x_n; \beta_K),$$

where $\theta = (\alpha^T \, \beta^T)^T$, with $\alpha = (\alpha_1 \dots \alpha_K)^T$ and $\beta = (\beta_1 \dots \beta_K)^T$, is the $2K \times 1$ vector of unknown parameters and $h(x_n; \beta_k)$ is nonlinear in the parameter β_k. Suppose that for estimating the parameters θ the weighted least squares method is chosen with symmetric and positive definite weighting matrix R. Show that the weighted least squares solution $\hat{b}$ for the parameters β may be obtained by minimizing the criterion

$$w^T \left[I - RH(H^T RH)^{-1} H^T R \right] w$$

with respect to $b = (b_1 \dots b_K)^T$, where I is the identity matrix of order N and H is an $N \times K$ matrix depending on b only and defined by

$$H = \begin{pmatrix} h_1(b_1) & \dots & h_1(b_K) \\ \vdots & & \vdots \\ h_N(b_1) & \dots & h_N(b_K) \end{pmatrix}$$

with $h_n(b_k) = h(x_n; b_k)$. Also show that

$$\hat{a} = (H^T RH)^{-1} H^T Rw\,,$$

with $\hat{b}$ substituted for b, is the weighted least squares solution for α.

5.16 In the iterative numerical computation of nonlinear least squares estimates using the Newton method both the gradient and the Hessian matrix of the least squares criterion with respect to the parameters are used. Derive an expression for the Hessian matrix of the general least squares criterion (5.173).

5.17 Suppose that the expectations of the observations $w = (w_0 \dots w_{N-1})^T$ are periodic with M and are described by

$$Ew_n = \sum_k \alpha_k \cos(2\pi kn/M) + \beta_k \sin(2\pi kn/M)$$

with $n = 0, \dots, JM - 1$, where α_k and β_k, $k = 1, \dots, K$ are the Fourier cosine and sine coefficients, respectively, and J is an integer.

Prove that the discrete Fourier transforms

$$\frac{2}{JM} \textstyle\sum_n w_n \cos(2\pi kn/N) \quad \text{and} \quad \frac{2}{JM} \textstyle\sum_n w_n \sin(2\pi kn/N)$$

with $n = 0, \dots, JM - 1$ are equivalent to the ordinary least squares estimators of α_k and β_k, respectively.

Remark:
The sequences $\cos(2\pi kn/M)$ and $\sin(2\pi \ell n/M)$ for all k and ℓ, $\cos(2\pi kn/M)$ and $\cos(2\pi \ell n/M)$ for $k \neq \ell$, and $\sin(2\pi kn/M)$ and $\sin(2\pi \ell n/M)$ for $k \neq \ell$ are orthogonal on an integer number J of periods M.

5.18 The observations $w = (w_1 \dots w_N)^T$ have a covariance matrix C and expectations $Ew = X\theta$ where X is a known $N \times K$ matrix and θ is a $K \times 1$ vector of unknown parameters. Furthermore, F is any $K \times N$ matrix such that the $K \times K$ matrix FX is nonsingular.

(a) Show that $\check{t} = (FX)^{-1}Fw$ is a linear unbiased estimator of θ.

(b) Compute the covariance matrix of $\check{t}$.

(c) For which matrix F is $\check{t}$ best linear unbiased?

5.19 The observations w_n have expectations $Ew_n = \alpha x_n$, where the $x_n \neq 0$ are known and α is an unknown scalar parameter. With each additional observation, the ordinary least squares estimate $\hat{a}_n$ of this parameter is computed from the last N observations $w_{n-N+1}, \ldots, w_n$. Consequently, earlier observations are disregarded. Suppose that $\alpha = \alpha'$ for $n < N'$ and $\alpha = \alpha''$ otherwise. Then, for $n < N'$ we have $E\hat{a}_n = \alpha'$, and for $n \geq N' + N - 1$ we have $E\hat{a}_n = \alpha''$. On the interval $N' \leq n < N' + N - 1$, the estimator uses observations with expectation $\alpha' x_n$ and observations with expectation $\alpha'' x_n$. Show that the transition of the expectation of the estimator $\hat{a}_n$ from α' to α'' on $N' - 1 \leq n \leq N' + N - 1$ is monotonic.

CHAPTER 6

NUMERICAL METHODS FOR PARAMETER ESTIMATION

6.1 INTRODUCTION

In Chapter 5, we have seen that maximizing likelihood functions or minimizing least squares criteria with respect to parameters of expectation models usually results in a nonlinear optimization problem that cannot be solved in closed form. Therefore, it has to be solved by iterative numerical optimization. In this chapter, numerical optimization methods are discussed suitable for or specialized to such optimization problems. The relevant literature is vast and an exhaustive discussion is outside the scope of this book. Therefore, the discussion will be limited to a relatively small number of methods that have been found to solve the majority of the relevant practical parameter estimation and optimal design problems. Since maximizing a function is equivalent to minimizing its additive inverse, minimization methods discussed below are equally suitable for maximization and the converse.

The outline of this chapter is as follows. In Section 6.2, mathematical concepts basic to optimization are presented and their use in numerical optimization is explained. Also, reference log-likelihood functions and reference least squares criteria used for software testing are introduced. Section 6.3 is devoted to the steepest descent method. This is a general function minimization method. It is not specialized to optimizing least squares criteria or likelihood functions. The method converges under general conditions, but its rate of convergence may be impractical. This is improved by the Newton minimization method discussed in Section 6.4. This is also a general function minimization method, but the conditions for convergence are less general than those for the steepest descent method. In Section 6.5, the Fisher scoring method is introduced. This method is an approximation to the Newton method when used for maximizing log-likelihood functions and is, therefore, a specialized method. In Section 6.6, an expression is derived for the Newton iteration step for maximizing the log-likelihood function for normally distributed observations. As a consequence of the particular form of the log-likelihood function concerned, this step is also the Newton step for minimizing the nonlinear least squares criterion for observations of any distribution. From the Newton step for normal observations, a much simpler approximate step is derived. The method using this step is called the Gauss–Newton method and is the subject of Section 6.7. The Newton steps for maximizing the Poisson and the multinomial log-likelihood functions are discussed in Section 6.8 and Section 6.9, respectively. In Section 6.10, an expression is derived for the Newton step for maximizing the log-likelihood function if the distribution of the observations is a linear exponential family. From this Newton step, a much simpler approximate step is derived that is used by the generalized Gauss–Newton method. This method is the subject of Section 6.11. In Section 6.12, the iteratively reweighted least squares method is described. It is shown that it is identical to the generalized Gauss–Newton method and with the Fisher scoring method if the assumed distribution of the observations is a linear exponential family. Like the Newton method, the Gauss–Newton method solves a system of linear equations in each step. The Levenberg–Marquardt method, discussed in Section 6.13, is a version of the Gauss–Newton method that can handle (near-)singularity of these equations that could occur during the iteration process. Section 6.14 summarizes the numerical optimization methods discussed in this chapter. Finally, Section 6.15 is devoted to the methodology of estimating parameters of expectation models. In it, consecutive steps are proposed to be made in the process starting with the choice of model of the observations and ending with actually estimating the parameters.

6.2 NUMERICAL OPTIMIZATION

6.2.1 Key notions in numerical optimization

In this section, a number of key notions in numerical optimization is summarized. This summary will be restricted to notions relevant to optimizing log-likelihood functions and least squares criteria used for estimating parameters of expectation models and to optimizing experimental designs. As in the optimization literature, the function to be optimized will be called *objective function*. Furthermore, it will be assumed throughout that the objective functions considered are twice continuously differentiable.

The most important characteristic of maxima and minima is that they are *stationary points* of the objective function. A stationary point is a point where the gradient vector of the function vanishes. If $f(x)$ is a function of the elements of $x = (x_1 \dots x_K)^T$, then x^* is a stationary point if

$$\left.\frac{\partial f}{\partial x}\right|_{x=x^*} = \begin{pmatrix} \dfrac{\partial f}{\partial x_1} \\[2ex] \dfrac{\partial f}{\partial x_2} \\ \vdots \\ \dfrac{\partial f}{\partial x_K} \end{pmatrix}_{x=x^*} = o, \tag{6.1}$$

where, for simplicity, the argument of $f(x)$ has been left out and o is the $K \times 1$ null vector. Stationarity is a *necessary* condition for a point to be a maximum or a minimum. A *sufficient* condition for a stationary point to be a minimum is that at that point the Hessian matrix of the function is positive definite. That is,

$$\left.\frac{\partial^2 f}{\partial x\,\partial x^T}\right|_{x=x^*} = \begin{pmatrix} \frac{\partial^2 f}{\partial x_1^2} & \frac{\partial^2 f}{\partial x_1 \partial x_2} & \cdots & \frac{\partial^2 f}{\partial x_1 \partial x_K} \\ \frac{\partial^2 f}{\partial x_1 \partial x_2} & \frac{\partial^2 f}{\partial x_2^2} & & \vdots \\ \vdots & & \ddots & \vdots \\ \frac{\partial^2 f}{\partial x_1 \partial x_K} & \cdots & \cdots & \frac{\partial^2 f}{\partial x_K^2} \end{pmatrix}_{x=x^*} \succ O, \tag{6.2}$$

where O is the $K \times K$ null matrix. Similarly, a sufficient condition for a stationary point to be a maximum is that the Hessian matrix is negative definite, or

$$\left.\frac{\partial^2 f}{\partial x\,\partial x^T}\right|_{x=x^*} \prec O. \tag{6.3}$$

Definiteness of symmetric matrices is the subject of Appendix C. By Theorem C.6, a necessary and sufficient condition for positive definiteness of a symmetric matrix is that all eigenvalues of the matrix are positive. Thus, in practical problems, the test if a stationary point is a minimum may be conducted by computing the eigenvalues of the Hessian matrix at the stationary point and checking their signs. The test for a maximum is analogous.

The direction of

$$-\frac{\partial f}{\partial x}, \tag{6.4}$$

that is, the direction opposed to that of the gradient, is, by definition, a direction in which the function decreases. The direction of any $K \times 1$ vector y is called a *descent direction* if the scalar

$$\boxed{y^T\left(-\frac{\partial f}{\partial x}\right) > 0}\,. \tag{6.5}$$

Theorem B.1 shows that this condition is met if and only if the vector $-\partial f/\partial x$ and the orthogonal projection of y on it point in the same direction. Analogously, the direction of y is called an *ascent direction* if

$$y^T \frac{\partial f}{\partial x} > 0. \tag{6.6}$$

In this chapter, use will be made of the multivariate Taylor expansion. Suppose that $f(x)$ has continuous derivatives up to order p and define

$$\Delta x = (\Delta x_1 \; \Delta x_2 \; \ldots \; \Delta x_K)^T. \tag{6.7}$$

Then, Taylor's theorem states that in a neighborhood of the point x_o:

$$f(x_o + \Delta x) = f(x_o) + \sum_{n=1}^{p-1} \frac{1}{n!}\left(\Delta x_1 \frac{\partial}{\partial x_1} + \Delta x_2 \frac{\partial}{\partial x_2} + \cdots + \Delta x_K \frac{\partial}{\partial x_K}\right)^n f(x) + R_p, \tag{6.8}$$

where all derivatives are evaluated at x_o and R_p is the *remainder* defined by

$$R_p = \frac{1}{p!} \left(\Delta x_1 \frac{\partial}{\partial x_1} + \Delta x_2 \frac{\partial}{\partial x_2} + \cdots + \Delta x_K \frac{\partial}{\partial x_K} \right)^p f(x) \tag{6.9}$$

at $x = x_o + \nu \Delta x$ with $0 < \nu < 1$.

The *linear Taylor polynomial* is obtained from (6.8) by leaving out all quadratic and higher–degree terms:

$$f(x_o) + \frac{\partial f}{\partial x_1}\Delta x_1 + \frac{\partial f}{\partial x_2}\Delta x_2 + \cdots + \frac{\partial f}{\partial x_K}\Delta x_K = f(x_o) + \frac{\partial f}{\partial x^T}\Delta x. \tag{6.10}$$

A *linear function* is a function with a constant gradient. It is fully represented by its linear Taylor polynomial. The linear Taylor polynomial reduces to the constant $f(x_o)$ if $\partial f / \partial x$ is equal to the null vector, that is, if x_o is a stationary point. Therefore, in a sufficiently small neighborhood of a point x_o, the linear Taylor polynomial (6.10) may be used as an approximation to $f(x_o + \Delta x)$ unless x_o is stationary.

Similarly, the *quadratic Taylor polynomial* is described by

$$f(x_o) + \frac{\partial f}{\partial x_1}\Delta x_1 + \frac{\partial f}{\partial x_2}\Delta x_2 + \cdots + \frac{\partial f}{\partial x_K}\Delta x_K + \frac{1}{2!}\sum_{k,\ell} \frac{\partial^2 f}{\partial x_k \partial x_\ell}\Delta x_k \Delta x_\ell$$

$$= f(x_o) + \frac{\partial f}{\partial x^T}\Delta x + \frac{1}{2!}\Delta x^T \frac{\partial^2 f}{\partial x\, \partial x^T}\Delta x\,, \tag{6.11}$$

where $k, \ell = 1, \ldots, K$. A *quadratic function* is defined as a function with a constant Hessian matrix $\partial^2 f / \partial x \partial x^T$. It is exactly represented by its quadratic Taylor polynomial. If x_o is a stationary point, the quadratic Taylor polynomial reduces to

$$f(x_o) + \frac{1}{2!}\Delta x^T \frac{\partial^2 f}{\partial x \partial x^T}\Delta x\,. \tag{6.12}$$

Therefore, in a sufficiently small neighborhood of the stationary point x_o, (6.12) may be used as an approximation to the function $f(x_o + \Delta x)$. Since a minimum is a stationary point, an objective function will in a sufficiently small neighborhood of the minimum behave like a quadratic function. *This is the reason why fast convergence for quadratic functions is generally considered as a minimum requirement to be met by numerical minimization methods.*

6.2.2 Reference log-likelihood functions and least squares criteria

By definition, a log-likelihood function $q(w; t)$ is a function of the elements of the parameter vector t and is parametric in the observations w. As a result, the location of the absolute maximum of the log-likelihood function depends on the particular realization of the observations used. Therefore, this location is, typically, unpredictable and can be determined by numerical optimization only. Similar considerations apply to the location of the absolute minimum of the nonlinear least squares criterion. This implies that simulated or actually measured statistical observations are not suitable for testing software for log-likelihood maximizing and nonlinear least squares minimizing since the outcome is unknown. To cope with this difficulty, we introduce two artificial but very practical concepts: (a) *exact observations* and (b) *reference log-likelihood functions* or *reference least squares criteria*.

Exact observations are defined as observations that are equal to their expectations:

$$w_n = Ew_n = g_n(\theta). \tag{6.13}$$

They need not exist. For example, if observations have a Poisson distribution, they are integers. However, their expectation $g_n(\theta)$ is, typically, not integer and can, therefore, not be an observation generated by the Poisson distribution. This is the reason why we called the concept exact observations *artificial*.

The definition of the reference log-likelihood function follows from the definition of the exact observations. It is the log-likelihood function $q(w;t)$ for the exact observations $w = g(\theta)$, that is, $q(g(\theta);t)$ with $g(\theta) = [g_1(\theta) \ \dots \ g_N(\theta)]^T$. The reference least squares criterion is defined similarly .

■ EXAMPLE 6.1

The reference ordinary least squares criterion

The ordinary nonlinear least squares criterion is defined as

$$J(t) = \sum_n [w_n - g_n(t)]^2 . \tag{6.14}$$

Substituting the exact observations $w_n = g_n(\theta)$ in this expression yields the reference least squares criterion

$$J(t) = \sum_n [g_n(\theta) - g_n(t)]^2 . \tag{6.15}$$

Then, $J(t)$ is absolutely minimum and equal to zero if $t = \theta$. ■

■ EXAMPLE 6.2

The reference log-likelihood function for Poisson distributed observations

For independent, Poisson distributed observations $w = (w_1 \dots w_N)^T$, the log-likelihood function is described by (5.112):

$$q(w;t) = \sum_n -g_n(t) + w_n \ln g_n(t) - \ln w_n! . \tag{6.16}$$

The version of this function parametric in continuous observations w_n is described by

$$q(w;t) = \sum_n -g_n(t) + w_n \ln g_n(t) - \ln \Gamma(w_n + 1) , \tag{6.17}$$

where $\Gamma(w_n + 1)$ is the *gamma function* which is defined for $w_n \geq 0$ and has the properties $\Gamma(w_n + 1) = w_n \Gamma(w_n)$ and $\Gamma(1) = 1$. Thus, (6.16) is consistent with (6.17) if w_n is integer since then $\Gamma(w_n + 1) = w_n!$. If, subsequently, the exact observations are substituted in (6.17), the reference log-likelihood function is obtained:

$$q(g(\theta);t)) = \sum_n -g_n(t) + g_n(\theta) \ln g_n(t) - \ln \Gamma(g_n(\theta) + 1) . \tag{6.18}$$

Elementary calculations show that this function is maximized by $t = \theta$. ■

Examples 6.1 and 6.2 show that in least squares and maximum likelihood problems exact observations may be used to test parameter estimation software since the solutions corresponding to these observations are the exact parameter values θ.

6.3 THE STEEPEST DESCENT METHOD

6.3.1 Definition of the steepest descent step

In the preceding section, the concept descent direction has been introduced but the question in which direction the objective function decreases most has not been posed. The answer is of course of great importance for numerical optimization and will be dealt with in this section. From here on, $f(t)$ will denote the objective function. Thus, in this book, $f(t)$ will, often but not always, be the log-likelihood function $q(w;t)$ or the least squares criterion $J(t)$.

The gradient of $f(t)$ with respect to the vector t at the *current point* $t = t_c$ is described by

$$\left.\frac{\partial f(t)}{\partial t}\right|_{t=t_c}. \tag{6.19}$$

Suppose that a vector of increments

$$\Delta t = (\Delta t_1 \cdots \Delta t_K)^T \tag{6.20}$$

is added to t_c and consider all Δt of length Δ defined by

$$\Delta^2 = \|\Delta t\|^2 = (\Delta t_1)^2 + \cdots + (\Delta t_K)^2 . \tag{6.21}$$

Then, if Δ is taken sufficiently small, $\Delta f = f(t_c + \Delta t) - f(t_c)$ may be approximated by

$$\overline{\Delta f} = \overline{f(t_c + \Delta t)} - f(t_c) = \frac{\partial f}{\partial t_1}\Delta t_1 + \cdots + \frac{\partial f}{\partial t_K}\Delta t_K , \tag{6.22}$$

where $\overline{f(t_c + \Delta t)}$ is defined as the linear Taylor polynomial

$$\overline{f(t_c + \Delta t)} = f(t_c) + \frac{\partial f}{\partial t_1}\Delta t_1 + \cdots + \frac{\partial f}{\partial t_K}\Delta t_K \tag{6.23}$$

and the derivatives of $f = f(t)$ are taken at $t = t_c$. The following question then arises: Which Δt produces the absolutely largest, negative $\overline{\Delta f}$ under the equality constraint (6.21)?

The solution must be a stationary point of the Lagrangian function

$$\varphi(\Delta t, \lambda) = \overline{\Delta f} + \lambda(\Delta^2 - \|\Delta t\|^2) . \tag{6.24}$$

where the scalar λ is the Lagrange multiplier. These stationary points satisfy

$$\frac{\partial \varphi}{\partial \Delta t_k} = \frac{\partial f}{\partial t_k} - 2\lambda \Delta t_k = 0 \tag{6.25}$$

with $k = 1, \ldots, K$, and

$$\frac{\partial \varphi}{\partial \lambda} = \Delta^2 - \|\Delta t\|^2 = 0 , \tag{6.26}$$

where the arguments of $\varphi(\Delta t, \lambda)$ have been omitted. Equation (6.25) shows that

$$\Delta t_k = \frac{1}{2\lambda}\frac{\partial f}{\partial t_k} \tag{6.27}$$

and, therefore,

$$\Delta t = \frac{1}{2\lambda}\frac{\partial f}{\partial t}. \tag{6.28}$$

Substituting this in (6.26) yields

$$\frac{1}{4\lambda^2}\left\|\frac{\partial f}{\partial t}\right\|^2 = \Delta^2. \tag{6.29}$$

By (6.5), the direction of Δt is a descent direction if and only if

$$-\frac{\partial f}{\partial t^T}\Delta t > 0. \tag{6.30}$$

Then, (6.28) shows that λ must be negative and, by (6.29), equal to

$$\lambda = -\frac{1}{2\Delta}\left\|\frac{\partial f}{\partial t}\right\|. \tag{6.31}$$

The corresponding step is the *steepest descent step* Δt_{SD}. It is, by (6.28), equal to

$$\boxed{\Delta t_{SD} = -\frac{\dfrac{\partial f}{\partial t}}{\left\|\dfrac{\partial f}{\partial t}\right\|}\Delta}, \tag{6.32}$$

where the derivatives of $f = f(t)$ are taken at $t = t_c$. The vector

$$\frac{\dfrac{\partial f}{\partial t}}{\left\|\dfrac{\partial f}{\partial t}\right\|} \tag{6.33}$$

is the normalized gradient. Therefore, the step defined by (6.32) has a step length Δ and a direction opposite to that of the gradient. The minimization method employing this step is called *steepest descent method.*

Then, one iteration of the steepest descent method in its most elementary form may consist of the following steps:

1. Test if the chosen conditions for convergence are met. If not, go to 2. Otherwise, stop and take t_c as the solution.

2. Compute the gradient of the objective function $f(t)$ at the point $t = t_c$.

3. From the gradient, compute the steepest descent step.

4. Compute $f(t_c + \Delta t_{SD})$. If $f(t_c + \Delta t_{SD}) < f(t_c)$, take $t_c + \Delta t_{SD}$ as new t_c, and go to 1. Else, reduce step length Δ, compute the corresponding Δt_{SD}, and repeat this step.

The test in Step 4 is needed since in the neighborhood of t_c the function $f(t)$ may be such that the current step length Δ is too large for the linear approximation of $f(t)$ to be valid.

The procedure shows that the computational effort in every iteration consists almost entirely of the computation of the gradient of the objective function.

The *steepest ascent step* Δt_{SA} used for numerically *maximizing* is the additive inverse of the steepest descent step:

$$\Delta t_{SA} = -\Delta t_{SD}. \tag{6.34}$$

For a number of frequently occurring log-likelihood functions, the gradient vectors, to be used in the steepest ascent method, are restated in the following examples. They give an impression of the computational effort involved.

■ EXAMPLE 6.3

Gradient vectors of normal and Poisson log-likelihood functions

The gradient vector of the normal log-likelihood function is described by (5.60):

$$s_t = \frac{\partial g^T(t)}{\partial t} C^{-1} d(t), \tag{6.35}$$

where the covariance matrix C is constant. The gradient vector of the Poisson log-likelihood function is described by (5.114) and (5.115):

$$s_t = \frac{\partial g^T(t)}{\partial t} C^{-1} d(t), \tag{6.36}$$

and

$$C = \text{diag}\, g(t), \tag{6.37}$$

which shows that the covariance matrix C depends on t. ■

Since the normal and the Poisson distribution are linear exponential families, (6.35) and (6.36) are special cases of the general expression for the gradient of log-likelihood functions for distributions that are linear exponential families. This gradient is restated in the following example.

■ EXAMPLE 6.4

The gradient vector of the log-likelihood function for a linear exponential family of distributions

The gradient vector of the log-likelihood function for a linear exponential family of distributions is described by (5.129):

$$s_t = \frac{\partial g^T(t)}{\partial t} C^{-1} d(t), \tag{6.38}$$

where, generally, the covariance matrix of the observations C is a function of the parameters t as in (6.36). ■

6.3.2 Properties of the steepest descent step

6.3.2.1 Convergence Unless t_c is a stationary point, the direction of the steepest descent step is a descent direction. This means that reducing $f(t)$ can always be achieved if a sufficiently small step length is chosen. Unfortunately, as will be shown below, this almost guaranteed convergence does not always imply a fast convergence rate.

6.3.2.2 Direction of the steepest descent step A *contour* is a collection of points of constant function value. In two dimensions, contours are equivalent to contour lines on a map. Suppose that a step Δt is made from t_c to a neighboring point on the contour $f(t) = f(t_c)$. Then, by definition, $\Delta f = f(t_c + \Delta t) - f(t_c) = 0$. Therefore,

$$\overline{\Delta f} = \overline{f(t_c + \Delta t)} - f(t_c) = \frac{\partial f}{\partial t_1}\Delta t_1 + \cdots + \frac{\partial f}{\partial t_K}\Delta t_K = \frac{\partial f}{\partial t^T}\,\Delta t \approx 0 \tag{6.39}$$

if the step length Δ is sufficiently small. The conclusion is that the contour $f(t) = f(t_c)$ and the gradient of $f(t)$ at $t = t_c$ are orthogonal. Therefore, the steepest descent step Δt_{SD} and the contour are orthogonal as well. This property is helpful in understanding the convergence properties of the steepest descent method as is illustrated in the following numerical examples.

■ **EXAMPLE 6.5**

Behavior of the steepest descent method for quadratic functions

A simple example of a quadratic function occurring in parameter estimation is the ordinary least squares criterion (5.223) for the straight-line model. The corresponding least squares solution is closed-form and is described by (5.228). Therefore, there is no need to

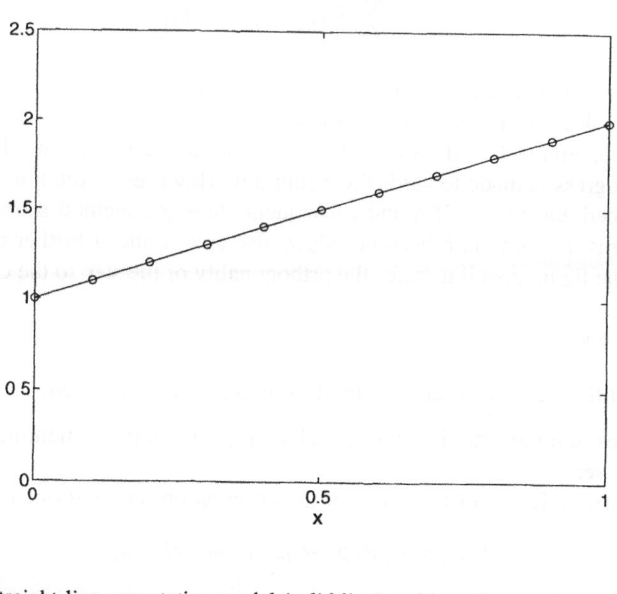

Figure 6.1. Straight-line expectation model (solid line) and its values at the measurement points (circles). (Example 6.5)

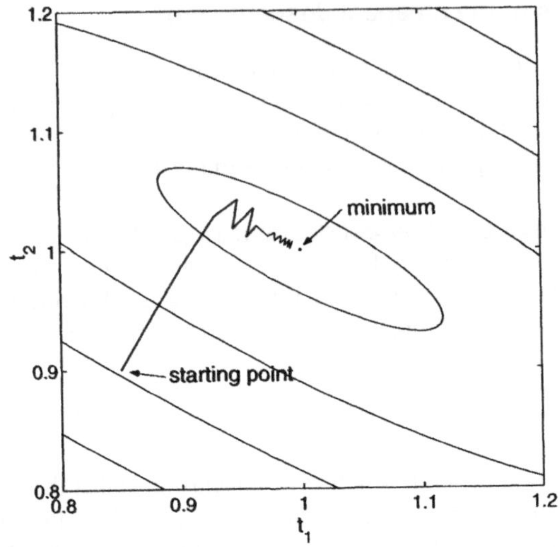

Figure 6.2. Contours of quadratic criterion and the first 20 iterations of the steepest descent method. (Example 6.5)

solve this problem iteratively, but this example is intended to investigate the performance of the steepest descent method if applied to quadratic functions.

In this example, the expectations of the observations are described by $Ew_n = \theta_1 x_n + \theta_2$ with $\theta_1 = \theta_2 = 1$ and $x_n = (n-1) \times 0.1$ with $n = 1, \ldots, 11$. They are shown in Fig. 6.1. Figure 6.2. shows contours of the reference least squares criterion

$$J(t) = \sum_n (w_n - t_1 x_n - t_2)^2 \tag{6.40}$$

with $w_n = Ew_n$ and the minimum located at $(t_1, t_2) = (\theta_1, \theta_2) = (1, 1)$.

The steepest descent minimization is started at the point $(t_1, \ t_2) = (0.85, 0.9)$ with a step length 0.05. Figure 6.2. shows its progress in the first 20 iterations. In the first three steps, good progress is made towards the minimum. However, in the fourth step, the step length is reduced and in the fifth and subsequent steps the method starts to zigzag and progress towards the minimum becomes slow, also as a result of further reduction of the step length. The figure also illustrates the orthogonality of the step to the contours. ■

■ **EXAMPLE 6.6**

Behavior of the steepest descent method for nonquadratic functions

As examples of nonquadratic nonlinear objective functions two nonlinear least squares criteria are chosen.

In the first example, the expectations of the observations are described by

$$Ew_n = \theta_1 \exp(-\theta_2 x_n) \cos(2\pi\theta_3 x_n). \tag{6.41}$$

Figure 6.3.(a) shows this model and its values at the measurement points. The latter are random numbers uniformly distributed on the interval $[0, 1.5]$ to emphasize that the obser-

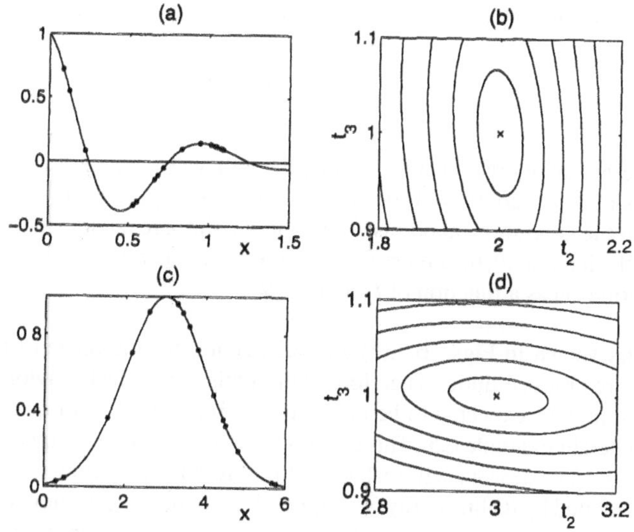

Figure 6.3. Expectation models and corresponding reference least squares criteria. (a) Exponentially damped cosine expectation model (solid line) and its values at the measurement points (circles). (b) Contours of corresponding reference nonlinear least squares criterion as a function of the decay constant and frequency with minimum (cross). (c) Gaussian peak expectation model (solid line) and its values at the measurement points (circles). (d) Contours of the corresponding reference nonlinear least squares criterion as a function of the location and half-width of the peak with minimum (cross). The linear parameters, the amplitude of the damped cosine, and the height of the peak have been eliminated from the criterion. (Example 6.6)

vations need not be equidistant. The parameters are the amplitude θ_1, the decay constant θ_2, and the frequency θ_3. They have been chosen as $\theta_1 = 1$, $\theta_2 = 2$, and $\theta_3 = 1$. The reference least squares criterion is described by

$$J(t) = \sum_n [w_n - t_1 \exp(-t_2 x_n) \cos(2\pi t_3 x_n)]^2 \tag{6.42}$$

with $w_n = Ew_n$. At the minimum,

$$\partial f(t)/\partial t_1 = 2 \sum_n [w_n - t_1 h_n(t_2, t_3)]\, h_n(t_2, t_3) = 0 \tag{6.43}$$

with

$$h_n(t_2, t_3) = \exp(-t_2 x_n) \cos(2\pi t_3 x_n). \tag{6.44}$$

Solving (6.43) for t_1 yields

$$t_1 = \frac{\sum_n w_n h_n}{\sum_n h_n^2}, \tag{6.45}$$

where $h_n = h_n(t_2, t_3)$. Substituting this expression for t_1 in (6.42) produces a least squares criterion that is a function of t_2 and t_3 only. Figure 6.3.(b) shows a number of contours of this criterion for a suitable range of (t_2, t_3) values. In the second example, a similar study is made for the Gaussian peak expectation model. Here, the expectations of the observations are described by

$$Ew_n = \theta_1 \exp\left(-\frac{1}{2}\left(\frac{x_n - \theta_2}{\theta_3}\right)^2\right), \tag{6.46}$$

where θ_1 is the height, θ_2 the location, and θ_3 the half-width parameter of the peak. The values of the parameters are $\theta_1 = 1$, $\theta_2 = 3$, and $\theta_3 = 1$. Figure 6.3.(c) shows this model and its values at the measurement points. Here, these points are random numbers uniformly distributed on the interval $[0, 6]$. Figure 6.3.(d) shows a number of contours of the reference least squares criterion as a function of t_2 and t_3. ■

The contours shown in Figs. 6.3.(b) and 6.3.(d) are, for the range of the parameters chosen, still more or less elliptic. Therefore, within this range, the behavior of the steepest descent method may be expected to be more or less similar to that for quadratic objective functions sketched in Example 6.5. This implies that, also for these nonquadratic functions, the method of steepest descent will make rapid progress as long as the objective function may be approximated by its linear Taylor polynomial around the current point. In any case, in the neighborhood of the minimum, this approximation is no longer valid and, as we have seen, an approximation by the quadratic Taylor polynomial should be used instead. This quadratic approximation is the basis of the *Newton optimization method* described in Section 6.4.

6.4 THE NEWTON METHOD

6.4.1 Definition of the Newton step

Like the steepest ascent and descent methods, the Newton method is a *general* numerical function optimization method. It has not been especially designed for minimizing least squares criteria or for maximizing likelihood functions. The principle of the Newton method is to approximate the objective function by its quadratic Taylor polynomial about he current point, to compute the stationary point of this quadratic approximation, and then to use this point as current point in the next step.

The Newton step Δt_{NE} is derived as follows. Suppose that t_c is the current point and that Δt is a vector of increments of t_c. Then, if Δt is taken sufficiently small, $f(t_c + \Delta t)$ may be approximated by its quadratic Taylor polynomial $\overline{f(t_c + \Delta t)}$ described by

$$\overline{f(t_c + \Delta t)} = f(t_c) + \frac{\partial f}{\partial t^T}\Delta t + \frac{1}{2!}\Delta t^T \frac{\partial^2 f}{\partial t \partial t^T}\Delta t, \tag{6.47}$$

where the derivatives of $f = f(t)$ are evaluated at $t = t_c$. Since the optimum to be found is a stationary point of $\overline{f(t_c + \Delta t)}$ with respect to Δt :

$$\frac{\partial \overline{f(t_c + \Delta t)}}{\partial(\Delta t)} = \frac{\partial f}{\partial t} + \frac{\partial^2 f}{\partial t \partial t^T}\Delta t = o, \tag{6.48}$$

where Lemma 5.1 and Lemma 5.2 have been used. Therefore,

$$\boxed{\Delta t_{NE} = -\left(\frac{\partial^2 f}{\partial t \partial t^T}\right)^{-1} \frac{\partial f}{\partial t}}, \tag{6.49}$$

where the derivatives of $f = f(t)$ are evaluated at $t = t_c$. Expression (6.49) shows that the Newton method requires in every iteration the computation of the gradient and the Hessian matrix of the objective function at $t = t_c$ and, in addition, the solution of a system of linear equations.

6.4.2 Properties of the Newton step

The derivation of the Newton step in Section 6.4.1 shows that all that can be said about Δt_{NE} is that it is a stationary point of $\overline{f(t_c + \Delta t)}$. It may be a maximum, a minimum, or a saddle point. To see this, consider the Hessian matrix of $\overline{f(t_c + \Delta t)}$ at $\Delta t = \Delta t_{NE}$. Applying Lemma 5.3 to (6.47) shows that it is described by

$$\frac{\partial^2 \overline{f(t_c + \Delta t)}}{\partial(\Delta t)\, \partial(\Delta t)^T} = \frac{\partial^2 f}{\partial t \partial t^T}. \tag{6.50}$$

If the Hessian matrix $\partial^2 f/\partial t \partial t^T$ in this expression is positive definite, the point Δt_{NE} is a minimum of $\overline{f(t_c + \Delta t)}$. If it is negative definite, it is a maximum. If it is indefinite, Δt_{NE} is a saddle point. If $f(t)$ *is* quadratic, $f(t_c + \Delta t)$ and $\overline{f(t_c + \Delta t)}$ coincide and the Newton method converges in one step to this minimum, maximum, or saddle point. Generally, the direction of the Newton step is a descent direction if

$$-\frac{\partial f}{\partial t^T} \Delta t_{NE} = \frac{\partial f}{\partial t^T} \left(\frac{\partial^2 f}{\partial t \partial t^T} \right)^{-1} \frac{\partial f}{\partial t} > 0\,. \tag{6.51}$$

This is true for all nonzero gradients $\partial f/\partial t$ if, at $t = t_c$, the matrix $(\partial^2 f/\partial t \partial t^T)^{-1}$ and, therefore, the Hessian matrix $\partial^2 f/\partial t \partial t^T$ are positive definite. Similarly, the direction of the Newton step is an ascent direction if the Hessian matrix is negative definite. It is either an ascent or a descent direction if the Hessian matrix is indefinite.

6.4.2.1 The univariate Newton step To simplify the discussion of relevant properties of the Newton method, first univariate Newton optimization is considered. The following elementary example describes such a univariate problem.

■ **EXAMPLE 6.7**

The Newton method applied to a quartic polynomial

In this example, the behavior of the Newton method is investigated if we would apply it to the quartic polynomial

$$f(t) = -0.25t^4 + 3.75t^2 - t + 20\,. \tag{6.52}$$

This function has been chosen since it is simple and illustrative. In practice, a quartic polynomial is not minimized by means of the Newton method. Its first-order derivative is a cubic polynomial. For the roots of cubic polynomials, closed-form expressions are available. The real roots among these are the stationary points of the quartic polynomial. The second order derivative evaluated at these stationary points reveals their nature.

The quartic polynomial is shown in Fig. 6.4.(a). It has three stationary points: an absolute maximum, a relative minimum, and a relative maximum located at $t = -2.8030$, $t = 0.1337$, and $t = 2.6693$, respectively. These are the zero–crossings of its first-order derivative shown in Fig. 6.4.(b). The function has two points of inflection located at

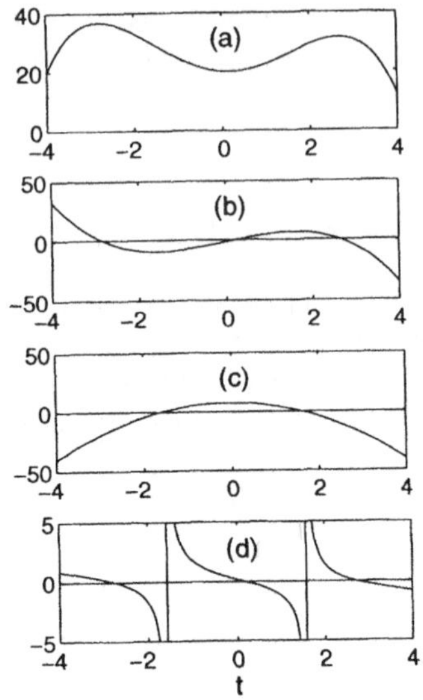

Figure 6.4. (a) Quartic polynomial with (b) its first-order derivative, (c) second-order derivative, and (d) the corresponding Newton step. (Example 6.7)

$t = 1.5811$ and $t = -1.5811$. These are the zero–crossings of the second-order derivative shown in Fig. 6.4.(c).

The Newton step is equal to

$$\Delta t_{NE} = -\frac{\dfrac{df(t)}{dt}}{\dfrac{d^2 f(t)}{dt^2}}, \tag{6.53}$$

which is the additive inverse of the ratio of the first-order derivative to the second-order derivative. It is shown in Fig. 6.4.(d). If it is positive, the step is made in the direction of the positive t-axis. A negative sign means the opposite direction. The figure also shows vertical asymptotes located at both points of inflection. If these points are approached, the Newton step goes to plus or minus infinity. Figure 6.4.(d) shows that the Newton step behaves differently on each of the three intervals separated by the points of inflection. On the left-hand and on the right-hand interval, the step is made in the direction of the absolute and the relative maximum, respectively. On the middle interval, it is made in the direction of the relative minimum. Thus, the Newton step is made in the direction of the stationary point located on the interval concerned. Hence, if the initial point is sufficiently close to one of these stationary points and, therefore, the step length is sufficiently small, the method will converge to that stationary point. However, on all intervals, the step length may become large at points close to a point of inflection. This has consequences for the behavior of the Newton method. For example, suppose that the initial point is located on the middle interval near one of both points of inflection. Then, the step may become so large that the method arrives in the first step at a point on the right-hand or the left-hand interval where the quadratic approximation used by the Newton method on the middle interval is no longer valid. Also, if the initial point is located on the left-hand or the right-hand interval near a point of inflection, the method may arrive in the first step in a point far away from the absolute or relative maximum instead of coming closer to these maxima. ■

The difficulties with the Newton method in Example 6.7 are caused by the occurrence of a large step length. As a remedy, the step length may be reduced. Since on the three different intervals, the Newton step is directed towards the maximum or minimum located on the interval concerned, convergence to these extrema may thus be expected. Then, on the middle interval the Newton method converges to the local minimum, and on the left-hand and right-hand interval it converges to the absolute and relative maximum, respectively. If the method is intended to *maximize* the function, this implies that it is successful in this respect on the left-hand and the right-hand interval only. This behavior is essentially different from that of the steepest ascent method. If this method had been applied to the maximization problem of Example 6.7, the plot of the first order derivative depicted by Fig. 6.4.(b) shows that it would have converged to the absolute maximum if the starting point had been chosen to the left of the relative minimum and to the relative maximum otherwise. That is, the steepest ascent method would always have converged to a maximum, be it not necessarily to the absolute maximum. These considerations suggest the solution of maximization problems like these by first applying the steepest ascent method until the function has sufficiently increased and then using the result of the steepest ascent method as initial value for the subsequent Newton method. In any case, "sufficiently increased" implies that the second-order derivative has become negative. If the second-order derivative does not stay negative after a Newton iteration, the iteration should be repeated using a part of the Newton step only. The next iteration should start using again a full Newton step to prevent the convergence from becoming slow. To increase the probability that the absolute maximum is found, this procedure has to be repeated from different starting points followed by selection of the solution corresponding to the largest objective function value.

6.4.2.2 The multivariate Newton step It will now be shown that the properties of the vector–valued Newton step are analogous to those of the univariate step just described. This will be done using an extensive example of Newton maximization of a log-likelihood function with respect to three parameters. For log-likelihood functions, the Newton step (6.49) is described by

$$\boxed{\Delta t_{NE} = -\left(\frac{\partial^2 q(w;t)}{\partial t \partial t^T}\right)^{-1} s_t} \tag{6.54}$$

with

$$s_t = \frac{\partial q(w;t)}{\partial t}. \tag{6.55}$$

■ **EXAMPLE 6.8**

Newton maximization of a likelihood function of three parameters.

Suppose that observations $w = (w_1 \ldots w_N)^T$ are available with expectations $Ew = g(\theta)$. Furthermore, suppose that the observations are independent and Poisson distributed. Then, their log-likelihood function is described by (5.112):

$$q(w;t) = \sum_n -g_n(t) + w_n \ln g_n(t) - \ln w_n!\,. \tag{6.56}$$

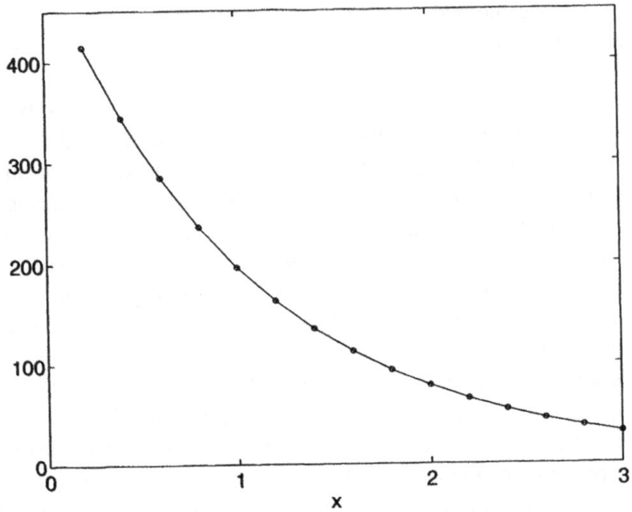

Figure 6.5. Biexponential expectation model (solid line) and its values at the measurement points (circles). (Example 6.8)

In this example, the expectations of the observations are described by

$$Ew_n = g_n(\theta) = \alpha[\rho \exp(-\beta_1 x_n) + (1-\rho)\exp(-\beta_2 x_n)] \tag{6.57}$$

with $\theta = (\alpha\ \beta_1\ \beta_2)^T$, where the amplitude α and the decay constants $\beta_1$and β_2 are the positive but otherwise unknown parameters. The known parameter ρ satisfies $0 < \rho < 1$ and distributes the amplitude α over the exponentials. Below, the following notation will be used:

$$g_n(\theta) = \alpha h_n(\beta_1, \beta_2) \tag{6.58}$$

with

$$h_n(\beta_1, \beta_2) = \rho \exp(-\beta_1 x_n) + (1-\rho)\exp(-\beta_2 x_n). \tag{6.59}$$

Then, substituting $a h_n(b_1, b_2)$ for $g_n(t)$ in (6.56) produces the log-likelihood function of $t = (a\ b_1\ b_2)^T$:

$$q(w;t) = -a\sum h_n + \ln a \sum w_n + \sum w_n \ln h_n - \sum \ln w_n!\,, \tag{6.60}$$

where all summations are over n and, for brevity, $h_n = h_n(b_1, b_2)$. Equation (6.60) shows that

$$s_a = -\sum h_n + \frac{1}{a}\sum w_n. \tag{6.61}$$

At the maximum of the log-likelihood function, $s_a = 0$ and, therefore,

$$a = \frac{\sum w_n}{\sum h_n}. \tag{6.62}$$

Substituting this result in (6.60) yields a function of b_1 and b_2 only:

$$\left[-1 + \ln(\sum w_n) - \ln(\sum h_n)\right]\sum w_n + \sum w_n \ln h_n - \sum \ln w_n!\,. \tag{6.63}$$

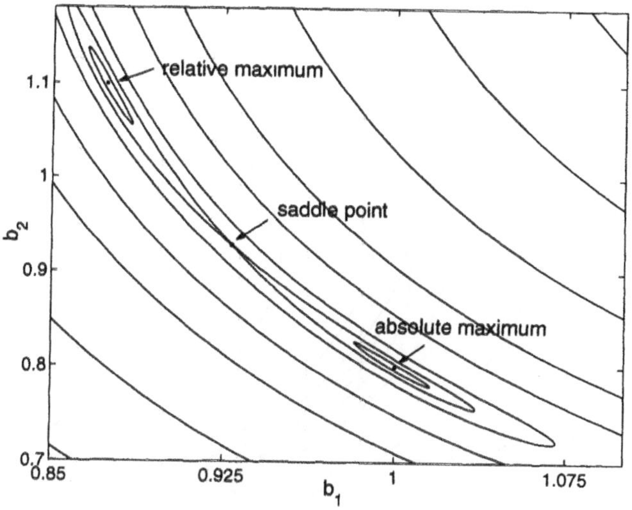

Figure 6.6. Contours of the Poisson log-likelihood function. (Example 6.8)

The terms of this expression dependent on b_1 and b_2 are

$$- \ln(\sum h_n) \sum w_n + \sum w_n \ln h_n. \tag{6.64}$$

The maximum likelihood estimator $(\tilde{b}_1, \tilde{b}_2)$ of (β_1, β_2) maximizes (6.63) while the corresponding maximum likelihood estimator $\tilde{a}$ of α is (6.62) with $(\tilde{b}_1, \tilde{b}_2)$ substituted for (b_1, b_2).

In this example, the reference log-likelihood function is chosen as log-likelihood function. Furthermore, the model is described by (6.58) and (6.59) with $\alpha = 500$, $\beta_1 = 1$, $\beta_2 = 0.8$. The known parameter ρ is 0.7. The measurement points are $x_n = n \times 0.2$ with $n = 1, \ldots, 15$. Figure 6.5. shows the biexponential model with these parameters and its values at the measurement points. Figure 6.6. shows contours of the log-likelihood function. Since the observations are exact, the absolute maximum is located at $(b_1, b_2) = (1, 0.8)$, which are the true parameter values. There is an additional, relative, maximum at $(b_1, b_2) = (0.8735, 1.0993)$. Its presence may be explained as follows. If the parameter ρ would be equal to 0.5 instead of 0.7, the parameters b_1 and b_2 would be interchangeable and the log-likelihood function would have two equivalent absolute maxima $(\tilde{b}_1, \tilde{b}_2)$ and $(\tilde{b}_2, \tilde{b}_1)$. If, subsequently, ρ is taken different from 0.5, the maxima continue to exist but an asymmetry is introduced which causes one of the maxima to become relative. Finally, there is a saddle point at $(0.9282, 0.9282)$. This point lies in the plane $b_1 = b_2$ where the model is a monoexponential $a \exp(-bx)$ and the log-likelihood function is the log-likelihood function for this model. The saddle point, therefore, represents the maximum $\tilde{b}$ of the log-likelihood function for the monoexponential model and is located at $(\tilde{b}, \tilde{b})$.

For an explanation of the behavior of the Newton method for log-likelihood functions like that of Fig. 6.6., the Hessian matrix of the log-likelihood function must be studied as a function of the parameters. Clearly, at both maxima the Hessian matrix is negative definite while at the saddle point it is indefinite. Apparently, there is a region of the (b_1, b_2) plane where the Hessian matrix is negative definite and a region where it is indefinite. The simplest

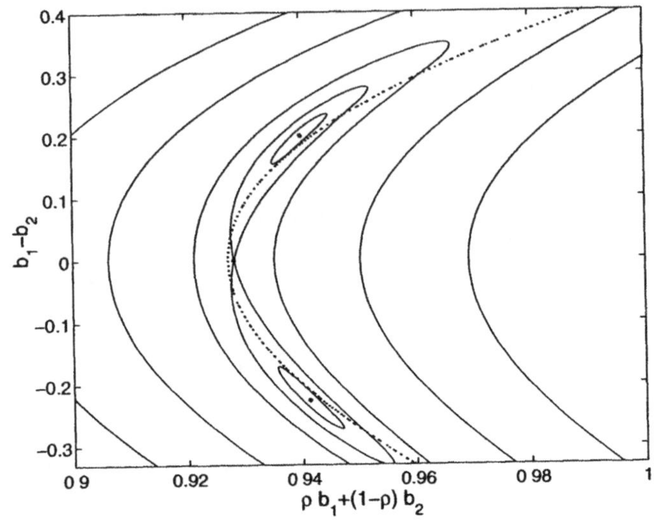

Figure 6.7. The contours of the Poisson log-likelihood function of Fig. 6.6. in transformed coordinates (solid lines) and the collection of points where the Hessian matrix is singular (dotted line). (Example 6.8)

way to find these regions is to compute their border. Since, in this two-dimensional example, negative definite is equivalent to two negative eigenvalues and indefinite is equivalent to one positive and one negative eigenvalue, the border may be found by computing the zero crossings of one of the eigenvalues, or, equivalently, those of the determinant of the Hessian matrix. The latter approach has been followed here. Figure 6.7. shows the results. The contours in this figure are the five innermost contours of Fig. 6.6. but they are displayed in the transformed coordinates $(\rho b_1 + (1-\rho)b_2, b_1 - b_2)$ instead of (b_1, b_2) to make the plot clearer. The dotted line in Fig. 6.7. is the collection of points where the determinant vanishes. To the left of this line the Hessian matrix of the log-likelihood function is negative definite, to the right it is indefinite. On the dotted line, the Hessian matrix is negative semidefinite and, therefore, singular. This implies that for points on the line the Newton step does not exist. The behavior of the Newton method in both regions will now be demonstrated by starting from three different points. Throughout, half the Newton step has been used to avoid steps so large that one of the eigenvalues changes sign. In Fig. 6.8., $(b_1, b_2) = (1.075, 1)$ is Starting Point 1. This point is located in the region where the Hessian matrix is indefinite. The dots indicate the current points in the subsequent iterations. The figure shows that the Newton method converges to the saddle point. If, on the other hand, the starting point is $(b_1, b_2) = (0.875, 0.975)$, which is Starting Point 2 located in the region where the Hessian matrix is negative definite, the Newton method converges to the relative maximum (squares). Finally, starting at $(b_1, b_2) = (0.875, 0.725)$, which is Starting Point 3 also located in the region where the Hessian matrix is negative definite, the Newton method converges to the absolute maximum (diamonds). It is concluded that the method converges to the saddle point if the initial point is chosen in the region where the Hessian matrix is indefinite and to the absolute or relative maximum if it is chosen in the region where this matrix is negative definite.

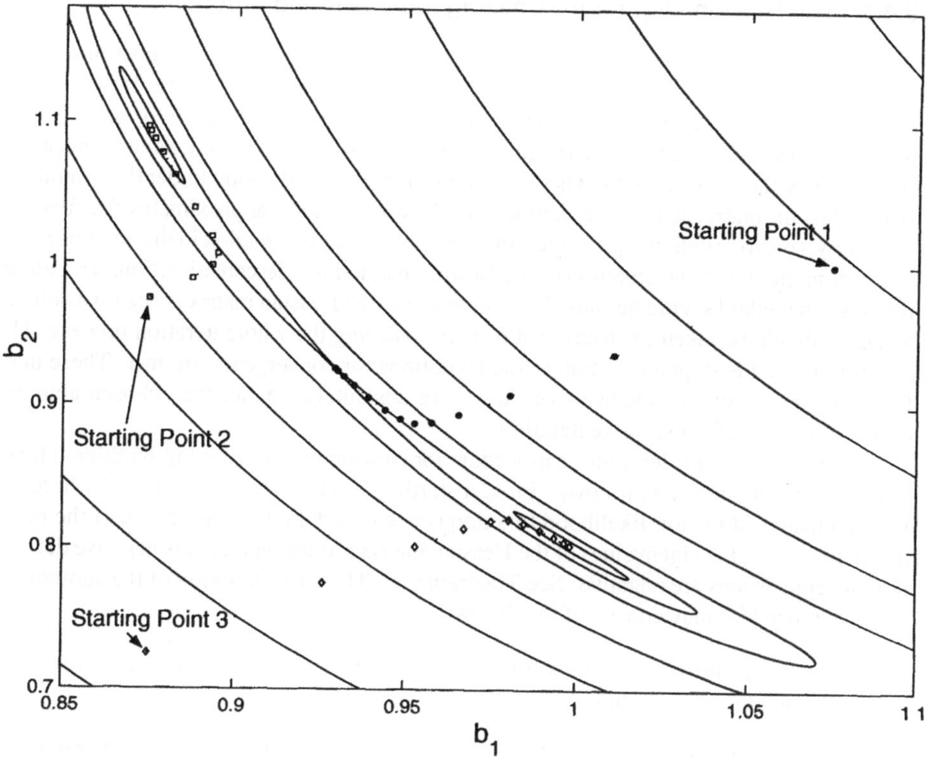

Figure 6.8. Paths of the Newton method applied to the log-likelihood function of Fig. 6.6. for different starting points. (Example 6.8)

A final observation is that in the neighborhood of the maxima the rate of convergence is fast as compared to that of the steepest ascent method. No zigzagging occurs. ■

Analogies of Example 6.7 and Example 6.8 are the following. Both the function of Example 6.7 and that of Example 6.8 have an absolute and a relative maximum. These maxima are characterized by a negative second-order derivative at the maxima in Example 6.7 and by a negative definite Hessian matrix in Example 6.8. Furthermore, both functions have a further stationary point which is not a maximum. In Example 6.7, this is a relative minimum where the second-order derivative is positive, while in Example 6.8 it is a saddle point where the Hessian matrix is indefinite. Finally, in Example 6.7, there are two points of inflection, where the Newton step does not exist since the second-order derivative vanishes. Each of these points separates a region where the second-order derivative is negative from a region where it is positive. In Example 6.8, the points where the Hessian matrix is singular constitute a curve separating a region where the Hessian matrix is negative definite from a region where it is indefinite.

6.4.3 The Newton step for maximizing log-likelihood functions

Example 6.8 shows that the Newton method for maximizing log-likelihood functions should start at a point where the Hessian matrix of this function is negative definite. If the Hessian matrix is indefinite, the Newton method locally approximates the function by an indefinite quadratic form, characteristic of a function in the neighborhood of a saddle point. Indeed, in Example 6.8, the Newton method is seen to converge to a saddle point under this condition. If the Hessian matrix is positive definite, the Newton method approximates the function locally by a positive definite quadratic form, characteristic of a function in the neighborhood of a minimum. Then, the direction of the Newton method is a descent direction. Therefore, the Newton method should be started at a point where the Hessian matrix is negative definite and care should be taken to ensure that it stays so during the whole iteration process. The Newton method is stopped as soon as the conditions for convergence are met. These may, for example, be that the elements of Δt_{NE} are absolutely smaller than chosen amounts during a number of consecutive iterations.

Based on these considerations, a procedure for Newton maximizing log-likelihood functions may be organized as follows. First, a starting point $t = t_{init}$ is selected where the Hessian matrix of the log-likelihood function is negative definite. For a check of the negative definiteness, the eigenvalues of the Hessian matrix are computed. It is negative definite if all its eigenvalues are negative. See Theorem C.6. Then, one iteration of the subsequent iterative procedure may consist of the following steps:

1. Test if the chosen conditions for convergence are met. If not, go to 2. Otherwise, stop and take t_c as solution.

2. Compute the Newton step Δt_{NE} and take κ that is the fraction of the Newton step used equal to one.

3. Compute the Hessian matrix and the value of the log-likelihood function $q(w;t)$ at $t = t_c + \kappa\, \Delta t_{NE}$. If the Hessian matrix is negative definite and $q(w; t_c + \kappa\, \Delta t_{NE}) > q(w; t_c)$, go to 4. Otherwise, reduce κ and repeat this step.

4. Take $t_c + \kappa \Delta t_{NE}$ as new t_c and go to 1.

This procedure guarantees that the Hessian matrix stays negative definite during the whole iteration process. Furthermore, at each iteration, first the full Newton step is tried because the Newton method is known to converge *quadratically* to $\hat{t}$ if t_c is sufficiently close to $\hat{t}$ and κ converges to one. Quadratic convergence implies that the number of correct decimals of t_c doubles at each iteration. This rate of convergence is often considered as the fastest realizable and, therefore, used as benchmark for other methods.

The computational effort in one iteration is as follows. First, the gradient of the log-likelihood function is computed and the system of linear equations is solved for the elements of Δt_{NE}. Then, the Hessian matrix, its eigenvalues, and the value of the log-likelihood function have to be computed one or more times. As we will see later in this chapter, computing the Hessian matrix usually requires all second-order derivatives of the expectation model with respect to the parameters to be computed at all measurement points.

The steepest ascent or descent method and the Newton method are *general* numerical function optimization methods. The optimization methods presented in the subsequent sections are methods *specialized* to maximizing log-likelihood functions or minimizing least squares criteria. They have in common that they are specialized versions of or approximations to the Newton method.

6.5 THE FISHER SCORING METHOD

6.5.1 Definition of the Fisher scoring step

The purpose of the iterative Fisher scoring method is computing maximum likelihood estimates, that is, maximizing the log-likelihood function with respect to the parameters. The Fisher scoring step Δt_{FS} is defined as

$$\boxed{\Delta t_{FS} = F_t^{-1}\, s_t} \tag{6.65}$$

at $t = t_c$, where F_t is the $K \times K$ Fisher information matrix F_θ as defined by (4.51) at $\theta = t$, and s_t is the gradient of the log-likelihood function $q(w;t)$ with respect to t. The expression (6.54) for Δt_{NE} shows that Δt_{FS} may be interpreted as a Newton step with the additive inverse of the Hessian matrix of the log-likelihood function replaced by the Fisher information matrix at $\theta = t_c$.

The ideas underlying the Fisher scoring method may be sketched as follows. Consider expression (6.54) for the multivariate Newton step for maximizing log-likelihood functions:

$$\Delta t_{NE} = -\left(\frac{\partial^2 q(w;t)}{\partial t\, \partial t^T}\right)^{-1} s_t. \tag{6.66}$$

Next, assume that t tends to θ and $-\partial^2 q(w;t)/\partial t\, \partial t^T$ to $F_\theta = -E\partial^2 q(w;\theta)/\partial\theta\, \partial\theta^T$ as the number of observations N increases. Then, Δt_{NE} and Δt_{FS} agree asymptotically. The assumption that $\partial^2 q(w;t)/\partial t\, \partial t^T$ converges to $E\partial^2 q(w;\theta)/\partial\theta\, \partial\theta^T$ may be relatively easily made plausible if the observations are independent and $q(w;t)$ is, consequently, described by (5.12). Then, the standard deviations of the elements of $\partial^2 q(w;t)/\partial t\, \partial t^T$ are roughly proportional to the square root of N since these elements are the sum of N independent stochastic variables. See Appendix A. On the other hand, the expectations of these elements are, absolutely, asymptotically roughly proportional to N since they are sums of N deterministic quantities. Then, in this sense, $\partial^2 q(w;t)/\partial t\, \partial t^T$ tends asymptotically to $E\partial^2 q(w;\theta)/\partial\theta\, \partial\theta^T$.

One iteration of the Fisher scoring method may be composed of the following steps:

1. Test if the chosen conditions for convergence are met. If not, go to 2. Otherwise stop and take t_c as solution.

2. Compute the Fisher scoring step Δt_{FS} and take κ that is the fraction of Δt_{FS} used equal to one.

3. Compute the value of the log-likelihood function $q(w; t_c + \kappa\Delta t_{FS})$. If $q(w; t_c + \kappa\Delta t_{FS}) > q(w; t_c)$ go to 4. Otherwise, reduce κ and repeat this step.

4. Take $t_c + \kappa\Delta t_{FS}$ as new t_c and go to 1.

6.5.2 Properties of the Fisher scoring step

The computational effort in every Fisher scoring iteration consists of computing all elements of the Fisher information matrix F_t and the gradient vector s_t at $t = t_c$, followed by solving the resulting system of linear equations for the elements of the step.

The replacement of the additive inverse of the Hessian matrix by the Fisher information matrix may have consequences for the optimization process. The main reason is that, by definition, the Fisher information matrix is a covariance matrix. See Section 4.4.1. Therefore, it is positive semidefinite. Since, in addition, the use of the inverse of F_t implies that it is supposed nonsingular, it is positive definite and so is F_t^{-1}. Then,

$$s_t^T \Delta t_{FS} = s_t^T F_t^{-1} s_t > 0 \tag{6.67}$$

if $s_t \neq o$ at $t = t_c$. Therefore, the direction of Δt_{FS} is an ascent direction. Since such a statement cannot always be made with respect to the direction of the Newton step, one might conclude that this property of the Fisher scoring method is an advantage over the Newton method. However, if at a particular point the Hessian matrix is not negative definite, the Fisher scoring method nevertheless replaces it by a negative definite matrix. This implies that the Fisher scoring method approximates the function locally by a negative definite quadratic form while the true quadratic approximation is not negative definite. Then, the sense of such an approximation may be seriously doubted and, strictly, there is no reason to prefer at such points the Fisher scoring step to any other step in an ascent direction. In particular, the steepest ascent step may then be preferred since it has, by definition, also an ascent direction but requires a much smaller computational effort and is steepest.

On the other hand, if the computational effort of the Fisher scoring step is no impediment, maximizing the log-likelihood function using the Fisher scoring method only from start till convergence is much simpler than starting with the steepest ascent method followed by switching to the Fisher scoring method when appropriate. Furthermore, near the maximum the full Fisher scoring step approximates the Newton step if the number of observations is not too small. Checking the eigenvalues of the Hessian matrix of the log-likelihood function is not needed. Therefore, in any case, the Fisher scoring method is much simpler and much less computationally demanding than the Newton method. The computational effort involved in the Fisher scoring method is further reduced by the fact that computing the Fisher information matrix requires *first-order* derivatives of the log-likelihood function with respect to the parameters only as (4.38) shows.

■ **EXAMPLE 6.9**

The Fisher scoring step for normally distributed observations

The Fisher information matrix F_t for normally distributed observations follows from (4.40):

$$F_t = \frac{\partial g^T(t)}{\partial t} C^{-1} \frac{\partial g(t)}{\partial t^T} \tag{6.68}$$

while s_t follows from (5.60):

$$s_t = \frac{\partial g^T(t)}{\partial t} C^{-1} d(t). \tag{6.69}$$

Then, the Fisher scoring step is equal to

$$\Delta t_{FS} = \left(X^T C^{-1} X\right)^{-1} X^T C^{-1} d(t), \tag{6.70}$$

where the $N \times K$ matrix X is defined as

$$X = \frac{\partial g(t)}{\partial t^T} \tag{6.71}$$

at $t = t_c$. ■

6.5.3 Fisher scoring step for exponential families

For linear exponential families of distributions, the expression for the Fisher information matrix F_t follows from (4.68):

$$F_t = \frac{\partial g^T(t)}{\partial t} C^{-1} \frac{\partial g(t)}{\partial t^T} \tag{6.72}$$

and that for s_t follows from (5.129):

$$s_t = \frac{\partial g^T(t)}{\partial t} C^{-1} d(t). \tag{6.73}$$

These expressions show that the Fisher scoring step for these distributions is described by

$$\Delta t_{FS} = (X^T C^{-1} X)^{-1} X^T C^{-1} d(t) \tag{6.74}$$

with

$$X = \frac{\partial g(t)}{\partial t^T} \tag{6.75}$$

at $t = t_c$, where the covariance matrix C depends, typically, on the parameters t.

6.6 THE NEWTON METHOD FOR NORMAL MAXIMUM LIKELIHOOD AND FOR NONLINEAR LEAST SQUARES

6.6.1 The Newton step for normal maximum likelihood

In this section, an expression is derived for the Newton step for maximizing the log-likelihood function for normally distributed observations. From this expression, the Newton step for minimizing the nonlinear least squares criterion follows directly. The latter expression is important since in practice nonlinear least squares estimation is used frequently and is applied to observations with all kinds of distributions.

The log-likelihood function for normally distributed observations is described by (5.59):

$$q(w;t) = -\frac{N}{2}\ln 2\pi - \frac{1}{2}\ln\det C - \frac{1}{2} d^T(t)\, C^{-1}\, d(t)\,. \tag{6.76}$$

It will be assumed that the covariance matrix C is independent of the parameters t. Then, the gradient of $q(w;t)$ thus defined is described by (5.60):

$$s_t = \frac{\partial g^T(t)}{\partial t}\, C^{-1}\, d(t). \tag{6.77}$$

Therefore, the kth element of the gradient is described by

$$\boxed{s_{t_k} = \frac{\partial g^T(t)}{\partial t_k}\, C^{-1}\, d(t)} \tag{6.78}$$

with $k = 1, \dots, K$. Differentiating this expression with respect to t_ℓ produces the (k, ℓ)th element of the Hessian matrix of the normal log-likelihood function

$$\boxed{\frac{\partial^2 q(w;t)}{\partial t_k \partial t_\ell} = -\frac{\partial g^T(t)}{\partial t_k}\, C^{-1}\, \frac{\partial g(t)}{\partial t_\ell} + \frac{\partial^2 g^T(t)}{\partial t_k \partial t_\ell}\, C^{-1}\, d(t)}\,, \tag{6.79}$$

where $k, \ell = 1, \ldots, K$. The equations (6.78) and (6.79) define the Newton step (6.54) for maximizing the log-likelihood function for normally distributed observations.

For completeness, we also present the log-likelihood function (6.76), its gradient (6.77), and the (k, ℓ)th element of its Hessian matrix (6.79) for two special cases.

First, let the w_n be uncorrelated. Then,

$$C = \text{diag}(\sigma_1^2 \ldots \sigma_N^2), \tag{6.80}$$

where $\sigma_n^2 = \text{var}\, w_n$. Then, the log-likelihood function assumes the special form

$$q(w; t) = -\frac{N}{2} \ln 2\pi - \sum_n \ln \sigma_n - \frac{1}{2} \sum_n \frac{d_n^2(t)}{\sigma_n^2}. \tag{6.81}$$

The corresponding kth element of the gradient is described by

$$s_{t_k} = \sum_n \frac{1}{\sigma_n^2} \frac{\partial g_n(t)}{\partial t_k} d_n(t) \tag{6.82}$$

and the (k, ℓ)th element of the Hessian matrix by

$$\frac{\partial^2 q(w; t)}{\partial t_k \partial t_\ell} = -\sum_n \frac{1}{\sigma_n^2} \frac{\partial g_n(t)}{\partial t_k} \frac{\partial g_n(t)}{\partial t_\ell} + \sum_n \frac{1}{\sigma_n^2} \frac{\partial^2 g_n(t)}{\partial t_k \partial t_\ell} d_n(t). \tag{6.83}$$

Next, let the w_n be uncorrelated and have an equal variance σ^2. Then,

$$C = \sigma^2 I, \tag{6.84}$$

where I is the identity matrix of order N. The log-likelihood function has the special form

$$q(w; t) = -\frac{N}{2} \ln 2\pi - N \ln \sigma - \frac{1}{2\sigma^2} \sum_n d_n^2(t)\,. \tag{6.85}$$

The corresponding kth element of the gradient and the (k, ℓ)th element of the Hessian matrix are described by

$$s_{t_k} = \frac{1}{\sigma^2} \sum_n \frac{\partial g_n(t)}{\partial t_k} d_n(t) \tag{6.86}$$

and

$$\frac{\partial^2 q(w; t)}{\partial t_k \partial t_\ell} = -\frac{1}{\sigma^2} \sum_n \frac{\partial g_n(t)}{\partial t_k} \frac{\partial g_n(t)}{\partial t_\ell} + \frac{1}{\sigma^2} \sum_n \frac{\partial^2 g_n(t)}{\partial t_k \partial t_\ell} d_n(t)\,. \tag{6.87}$$

Returning to the general expression for the (k, ℓ)th element of the Hessian matrix of the log-likelihood function $q(w; t)$ for normally distributed observations (6.79), we see that the first term of the matrix description of the Hessian matrix is

$$-\frac{\partial g^T(t)}{\partial t} C^{-1} \frac{\partial g(t)}{\partial t^T}. \tag{6.88}$$

This is a $K \times K$ negative definite matrix if the $N \times K$ Jacobian matrix $\partial g(t)/\partial t^T$ is nonsingular. See Theorem C.4. Therefore, if the Hessian matrix $\partial^2 q(w; t)/\partial t\, \partial t^T$ is not negative definite, this must have been caused by its second term.

6.6.2 The Newton step for nonlinear least squares

The parameter dependent part of the log-likelihood function (6.76) for normally distributed observations is equal to half the additive inverse of the least squares criterion

$$J(t) = d^T(t)\, C^{-1}\, d(t). \tag{6.89}$$

Then, (6.49) shows that the Newton step for maximizing the log-likelihood function (6.76) for normally distributed observations and the Newton step for minimizing the least squares criterion (6.89) are identical. Furthermore, the Newton step for minimizing the least squares criterion with arbitrary symmetric and positive definite weighting matrix R:

$$J(t) = d^T(t)\, R\, d(t) \tag{6.90}$$

is seen to be defined by the kth element of the gradient

$$\boxed{\frac{\partial J(t)}{\partial t_k} = -2\frac{\partial g^T(t)}{\partial t_k} R\, d(t)} \tag{6.91}$$

and the (k, ℓ)th element of the Hessian matrix

$$\boxed{\frac{\partial^2 J(t)}{\partial t_k \partial t_\ell} = 2\frac{\partial g^T(t)}{\partial t_k} R \frac{\partial g(t)}{\partial t_\ell} - 2\frac{\partial^2 g^T(t)}{\partial t_k \partial t_\ell} R\, d(t)}\,. \tag{6.92}$$

As in maximum likelihood estimation from normally distributed observations, there are two important special cases.

First, suppose that R is the diagonal matrix

$$\text{diag}(r_{11} \ldots r_{NN}). \tag{6.93}$$

Then, the kth element of the gradient (6.91) and the (k, ℓ)th element of the Hessian matrix (6.92) are described by

$$\frac{\partial J(t)}{\partial t_k} = -2\sum_n r_{nn} \frac{\partial g_n(t)}{\partial t_k}\, d_n(t) \tag{6.94}$$

and

$$\frac{\partial^2 J(t)}{\partial t_k \partial t_\ell} = 2\sum_n r_{nn} \frac{\partial g_n(t)}{\partial t_k}\frac{\partial g_n(t)}{\partial t_\ell} - 2\sum_n r_{nn} \frac{\partial^2 g_n(t)}{\partial t_k \partial t_\ell}\, d_n(t)\,. \tag{6.95}$$

Next, let the weighting matrix R be the identity matrix of order N. Therefore, the least squares criterion is the ordinary least squares criterion described by

$$J(t) = d^T(t)\, d(t) = \sum_n d_n^2(t)\,. \tag{6.96}$$

Then, the kth element of the gradient and the (k, ℓ)th element of the Hessian matrix simplify to

$$\frac{\partial J(t)}{\partial t_k} = -2\sum_n \frac{\partial g_n(t)}{\partial t_k}\, d_n(t) \tag{6.97}$$

and

$$\frac{\partial^2 J(t)}{\partial t_k \partial t_\ell} = 2\sum_n \frac{\partial g_n(t)}{\partial t_k}\frac{\partial g_n(t)}{\partial t_\ell} - 2\sum_n \frac{\partial^2 g_n(t)}{\partial t_k \partial t_\ell}\, d_n(t)\,. \tag{6.98}$$

6.7 THE GAUSS–NEWTON METHOD

6.7.1 Definition of the Gauss–Newton step

The Newton step for maximizing the log-likelihood function for normally distributed observations is defined by (6.78) and (6.79). The latter expression shows that computing the Hessian matrix requires computing the second-order derivatives of the expectation model with respect to the parameters at all measurement points. These are $NK(K+1)/2$ derivatives. Furthermore, in many applications, the expressions for the second-order derivatives are much more complicated than those for the first-order derivatives. As a result, the evaluation of the second-order derivatives may constitute a considerable computational burden. The expression for the Hessian matrix (6.92) shows that similar considerations apply to computing the Newton step for minimizing the nonlinear least squares criterion. Avoiding the computation of the second-order derivatives and thus reducing the computational burden has been the principal motive for introducing the *Gauss–Newton step* as an approximation to the Newton step. As a consequence of the particular form of the normal log-likelihood function, the Gauss–Newton step may also be used for minimizing nonlinear least squares criteria with arbitrary weighting matrix.

The Gauss–Newton step Δt_{GN} is obtained from the Newton step by leaving out the second term of the Hessian matrix (6.79). Thus,

$$\boxed{\Delta t_{GN} = \left(X^T C^{-1} X\right)^{-1} X^T C^{-1} d(t)} \tag{6.99}$$

with $X = \partial g(t)/\partial t^T$ evaluated at $t = t_c$. Therefore, Δt_{GN} is identical to the Fisher scoring step (6.70) for normally distributed observations.

One iteration of the Gauss–Newton method may be composed of the following steps:

1. Test if the chosen conditions for convergence are met. If not, go to 2. Otherwise stop and take t_c as solution.

2. Compute the Gauss–Newton step Δt_{GN} and take κ that is the fraction of Δt_{GN} to be used equal to one.

3. Compute the value of the log-likelihood function $q(w; t_c + \kappa\Delta t_{GN})$. If $q(w; t_c + \kappa\Delta t_{GN}) > q(w; t_c)$ go to 4. Otherwise, reduce κ and repeat this step.

4. Take $t_c + \kappa\Delta t_{GN}$ as new t_c and go to 1.

6.7.2 Properties of the Gauss–Newton step

Leaving out the second term of the Hessian matrix used in the Newton step may be justified if, in a suitable sense, the first term is large as compared with the second term. The expression (6.79) shows that this condition is met if either the *second-order derivatives* of the expectation model or the *deviations* $d(t)$ make the second term sufficiently small.

6.7.2.1 The second-order derivatives First, consider the second-order derivatives. In any case, these vanish if $g_n(t)$ is linear in all elements of t. Then, the Gauss–Newton step is identical to the Newton step. Furthermore, they are small if, for the values of t considered, the remainder R_2 of the Taylor expansion of $g_n(t)$ is sufficiently small.

6.7.2.2 The deviations Next consider the deviations. These may be decomposed as follows:

$$d_n(t) = w_n - g_n(t) = w_n - Ew_n + Ew_n - g_n(t)\,. \tag{6.100}$$

If the expectation model $g(x;t)$ is correct, this is equivalent to

$$d_n(t) = d_n(\theta) + e_n(t) \tag{6.101}$$

with

$$d_n(\theta) = w_n - Ew_n = w_n - g_n(\theta) \tag{6.102}$$

and

$$e_n(t) = g_n(\theta) - g_n(t). \tag{6.103}$$

Thus, $d_n(\theta)$ is the fluctuation of w_n while $e_n(t)$ is the deviation of $g_n(t)$ at the current point from its true value $g_n(\theta)$. For exact observations, the $d_n(\theta)$ vanish for all t while the $e_n(t)$ vanish for $t = \theta$ only. Also, the difference $e_n(t)$ will never vanish if the model $g_n(t)$ is not correct. However, if it is, the $e_n(t)$ will for continuous $g_n(t)$ become arbitrarily small if t approaches θ.

These considerations show that the second term of (6.79) may be split as follows:

$$r^T(t)\,d(\theta) + r^T(t)\,e(t), \tag{6.104}$$

where $d(\theta) = [d_1(\theta)\ldots d_N(\theta)]^T$, $e(t) = [e_1(t)\ldots e_N(t)]^T$ and $r(t)$ is the $N \times 1$ vector

$$r(t) = C^{-1}\frac{\partial^2 g(t)}{\partial t_k\,\partial t_\ell}\,. \tag{6.105}$$

Thus, $r^T(t)\,d(\theta)$ is a linear combination of fluctuations $d_n(\theta)$ that are by definition stochastic variables with expectation zero:

$$r_1(t)\,d_1(\theta) + \ldots + r_N(t)\,d_N(\theta). \tag{6.106}$$

The expectation of such a combination is also equal to zero. Its standard deviation is, characteristically, roughly proportional to the square root of N. On the other hand, the first term of (6.79) is deterministic. It will, absolutely, increase more or less proportionally to N. Then, the quantity (6.106) is a stochastic variable with expectation zero and a standard deviation that is asymptotically small compared with the first term of (6.79).

On the other hand, the quantity $r^T(t)\,e(t)$ in (6.104) becomes small in the neighborhood of $t = \theta$. Since the solution $\tilde{t}$ tends asymptotically to θ, the elements of $e(t)$ and, therefore, the quantity $r^T(t)\,e(t)$ decrease as the Newton method converges. Thus, in summary, the Gauss–Newton step becomes an increasingly accurate approximation of the Newton step as the asymptotic solution is approached.

6.7.2.3 An alternative interpretation of the Gauss–Newton step The fact that the Gauss–Newton step is identical to the Newton step for linear models underlies the following interpretation of it. Suppose that the model $g(t)$ is linearized about $t = t_c$. That is,

$$g(t) = g(t_c + \Delta t) = g(t_c) + \frac{\partial g(t)}{\partial t^T}\Delta t\,, \tag{6.107}$$

where the derivatives are taken at $t = t_c$, the elements of Δt are the parameters of the model, and $\Delta t = o$ is the current value of Δt. Equation (6.78) shows that the kth element of the gradient of the normal log-likelihood function of these parameters is described by

$$\frac{\partial q(w; t_c + \Delta t)}{\partial \Delta t_k} = \frac{\partial g^T(t_c + \Delta t)}{\partial \Delta t_k}\,C^{-1}\,[w - g(t_c + \Delta t)]\,. \tag{6.108}$$

Then, by (6.107),

$$\frac{\partial q(w;t_c+\Delta t)}{\partial \Delta t_k} = \frac{\partial g^T(t)}{\partial t_k} C^{-1} \left(w - g(t_c) - \frac{\partial g(t)}{\partial t^T} \Delta t \right), \tag{6.109}$$

where the derivatives $\partial g_n(t)/\partial t_k$ are taken at $t = t_c$. Therefore, at the current point $\Delta t = o$,

$$\frac{\partial q(w;t_c+\Delta t)}{\partial \Delta t_k} = \frac{\partial g^T(t)}{\partial t_k} C^{-1}\, d(t_c), \tag{6.110}$$

where $d(t_c) = w - g(t_c)$. This expression shows that the corresponding gradient vector is described by

$$s_{\Delta t} = X^T C^{-1}\, d(t_c) \tag{6.111}$$

with $X = \partial g(t)/\partial t^T$ evaluated at $t = t_c$.

Furthermore, (6.79) shows that the Hessian matrix of the normal log-likelihood function of the parameters Δt is described by

$$\begin{aligned}\frac{\partial^2 q(w;t_c+\Delta t)}{\partial \Delta t_k \partial \Delta t_\ell} &= -\frac{\partial g^T(t_c+\Delta t)}{\partial \Delta t_k} C^{-1} \frac{\partial g(t_c+\Delta t)}{\partial \Delta t_\ell} \\ &\quad + \frac{\partial^2 g^T(t_c+\Delta t)}{\partial \Delta t_k \partial \Delta t_\ell} C^{-1} \left[w - g(t_c+\Delta t) \right].\end{aligned} \tag{6.112}$$

Then, the second term of this expression is equal to zero since, by (6.107), the elements $g_n(t_c + \Delta t)$ are linear in the elements of Δt and hence at the current point $\Delta t = o$:

$$\frac{\partial^2 q(w;t_c+\Delta t)}{\partial \Delta t_k \partial \Delta t_\ell} = -\frac{\partial g^T(t)}{\partial t_k} C^{-1} \frac{\partial g(t)}{\partial t_\ell}, \tag{6.113}$$

where the derivatives $\partial g_n(t)/\partial t_k$ are taken at $t = t_c$. Therefore, the Newton step for maximizing the log-likelihood function of the parameters of the linearized model for normally distributed observations is equal to

$$\Delta t_{NE} = \left(X^T C^{-1} X\right)^{-1} X^T C^{-1}\, d(t) \tag{6.114}$$

evaluated at $t = t_c$.

The equality of the Newton step (6.114) to the Gauss–Newton step (6.99) shows that the latter is the Newton step for maximum likelihood estimation of the parameters of a linearized model from normally distributed observations. As shown in Section 5.15.2, the maximum likelihood estimator of the parameters of linear expectation models from normally distributed observations is the best linear unbiased estimator. This is the linear least squares estimator with the inverse covariance matrix of the observations as weighting matrix. Therefore, in every Gauss–Newton iteration a best linear unbiased least squares problem is solved with the matrix X and the observations $w - g(t_c)$ varying from step to step. Solving for best linear unbiased estimates is a standard numerical problem that can be dealt with by using specialized methods.

6.7.2.4 *The Gauss–Newton step for nonlinear least squares* The Gauss–Newton step for minimizing the nonlinear least squares criterion is obtained analogously by leaving out the second term of the expression for the Hessian matrix (6.92). The result is

$$\boxed{\Delta t_{GN} = \left(X^T R X\right)^{-1} X^T R\, d(t)} \tag{6.115}$$

at $t = t_c$ with $X = \partial g(t)/\partial t^T$. Of course, this step may also be interpreted as the solution of a weighted linear least squares problem in every step, be it this time with weighting matrix R.

6.7.2.5 Computational effort Comparing the Newton step for maximizing the normal log-likelihood function and for minimizing the nonlinear least squares criterion with the Gauss–Newton step for the same purposes shows that the latter step is by far computationally least demanding. The reason is that the Gauss–Newton method avoids the computing of $NK(K+1)/2$ second-order derivatives of the expectation model. However, since the Gauss–Newton method solves a linear least squares problem in every iteration, it requires many more operations than the steepest ascent or descent method.

6.7.2.6 Further properties The Gauss–Newton step (6.99) is identical to the Fisher scoring step for normal maximum likelihood (6.70). Therefore, its numerical properties are similar to those of the Fisher scoring method. Thus, the direction of the Gauss–Newton step for normal maximum likelihood is an ascent direction. The direction of the Gauss–Newton step for the corresponding nonlinear least squares estimation problem is a descent direction because the gradient of the normal log-likelihood function and that of the corresponding least squares criterion have opposite directions.

6.8 THE NEWTON METHOD FOR POISSON MAXIMUM LIKELIHOOD

In this section, the Newton step for maximizing the log-likelihood function for independent Poisson distributed observations is derived. This log-likelihood function is described by (5.112):

$$q(w;t) = \sum_n -g_n(t) + w_n \ln g_n(t) - \ln w_n! \,. \tag{6.116}$$

The kth element of the gradient of $q(w;t)$ thus defined with respect to t is described by (5.113):

$$s_{t_k} = \sum_n \frac{\partial g_n(t)}{\partial t_k} \frac{1}{g_n(t)} d_n(t) \tag{6.117}$$

or, alternatively, by (5.114):

$$\boxed{s_{t_k} = \frac{\partial g^T(t)}{\partial t_k} C^{-1} d(t)} \tag{6.118}$$

with $C = \operatorname{diag} g(t)$. Differentiating (6.117) once more produces the (k,ℓ)th element of the Hessian matrix

$$\begin{aligned} \frac{\partial^2 q(w;t)}{\partial t_k \partial t_\ell} &= -\sum_n \frac{\partial g_n(t)}{\partial t_k} \frac{1}{g_n(t)} \frac{\partial g_n(t)}{\partial t_\ell} \\ &+ \sum_n \left(\frac{1}{g_n(t)} \frac{\partial^2 g_n(t)}{\partial t_k \partial t_\ell} - \frac{1}{g_n^2(t)} \frac{\partial g_n(t)}{\partial t_k} \frac{\partial g_n(t)}{\partial t_\ell} \right) d_n(t) \,. \end{aligned} \tag{6.119}$$

Differentiating (6.118) instead shows that an alternative expression is

$$\boxed{\begin{aligned} \frac{\partial^2 q(w;t)}{\partial t_k \partial t_\ell} &= -\frac{\partial g^T(t)}{\partial t_k} C^{-1} \frac{\partial g(t)}{\partial t_\ell} \\ &+ \left(\frac{\partial^2 g^T(t)}{\partial t_k \partial t_\ell} C^{-1} + \frac{\partial g^T(t)}{\partial t_k} \frac{\partial C^{-1}}{\partial t_\ell} \right) d(t) \end{aligned}} \,. \tag{6.120}$$

Detailed calculations show that (6.119) and (6.120) are identical.

The expressions (6.118) for the gradient and (6.120) for the Hessian matrix define the Newton step for maximizing the Poisson log-likelihood function. Comparing these with the corresponding expressions (6.78) and (6.79) for the gradient and Hessian matrix of the normal log-likelihood function shows that they are similar in many respects. However, a difference is the presence of the term

$$\frac{\partial g^T(t)}{\partial t_k} \frac{\partial C^{-1}}{\partial t_\ell} d(t) \tag{6.121}$$

in the Hessian matrix of the Poisson log-likelihood function. The consequence for the Newton step is that the second term of (6.120) does not vanish if the model $g_n(t)$ is linear in the elements of t. Therefore, (near-)linearity of $g_n(t)$ is no justification for leaving out the second term of (6.120), as is done in the Gauss–Newton step for maximizing the normal log-likelihood function or for minimizing the least squares criterion. On the other hand, asymptotically, the second term of (6.120) will become small compared with the first term as the method converges and, therefore, become negligible as has been explained in Subsection 6.7.2.2.

6.9 THE NEWTON METHOD FOR MULTINOMIAL MAXIMUM LIKELIHOOD

In this section, the expression is derived for the Newton step for maximizing the multinomial log-likelihood function (5.119):

$$q(w;t) = \ln M! - M \ln M - \sum_{n=1}^{N} \ln w_n! + \sum_{n=1}^{N} w_n \ln g_n(t) \tag{6.122}$$

with $w_N = M - \sum_{n=1}^{N-1} w_n$ and $g_N(t) = M - \sum_{n=1}^{N-1} g_n(t)$. The kth element of the gradient of this log-likelihood function is described by (5.120):

$$s_{t_k} = \sum_{n=1}^{N-1} \frac{\partial g_n(t)}{\partial t_k} \left(\frac{w_n}{g_n(t)} - \frac{w_N}{g_N(t)} \right) \tag{6.123}$$

or, alternatively, by (5.121):

$$\boxed{s_{t_k} = \frac{\partial g^T(t)}{\partial t_k} C^{-1} d(t)} \tag{6.124}$$

with C defined by (5.122). Differentiating (6.123) once more produces the (k, ℓ)th element of the Hessian matrix

$$\begin{aligned} \frac{\partial^2 q(w;t)}{\partial t_k \, \partial t_\ell} &= -\sum_n \frac{w_n}{g_n^2(t)} \frac{\partial g_n(t)}{\partial t_k} \frac{\partial g_n(t)}{\partial t_\ell} \\ &+ \sum_n \frac{\partial^2 g_n(t)}{\partial t_k \, \partial t_\ell} \left(\frac{w_n}{g_n(t)} - \frac{w_N}{g_N(t)} \right) + \frac{w_N}{g_N^2(t)} \sum_n \frac{\partial g_n(t)}{\partial t_k} \sum_n \frac{\partial g_n(t)}{\partial t_\ell} \end{aligned} \tag{6.125}$$

where $n = 1, \ldots, N-1$. Differentiating (6.124) instead shows that an alternative expression is

$$\boxed{\begin{aligned}\frac{\partial^2 q(w;t)}{\partial t_k \partial t_\ell} &= -\frac{\partial g^T(t)}{\partial t_k} C^{-1} \frac{\partial g(t)}{\partial t_\ell} \\ &\quad + \left(\frac{\partial^2 g^T(t)}{\partial t_k \partial t_\ell} C^{-1} + \frac{\partial g^T(t)}{\partial t_k} \frac{\partial C^{-1}}{\partial t_\ell} \right) d(t)\end{aligned}} \,. \tag{6.126}$$

For the computation of C^{-1} and $\partial C^{-1}/\partial t_\ell$ in (6.126), the closed-form expression (5.123) for C^{-1} may be instrumental. Detailed calculations show that (6.125) and (6.126) are identical.

The expressions for the gradient and the Hessian matrix (6.124) and (6.126) define the Newton step for maximizing the log-likelihood function for multinomially distributed observations. They are similar in many respects to the corresponding expressions for the normal and the Poisson distribution. As with Poisson distributed observations, linearity of the expectation model in all parameters is not enough to reduce the expressions (6.125) or (6.126) to their first term.

6.10 THE NEWTON METHOD FOR EXPONENTIAL FAMILY MAXIMUM LIKELIHOOD

In this section, expressions will be derived for the Newton step for maximizing the log-likelihood function of the parameters of the expectation model if the probability (density) function of the observations is a linear exponential family.

The gradient s_t of the log-likelihood function for linear exponential families of distributions is described by (5.129):

$$s_t = \frac{\partial g^T(t)}{\partial t} C^{-1} d(t) \tag{6.127}$$

with kth element:

$$\boxed{s_{t_k} = \frac{\partial g^T(t)}{\partial t_k} C^{-1} d(t)} \,. \tag{6.128}$$

Consequently, the (k, ℓ)th element of the Hessian matrix of $q(\omega; t)$ is described by

$$\boxed{\begin{aligned}\frac{\partial^2 q(w;t)}{\partial t_k \partial t_\ell} &= -\frac{\partial g^T(t)}{\partial t_k} C^{-1} \frac{\partial g(t)}{\partial t_\ell} \\ &\quad + \left(\frac{\partial^2 g^T(t)}{\partial t_k \partial t_\ell} C^{-1} + \frac{\partial g^T(t)}{\partial t_k} \frac{\partial C^{-1}}{\partial t_\ell} \right) d(t)\end{aligned}} \,. \tag{6.129}$$

For the differentiation of C^{-1}, use may be made of (D.13):

$$\frac{\partial C^{-1}}{\partial t_\ell} = -C^{-1} \frac{\partial C}{\partial t_\ell} C^{-1} . \tag{6.130}$$

The expressions (6.128) for the gradient and the expression (6.129) for the Hessian matrix define the Newton step for maximizing the log-likelihood function if the probability (density) function of the observations is a linear exponential family.

Alternative expressions for the gradient and the Hessian matrix may be obtained as follows. Equation (5.131) shows that the relevant gradient vector is also described by

$$s_t = \frac{\partial \gamma^T(t)}{\partial t} d(t). \tag{6.131}$$

The kth element of this gradient is

$$\boxed{s_{t_k} = \frac{\partial \gamma^T(t)}{\partial t_k} d(t)}\ . \tag{6.132}$$

The (k, ℓ)th element of the Hessian matrix of the log-likelihood function follows from this expression and (3.109):

$$\boxed{\frac{\partial^2 q(w;t)}{\partial t_k \partial t_\ell} = -\frac{\partial \gamma^T(t)}{\partial t_k} C \frac{\partial \gamma(t)}{\partial t_\ell} + \frac{\partial^2 \gamma^T(t)}{\partial t_k \partial t_\ell} d(t)}\ , \tag{6.133}$$

where each of both terms is equal to the corresponding term of (6.129). An advantage of the alternative expressions (6.132) and (6.133) is their relative simplicity. Like the equations (6.128) and (6.129), the equations (6.132) and (6.133) define the Newton step.

Finally, (6.128) and (6.129) show that the expressions for the gradient and the Hessian matrix of $q(w; t)$ are the same for all linear exponential families. This is illustrated by the expressions (6.78) and (6.79), (6.118) and (6.120), and (6.124) and (6.126) that describe the gradient and the Hessian matrix of $q(w; t)$ for normal, Poisson, and multinomial observations, respectively. However, the functional dependence of the covariance matrices C on the parameters differs between families.

6.11 THE GENERALIZED GAUSS–NEWTON METHOD FOR EXPONENTIAL FAMILY MAXIMUM LIKELIHOOD

6.11.1 Definition of the generalized Gauss–Newton step

The *generalized Gauss–Newton step* Δt_{GGN} is obtained from the pure Newton step for maximizing log-likelihood functions for linear exponential families of distributions by omitting the second term in the expression (6.129) for the Hessian matrix concerned. This produces

$$\boxed{\Delta t_{GGN} = \left(X^T C^{-1} X\right)^{-1} X^T C^{-1} d(t)} \tag{6.134}$$

with $X = \partial g(t)/\partial t^T$.

Alternatively, from (6.132) and (6.133),

$$\boxed{\Delta t_{GGN} = \left(\frac{\partial \gamma^T(t)}{\partial t} C \frac{\partial \gamma(t)}{\partial t^T}\right)^{-1} \frac{\partial \gamma^T(t)}{\partial t} d(t)}\ . \tag{6.135}$$

The generalized Gauss–Newton step thus defined is a generalization of the conventional Gauss–Newton step since it applies to the log-likelihood function for any linear exponential family of distributions while, strictly, the conventional Gauss–Newton step applies to the normal log-likelihood function only. As opposed to the covariance matrix C appearing in the expression for the conventional Gauss–Newton step (6.99), the covariance matrix C appearing in the expressions (6.134) and (6.135) for Δt_{GGN} depends generally on the parameters t and, therefore, changes from iteration to iteration.

The generalized Gauss–Newton step (6.134) is identical to the Fisher scoring step (6.74) for distributions that are linear exponential families. However, the Fisher scoring method is more general than the generalized Gauss–Newton method since it also applies to distributions that are *not* linear exponential families.

6.11.2 Properties of the generalized Gauss–Newton method

Analogous to the Gauss–Newton step discussed in Section 6.7, the use of the generalized Gauss–Newton step as an approximation to the pure Newton step is justified if the second term of the Hessian matrix is, in a suitable sense, small compared with the first one. The expression (6.133) shows that this condition may be met if either the second-order derivatives $\partial^2\gamma_n(t)/\partial t_k \partial t_\ell$ or the deviations $d_n(t)$ are sufficiently small.

6.11.2.1 The second-order derivatives First, consider the second-order derivatives $\partial^2\gamma_n(t)/\partial t_k \partial t_\ell$. These are small if, for the values of t considered, the remainder R_2 of the Taylor expansion of the $\gamma_n(t)$ is sufficiently small. They vanish if the $\gamma_n(\theta)$ are linear in θ. Equation (3.109) shows how $\gamma(\theta)$ depends on the elements of $g(\theta)$ and those of the covariance matrix C of the observations. Typically, the latter also depend on θ in a way characteristic of the pertinent distribution of the observations. If, for a particular form of the covariance matrix, $g_n(\theta)$ produces a $\gamma_n(\theta)$ linear in the elements of θ, the corresponding expectation model will be called *generalized linear expectation model*. As an illustration, the generalized linear expectation models will be derived for the normal, the Poisson, and the multinomial distribution. The resulting generalized linear expectation models are of more than theoretical value since they are all used in practice.

■ **EXAMPLE 6.10**

The generalized linear expectation model for normally distributed observations

For normally distributed observations, the vector $\gamma(\theta)$ is described by (3.82):

$$\gamma(\theta) = C^{-1}g(\theta), \tag{6.136}$$

where the covariance matrix C is supposed independent of θ. Thus, the linearity condition is met if $g(\theta) = X\theta$. Therefore, for normally distributed observations with constant covariance matrix, the conventional linear expectation model is also the generalized linear expectation model. ■

■ **EXAMPLE 6.11**

The generalized linear expectation model for Poisson distributed observations

For Poisson distributed observations, the vector $\gamma(\theta)$ is described by (3.87):

$$\gamma(\theta) = [\ln g_1(\theta) \ldots \ln g_N(\theta)]^T. \tag{6.137}$$

Therefore, the elements of $\gamma(\theta)$ are linear in the elements of θ if

$$g_n(\theta) = \exp(x_n^T \theta). \tag{6.138}$$

For Poisson distributed observations, the corresponding expectation model is the generalized linear expectation model. In the literature, it is called *log-linear Poisson model.* ■

■ **EXAMPLE 6.12**

The generalized linear expectation model for multinomially distributed observations

For multinomially distributed observations, the vector $\gamma(\theta)$ is described by (3.93):

$$\gamma(\theta) = \left(\ln \frac{g_1(\theta)}{g_N(\theta)} \ldots \ln \frac{g_{N-1}(\theta)}{g_N(\theta)} \right)^T \tag{6.139}$$

with $g_N(\theta) = M - \sum_{n=1}^{N-1} g_n(\theta)$. Therefore, the nth element of the vector $\gamma(\theta)$ is linear in θ if

$$\gamma_n(\theta) = \ln \frac{g_n(\theta)}{g_N(\theta)} = x_n^T \theta, \tag{6.140}$$

that is, if $g_n(\theta) = g_N(\theta) \exp(x_n^T \theta)$. Then, by definition,

$$g_N(\theta) = M - g_N(\theta) \sum_{n=1}^{N-1} \exp(x_n^T \theta). \tag{6.141}$$

Therefore,

$$g_n(\theta) = \frac{M \exp(x_n^T \theta)}{1 + \sum_{n=1}^{N-1} \exp(x_n^T \theta)} \quad n = 1, \ldots, N-1 \tag{6.142}$$

and

$$g_N(\theta) = \frac{M}{1 + \sum_{n=1}^{N-1} \exp(x_n^T \theta)} \,. \tag{6.143}$$

For multinomial observations, the corresponding expectation model is the generalized linear expectation model. It is an example of a *logistic model.* ■

This concludes the discussion of the influence of the second-order derivatives of $\gamma_n(t)$ on the Hessian matrix (6.133).

6.11.2.2 The deviations Next, consider the deviations $d_n(t) = w_n - g_n(t)$. The discussion of their influence on the Hessian matrix (6.133) is similar to the discussion presented in Subsection 6.7.2.2 if $r(t)$ defined by (6.105) is replaced by

$$r(t) = \frac{\partial^2 \gamma(t)}{\partial t_k \, \partial t_\ell} \,. \tag{6.144}$$

As in Section 6.7.2.2, the conclusion is that, asymptotically, the second term of the Hessian matrix (6.133) is generally small as compared with the first term as the method converges to the solution. This is the justification of omitting it in the generalized Gauss–Newton step.

6.11.2.3 Further properties of the generalized Gauss–Newton step Since the generalized Gauss–Newton method is identical to the Fisher scoring method for distributions that are linear exponential families, it has the same numerical properties as the latter method. These may be summarized as follows. The direction of the generalized Gauss–Newton step is an ascent direction. Furthermore, the generalized Gauss–Newton step is not always a sensible approximation to the Newton step but as it converges to the maximum, it approximates the Newton step if the number of observations used is sufficiently large.

6.12 THE ITERATIVELY REWEIGHTED LEAST SQUARES METHOD

The *iterratively reweighted least squares step* Δt_{IRLS} for estimating the parameters of the expectation model supposes the elements of the covariance matrix C of the observations to be known functions of the parameters θ. This known dependence is used in the step

$$\boxed{\Delta t_{IRLS} = \left(X^T C^{-1} X\right)^{-1} X^T C^{-1} d}, \qquad (6.145)$$

where $X = \partial g(t)/\partial t^T$ and $d = w - g(t)$. The name of the step derives from the fact that the weighting matrix C^{-1} is updated in every step of the iteration and that Δt_{IRLS} is a weighted least squares estimator of the parameters Δt of the model

$$\left.\frac{\partial g(t)}{\partial t^T}\right|_{t=t_c} \Delta t \qquad (6.146)$$

linearized around $t = t_c$ from observations $w - g(t_c)$. The iteratively reweighted least squares step is not derived from a particular distribution or family of distributions. A difficulty with the method is that it is not clear what particular objective function it optimizes. Since the objective function is unknown, its gradient is unknown. Therefore, it is not known if the direction of Δt_{IRLS} is an ascent direction. This difficulty is removed if it may be assumed that the distribution of the observations is a linear exponential family parametric in θ. Then, Δt_{IRLS} is identical to the generalized Gauss–Newton step Δt_{GGN} intended for maximizing the log-likelihood function for linear exponential family distributed observations.

6.13 THE LEVENBERG–MARQUARDT METHOD

6.13.1 Definition of the Levenberg–Marquardt step

As the Gauss–Newton method, the Levenberg–Marquardt method is intended for minimizing the nonlinear least squares criterion. However, it modifies the system of linear equations (6.99) defining Gauss–Newton step so that this system cannot become singular or near-singular. In Section 6.7.2.3, the Gauss–Newton step for maximizing the normal log-likelihood function was shown to be equal to the weighted linear least squares solution for the parameters Δt of the linearized model $X\Delta t$ from observations $d(t_c) = w - g(t_c)$. In this model, $X = \partial g(t)/\partial t^T$ evaluated at $t = t_c$. The weighting matrix is the inverse C^{-1} of the covariance matrix of the observations. Thus, the Gauss–Newton step minimizes

$$J(t_c + \Delta t) = [d(t_c) - X\Delta t]^T C^{-1} [d(t_c) - X\Delta t] \qquad (6.147)$$

with respect to Δt and is described by (6.99):

$$\Delta t_{GN} = \left(X^T C^{-1} X\right)^{-1} X^T C^{-1} d(t) \tag{6.148}$$

evaluated at $t = t_c$. The assumption in the Levenberg–Marquardt method is that during the iteration process the matrix $X^T C^{-1} X$ may become singular or nearly singular. By Theorem C.4, this must be a consequence of the $N \times K$ matrix X becoming singular or nearly so. It will be shown in this subsection that this possible singularity is cured by minimizing (6.147) under the equality constraint

$$\|\Delta t\|^2 = (\Delta t_1)^2 + \cdots + (\Delta t_K)^2 = \Delta^2, \tag{6.149}$$

where Δ is a positive scalar. This means that $J(t)$ is minimized on the sphere with t_c as center and Δ as radius. The Lagrangian function for minimizing (6.147) under the equality constraint (6.149) is described by

$$J(t_c + \Delta t) + \lambda(\|\Delta t\|^2 - \Delta^2), \tag{6.150}$$

where the scalar λ is the Lagrange multiplier. Then, the solution for Δt is found among the stationary points of (6.150) with respect to Δt and λ. These are the solutions for Δt and λ of the equations

$$\frac{\partial J(t_c + \Delta t)}{\partial \Delta t} + 2\lambda \Delta t = o \tag{6.151}$$

and

$$\|\Delta t\|^2 - \Delta^2 = 0. \tag{6.152}$$

Computing the gradient of $J(t_c + \Delta t)$ in (6.151) using (5.245) and substituting the result yields

$$-2X^T C^{-1}[d(t) - X\Delta t] + 2\lambda \Delta t = o \tag{6.153}$$

and hence

$$\boxed{\Delta t_{LM} = \left(X^T C^{-1} X + \lambda I\right)^{-1} X^T C^{-1} d(t)} \tag{6.154}$$

evaluated at $t = t_c$ with I the identity matrix of order K. Δt_{LM} is the *Levenberg–Marquardt step* for maximizing the log-likelihood function for normally distributed observations with covariance matrix C or, equivalently, minimizing the nonlinear least squares criterion weighted with C^{-1} for observations of any distribution. Then, the expression for the Levenberg–Marquardt step for minimizing the weighted nonlinear least squares criterion with an arbitrary symmetric and positive definite weighting matrix R is

$$\boxed{\Delta t_{LM} = \left(X^T R X + \lambda I\right)^{-1} X^T R\, d(t)} \tag{6.155}$$

or, equivalently,

$$\Delta t_{LM} = -\frac{1}{2}\left(X^T R X + \lambda I\right)^{-1} \frac{\partial J(t)}{\partial t}, \tag{6.156}$$

where (5.179) has been used and $J(t)$ is the weighted nonlinear least squares criterion

$$J(t) = d^T(t)\, R\, d(t). \tag{6.157}$$

The Levenberg–Marquardt step most often used in practice is

$$\boxed{\Delta t_{LM} = \left(X^T X + \lambda I\right)^{-1} X^T d(t)}. \tag{6.158}$$

It minimizes the ordinary nonlinear least squares criterion

$$J(t) = d^T(t)\, d(t) = \sum_n d_n^2(t). \tag{6.159}$$

All expressions for the Levenberg–Marquardt step Δt_{LM} mentioned show that it approaches the Gauss–Newton step (6.99) if $\lambda \to 0$. Furthermore, if $\lambda \to \infty$, the inverse matrix in the right–hand members approaches

$$\frac{1}{\lambda} I \tag{6.160}$$

and, consequently, the Levenberg–Marquardt step approaches a steepest ascent step for log-likelihood function maximizing or steepest descent step for least squares minimizing with a step length approaching zero. It can be shown that the length of the Levenberg–Marquardt step is a continuous and monotonically decreasing function of λ.

Using (6.155), the procedure in a Levenberg–Marquardt iteration may be as follows. Suppose that the scalar $\nu > 1$. Let t_c and λ_c be the values of t and λ produced in the previous iteration. Then,

1. Test if the chosen conditions for convergence are met. If not, go to 2. Otherwise, stop and take t_c as solution.

2. Compute Δt_{LM} from (6.155) for $\lambda = \lambda_c/\nu$. Next, compute $J(t_c + \Delta t_{LM})$. If $J(t_c + \Delta t_{LM}) < J(t_c)$, take $t_c + \Delta t_{LM}$ and λ_c/ν as new t_c and λ_c, and go to 1. Else, leave λ_c unaltered and go to 3.

3. Compute Δt_{LM} from (6.155) for $\lambda = \lambda_c$. Next, compute $J(t_c + \Delta t_{LM})$. If $J(t_c + \Delta t_{LM}) < J(t_c)$, take $t_c + \Delta t_{LM}$ as new t_c, leave λ_c unaltered, and go to 1. Else, go to 4.

4. Compute Δt_{LM} from (6.155) for $\lambda = \nu\lambda_c$. Next, compute $J(t_c + \Delta t_{LM})$. If $J(t_c + \Delta t_{LM}) < J(t_c)$, take $t_c + \Delta t_{LM}$ and $\nu\lambda_c$ as new t_c and λ_c, and go to 1. Else, take $\nu\lambda_c$ as new λ_c, leave t_c unaltered, and repeat this step.

This procedure reduces λ whenever possible. This means that the step is made as Gauss–Newtonlike as possible. The value of λ is increased only if the least squares criterion does not decrease if λ_c is reduced or kept the same. Increasing λ makes the step more steepest-descentlike. The underlying idea is to make use of the fact that, if the Gauss–Newton method converges, it does so quickly. On the other hand, if it fails to converge, the alternative is the steepest descent step which always reduces the least squares criterion to some extent.

6.13.2 Properties of the Levenberg–Marquardt step

Theorem C.4 shows that the matrix $X^T R X$ in (6.155) is positive semidefinite since R is positive definite. Therefore, the matrix

$$X^T R X + \lambda I \tag{6.161}$$

and its inverse, also appearing in (6.155), are positive definite because $\lambda > 0$. Then, this matrix is also nonsingular. Furthermore, (6.156) shows that

$$(\Delta t_{LM})^T \left(-\frac{\partial J(t)}{\partial t} \right) = \frac{1}{2} \frac{\partial J(t)}{\partial t^T} \left(X^T R X + \lambda I \right)^{-1} \frac{\partial J(t)}{\partial t} > 0\,, \tag{6.162}$$

which proves that the direction of Δt_{LM} is a descent direction.

The description of the Levenberg–Marquardt procedure in Section 6.13.1 shows that the computational effort involved is substantial. An iteration may require the solving of (6.155) for several values of λ. In addition, this system of equations does not have a form making it suitable for treatment as a standard linear least squares problem in every step as opposed to the system of equations to be solved for the Gauss–Newton step or for the generalized Gauss–Newton step for which efficient specialized methods exist.

These considerations show that the Levenberg–Marquardt step will always decrease the least squares criterion to some extent as a consequence of its descent direction. Also, it can cope with (near-)singularity of the matrix $X^T RX$ without interference of the experimenter. The method may, therefore, be characterized as reliable, in the sense of usually converging, but the computational effort in each step is greater than that in the Gauss–Newton step. Therefore, the Levenberg–Marquardt step is preferable to the Gauss–Newton step only if (near-)singularity of the matrix $X^T RX$ is to be expected.

6.14 SUMMARY OF THE DESCRIBED NUMERICAL OPTIMIZATION METHODS

6.14.1 Introduction

In this section, the properties of the discussed numerical methods for maximizing the log-likelihood function or minimizing the least squares criterion are summarized and compared. Two general numerical function optimization methods have been the starting point in this discussion: the steepest ascent (descent) method and the Newton method. The steepest ascent (descent) step is derived from the linear Taylor polynomial of the objective function. The Newton step is derived from the quadratic Taylor polynomial. All further methods discussed in this chapter are approximations of the Newton method since they use approximations of the Hessian matrix.

6.14.2 The steepest ascent (descent) method

The direction of the steepest ascent (descent) step is, by definition, an ascent (descent) direction. The method is reliable, in the sense that convergence is almost guaranteed. As compared with the Newton method and related methods, the steepest ascent (descent) method may become slow as the optimum is approached.

The computational effort is modest since it consists of computing the gradient only. As a consequence, the steepest ascent (descent) method requires the values of the first-order partial derivatives of the expectation model with respect to the parameters at all measurement points only.

The step length of the steepest ascent (descent) method must be specified by the user.

6.14.3 The Newton method

If the Newton method converges, it converges to a stationary point of the objective function. To guarantee that this is a local maximum (minimum), the Hessian matrix of the objective function should be negative (positive) definite at the starting point and remain so until convergence. The most important property of the Newton method is its quadratic convergence.

The computational effort consists, in the first place, of the computation of the Hessian matrix and the gradient of the objective function. If the objective function is the log-likelihood function of the expectation model parameters, this requires the values of both the first-order and the second-order partial derivatives of the expectation model with respect to the parameters at all measurement points. Furthermore, the method requires the values of the eigenvalues of the Hessian matrix to check whether this matrix is negative (positive) definite if the purpose is maximizing (minimizing) the objective function. Finally, the Newton method requires the solution of a system of linear equations in the elements of the step vector. Whether these computational requirements are an impediment to the use of the Newton method, is problem dependent.

The step length is defined by the method.

6.14.4 The Fisher scoring method

The Fisher scoring method is specialized to maximizing the log-likelihood function. In it, the additive inverse of the Hessian matrix used in the Newton step is approximated by the supposedly nonsingular Fisher information matrix at the current point. As a result of this approximation, the direction of the Fisher scoring step is always an ascent direction while such a guarantee cannot be given with respect to the direction of the Newton step. Furthermore, for a sufficiently large number of observations, the method converges to the Newton step in the neighborhood of the maximum.

The computational effort required by the Fisher scoring method consists of computing the Fisher information matrix and the gradient at the current point. These are the coefficient matrix and right-hand member of a system of linear equations to be subsequently solved for the elements of the Fisher scoring step. Computing the Fisher information matrix requires the values of the first-order partial derivatives of the expectation model with respect to the parameters only. The Fisher scoring method does not require eigenvalue analysis as the Newton method does.

Far from the maximum, the only relevant property of the Fisher scoring step is its ascent direction. Then, the steepest ascent method might be preferred being computationally much less expensive and, by definition, steepest. However, if the larger amount of computation is no impediment, using the Fisher scoring method from start till convergence involves a simpler program structure than starting with the steepest ascent method followed by switching to the Fisher scoring method when appropriate.

The Fisher scoring step for linear exponential families of distributions is identical to the generalized Gauss–Newton step. Therefore, the Fisher scoring step for the normal distribution and the conventional Gauss–Newton step also coincide. The Fisher scoring step for linear exponential families is also identical to the iteratively reweighted least squares step. However, the Fisher scoring method is more general than the generalized Gauss–Newton method or the iteratively reweighted least squares method since it provides a step converging to the Newton step for any distribution for which a Fisher information matrix is defined.

6.14.5 The Gauss–Newton method

The conventional Gauss–Newton method is specialized to maximizing the log-likelihood function of the parameters of an expectation model for the normal distribution with a covariance matrix that is independent of the parameters and known. Since the Gauss–Newton step is identical to the Fisher scoring step for maximizing normal log-likelihood functions, its properties are those of the Fisher scoring step described above. As a consequence of the particular form of the normal log-likelihood function, the Gauss–Newton method is also suitable for minimizing nonlinear least squares criteria.

6.14.6 The generalized Gauss–Newton method

The generalized Gauss–Newton step is specialized to maximizing log-likelihood functions for linear exponential families. Since the generalized Gauss–Newton step is identical to the Fisher scoring step for linear exponential families, its properties are those of the Fisher scoring step described above. The method includes the conventional Gauss–Newton method that is intended for maximizing the normal log-likelihood function as a special case. For generalized linear expectation models, the generalized Gauss–Newton step is identical to the Newton step. Then, the Newton step, the Fisher scoring step, the generalized Gauss–Newton step, and the iteratively reweighted least squares step for maximizing the log-likelihood function are identical.

6.14.7 The iteratively reweighted least squares method

The iteratively reweighted least squares step has not been derived for maximizing a particular log-likelihood function. It requires an expectation model and an expression for the elements of the covariance matrix of the observations as a function of the unknown parameters. In each step, it computes a weighted least squares estimate of the parameters of a model linearized around the current parameter values using the deviations of the current model from the observations as observations. As weighting matrix, the inverse of the covariance matrix of the observations is used, evaluated at the current point. Therefore, this weighting matrix changes from step to step. This is recognized as a conventional Gauss–Newton step for minimizing a weighted nonlinear least squares criterion with a different weighting matrix in every step which is the same as a generalized Gauss–Newton step. Therefore, if the probability (density) function of the observations is a linear exponential family, the objective function is the log-likelihood function and the method produces maximum likelihood estimates. Under these conditions, the method is also identical to the Fisher scoring method.

6.14.8 The Levenberg–Marquardt method

The purpose of the Levenberg–Marquardt method as presented in this book is numerically minimizing the weighted nonlinear least squares criterion. The matrix used to approximate the Hessian matrix is that used in the Gauss–Newton method but with the same positive quantity added to each of its diagonal elements. The purpose is to prevent the matrix from becoming (near-)singular. The computational effort is greater than that required by the Gauss–Newton method.

6.14.9 Conclusions

For maximum likelihood estimation of the parameters of expectation models, the Fisher scoring step has a unique combination of advantages over other methods:

- It produces maximum likelihood estimates when the Fisher information matrix used corresponds to the distribution of the observations.

- The direction of the Fisher scoring step is an ascent direction.
- The method defines its own step length.
- If the number of observations is sufficiently large, the full Fisher scoring step converges to the Newton step in the neighborhood of the maximum.
- Computing the gradient of the log-likelihood function and the Fisher information matrix requires the computation of only the first-order partial derivatives of the expectation model at all measurement points.

A disadvantage of the method is that, far from the maximum, the only property of the Fisher scoring step used is its ascent direction. Therefore, the steepest ascent step would then, as a rule, be computationally cheaper and more effective.

6.15 PARAMETER ESTIMATION METHODOLOGY

6.15.1 Introduction

In this section, steps are described recommended in the process starting with choosing the model of the observations and ending with actually estimating the model parameters. Each of the steps will be illustrative of concepts developed in the Chapters 4–6. It will be seen that in this process numerical experiments and simulation are essential. In the first place, these enable the experimenter to find out if the intended experiment is suitable for the intended purpose. This is discussed in Subsection 6.15.2. Furthermore, they enable the experimenter to check the mathematical expressions and the software used for estimating the parameters. This is discussed in Subsection 6.15.3. Finally, they enable the experimenter to get used to aspects of the intended estimation experiment such as the properties of the log-likelihood function to be maximized and the convergence properties of the numerical optimization method chosen. This is also discussed in the Subsections 6.15.2 and 6.15.3.

In the following subsections, three assumptions are made. The first is that the choice of expectation model has already been made. Furthermore, it is assumed that a distribution has been chosen for the observations. Finally, it is assumed that a preliminary choice of experimental design has been made. The concept experimental design has been introduced in Subsection 4.10.1. The choice of expectation model, distribution, and design are the domain and responsibility of the experimenter since they require his or her expert knowledge.

6.15.2 Investigating the feasibility of the observations

The fact that the expectation model, the distribution of the observations, and the experimental design are supposed to be known makes it possible to compute the Fisher information matrix and the corresponding Cramér–Rao lower bound matrix. These matrices and their meaning have extensively been discussed in Chapter 4.

First, it is investigated if the Fisher information matrix is nonsingular. If not, the parameters to be estimated are not identifiable. Identifiability has been discussed in Section 4.9. If the parameters are identifiable, the Cramér–Rao standard deviations are computed next. If these are not small enough to reach the intended conclusions about the true numerical values of the parameters, it is decided that the observations are not feasible. The reason is that these standard deviations are the smallest attainable by unbiased estimators. Then, the experimenter's only option is to change the experimental design or the measurement method or instrument used, which, unfortunately, for practical reasons is not always possible. In any case, investigating the feasibility of the observations is a valuable tool to avoid useless attempts to unbiasedly estimate the parameters with a specified standard deviation from the available observations. From here on, it will be assumed that it has been established that the Cramér–Rao standard deviations produced by the existing or newly developed experimental design meet the experimenter's demands.

The Cramér–Rao lower bound matrix depends on the exact values of the parameters as the expressions presented in Section 4.5 show. Since these exact parameters are the quantities to be estimated, they are, of course, unknown. Fortunately, in science and engineering, experimenters have usually a reasonably accurate idea of the magnitude or order of magnitude of the parameters to be measured. Therefore, in practice, the Cramér–Rao lower bound is computed for *nominal* values of the parameters. The use of these values will provide the experimenter with a first quantitative impression of the limits to the precision of the parameters measured. Later, when estimates of the parameters have become available, the nominal values may be modified if needed.

6.15.3 Preliminary simulation experiments

If the feasibility of the observations with the chosen design has been established, the actual estimation of the parameters has to be carefully investigated by numerical simulation experiments. This requires a number of steps to be made described in this subsection.

6.15.3.1 Choice of optimization method First, a numerical optimization method must be chosen. We emphasize that *all* derivatives occurring in the gradient vector, in the Hessian matrix, or in Jacobian matrices used must be carefully tested by means of finite difference approximations. In practice, deriving analytical expressions for the derivatives and programming these expressions are notorious sources of error.

6.15.3.2 Generation of exact observations Using the chosen expectation model and experimental design, we generate exact observations as described in Section 6.2.2. As numerical values for the exact parameters, the nominal values may be chosen that were used earlier in the feasibility study of the observations described in Section 6.15.2. Then, substituting these exact observations in the log-likelihood function yields the reference log-likelihood function defined in Section 6.2.2.

6.15.3.3 Maximizing the reference log-likelihood function Next, the chosen numerical optimization procedure is applied to the reference log-likelihood function using three different types of starting points.

First, the optimization procedure is started taking the chosen exact values of the parameters as starting point. This means that the location of the maximum sought for is taken as starting point. One of the purposes of this experiment is to check if the gradient is equal to zero indeed. This is, however, a necessary and not a sufficient condition for the gradient to be correct. For example, if in (6.127) the deviations $d(t)$ of the expectation model from the observations vanish, the resulting gradient will vanish even if there are analytical or programming errors in the rest of the expression. This experiment also produces the maximum value of the reference log-likelihood function to be used later for comparison.

Next, the optimization procedure is started from slightly modified values of the exact parameters. The purpose is to verify if the procedure converges to the exact parameter values and, if so, how fast.

Finally, the optimization procedure is carried out a number of times with starting points produced by a random number generator. They may, for example, be uniformly distributed and have expectations equal to the exact parameter values while the interval over which they are distributed reflects the uncertainty of the experimenter about their value. One of the purposes of these experiments is to find out if there are relative maxima. Since the reference log-likelihood function is used, the maximum likelihood estimates should still be equal to the exact parameter values.

6.15.3.4 Maximizing the log-likelihood function for statistical observations

If the optimization experiments with the reference log-likelihood function have been concluded successfully, the next step is applying the procedure to computer generated *statistical* observations. The expectations of these observations are taken as the values of the expectation model at the measurement points. The distribution of the observations around these values should be the assumed distribution.

Next, using the exact parameters as starting point, we carry out a number of trial experiments. The purpose is to get an impression of the number of iterations needed. The optimization process may be stopped if, in a number of consecutive steps, each of the elements of the step becomes absolutely smaller than specified amounts.

At this stage, procedures for plotting the results of the optimization procedure may be chosen and tested. In any case, the plots produced should show in one figure: the observations, the estimated expectation model which is the expectation model with the estimated parameters as parameters, and the residuals, which are the deviations of the observations from the estimated expectation model at the measurement points. An example is Fig. 5.3. of Example 5.4. It shows the Poisson distributed observations of an exponential decay model, the estimated decay model with as parameters the maximum likelihood estimates of the amplitude and decay constant, and the residuals. Later, when parameters are estimated from experimental, nonsimulated observations, inspecting and testing the residuals may be of great importance for the interpretation of the results.

The next step is to repeat the simulation experiments a substantial number of times. This number must be large enough to allow the experimenter to draw conclusions about the average value and the standard deviation of the parameter estimates thus obtained. In each experiment, a set of observations is generated and the parameter estimates, the final gradient, the eigenvalues of the Hessian matrix, and the maximum values of the likelihood function are computed and stored. Next, these quantities are inspected for each experiment separately. If they look acceptable, the average and the sample variance of the parameter estimates is computed. Since the exact values of the parameters are known in these simulation experiments, the bias of the maximum likelihood estimator may be estimated from the average of the estimated parameters and the exact value. Example 5.4 illustrates these operations.

The mean squared error of the parameter estimates, discussed in Section 4.2, is computed next and, from it, the efficiency, that is, the ratio of the Cramér–Rao variance to the mean squared error. An efficiency significantly lower than one indicates that the design chosen prevents the maximum likelihood estimator used from having, or almost having, its desirable optimal asymptotic properties. This may mean that, possibly, a different experimental design should be chosen.

By this time, the experimenter has gained vast experience with the statistics and the numerical properties of the parameter estimation problem at hand under almost perfectly controlled conditions. Moreover, his or her experimental design, the derived mathematical expressions, and the software have been put to severe tests. Therefore, applying the estimation procedure to experimental nonsimulated observations may now be faced with much more confidence than without these careful preparations. The ensuing actual estimation of parameters from experimental data is concluded by a model test such as described in Section 5.8 to find out whether or not the model used is acceptable.

6.16 COMMENTS AND REFERENCES

The book by Gill, Murray and Wright [12] is an excellent general reference on numerical optimization. The importance of the book for our purposes is that it discusses the necessary key notions in numerical optimization. The description of methods in this chapter shows that, with the exception of the steepest descent method, all the methods require is a routine for the numerical solution of a system of linear equations. The Newton method also requires a routine for computing eigenvalues. Both routines are standard. Additional routines are needed for the computation of the gradient vectors and relevant Jacobian and Hessian matrices. These are problem dependent and have to be supplied by the experimenter.

The Fisher scoring method discussed in Section 6.5 originates from [10]. Our description of the Levenberg–Marquardt method largely follows [23].

6.17 PROBLEMS

6.1 Show that the reference log-likelihood function for Poisson distributed observations is maximized by the exact parameters.

6.2 Suppose that the observations $w = (w_1 \ldots w_N)^T$ are independent and binomially distributed as in Problem 3.3(b).

(a) Find an expression for the reference log-likelihood function of the parameters of the expectation model of the observations.

(b) Show that $t = \theta$ is a stationary point of the reference log-likelihood function and that this stationary point is a maximum.

(c) Do exact binomial observations exist?

6.3 Suppose that the observations $w = (w_1 \ldots w_N)^T$ are independent and exponentially distributed as in Problem 3.7(b).

(a) Find an expression for the reference log-likelihood function of the parameters of the expectation model of the observations.

(b) Show that $t = \theta$ is a stationary point of the reference log-likelihood function and that this stationary point is a maximum.

6.4 Suppose that the distribution of the observations $w = (w_1 \ldots w_N)^T$ is a linear exponential family. Show that $t = \theta$ is a stationary point of the reference log-likelihood function and that this stationary point is a maximum.

6.5 In an experiment, the observations w_n are independent and normally distributed around expectation values described by

$$Ew_n = g_n(\theta) = \alpha_1 \exp(-\beta_1 x_n) + \alpha_2 \exp(-\beta_2 x_n),$$

where $x_n \geq 0$ is the nth measurement point and $\theta = (\alpha_1\ \alpha_2\ \beta_1\ \beta_2)^T$ is the vector of parameters. The variance of all observations is equal to σ^2.

(a) Suppose that $\theta = (0.7\ 0.3\ 1\ 0.8)^T$ and σ is arbitrary. Use the steepest descent method to numerically compute the optimal experimental design in the sense of the criterion Q defined by (4.247) with weights $\lambda_1 = 0.1299$, $\lambda_2 = 0.7071$, $\lambda_3 = 0.0636$, and $\lambda_4 = 0.0994$ corresponding to (4.248) and under the restriction that only 10 measurement points $x_n \geq 0$ may be used. Repeat the optimization from different initial x. Check if the solution is a minimum of Q.

Remarks:

To keep the program suitable for other expectation models being linear combinations of nonlinearly parametric functions, derive the expressions for the gradient and Hessian matrix of Q for the general model $g(x_n;\theta) = \alpha_1 h(x_n;\beta_1) + \ldots + \alpha_K h(x_n;\beta_K)$ with respect to the x_n and use function subroutines for the computation of the required derivatives of $\alpha_k h(x_n;\beta_k)$ with respect to x_n, α_k and β_k. Next specialize to the exponential model used in this example.

To prevent the x_n from becoming negative, introduce a new variable $r_n^2 = x_n$ and use the fact that $\partial Q/\partial r_n = 2r_n \partial Q/\partial x_n$ and that $\partial^2 Q/\partial r_n \partial r_m = 2\delta_{n,m}\,\partial Q/\partial x_n + 4r_n r_m \partial^2 Q/\partial x_n \partial x_m$.

(b) Compare the Cramér–Rao variances for the computed optimal design to those for the uniform design $x_n = (n-1)/3$, $n = 1, \ldots, 10$.

(c) Specify the approximate standard deviation of the observations needed for a maximum 10% relative Cramér–Rao standard deviation of all parameters for the optimal and for the uniform design.

6.6 The function $f(x) = x_1^2 x_2^2 - 4x_1^2 x_2 - 2x_1 x_2^2 + 5x_1^2 + 3x_2^2 + 8x_1 x_2 - 10x_1 - 12x_2 + 15$ is minimized with respect to $x = (x_1\ x_2)^T$ by means of the Newton method.

(a) Numerically compute the points reached in the first and the second step for the starting points $x = (2\ 3)^T$, $x = (2\ 4)^T$, and $x = (2\ \ 2 + \sqrt{3})^T$, respectively. If a stationary point is reached, what is its nature?

(b) Plot contours of $f(x)$ and the collection of all points where the determinant of the Hessian matrix $\partial^2 f(x)/\partial x\, \partial x^T$ vanishes on $-5 \leq x_1, x_2 \leq 5$. Use the plot to find the collection of points suitable as starting points for the Newton method.

6.7

(a) Derive an expression for the generalized linear expectation model for observations that are independent and binomially distributed as in Problem 3.3(b).

(b) Derive an expression for the generalized linear expectation model for observations that are independent and exponentially distributed as in Problem 3.7(b).

6.8 Show that the system of linear equations defining the Levenberg–Marquardt step cannot be singular.

6.9 The function

$$f(x) = 100(x_2 - x_1^2)^2 + (1 - x_1)^2$$

with $x = (x_1\, x_2)^T$ is called *Rosenbrock's function*. It is used for comparing the performance of numerical minimization methods.

(a) Show that $(1, 1)$ is the location of the absolute minimum.

(b) Plot contours of $f(x)$ for $-1.5 \leq x_1 \leq 1.5$ and $-0.5 \leq x_2 \leq 1.5$. Derive an expression for all points (x_1, x_2) where the determinant of the Hessian matrix of $f(x)$ vanishes and plot these points in a contour plot of $f(x)$.

(c) Which points are suitable starting points for the Newton method?

6.10 Suppose that it is known that the distribution of a set of observations is a linear exponential family. Show that then the direction of the Newton step for maximizing the log-likelihood function of the parameters of the corresponding generalized linear expectation model is an ascent direction.

6.11 The *Newton–Raphson method* is an iterative method for the numerical solution of a system of K nonlinear equations in K unknowns. It consists of linearizing the equations around the current point, solving the system of K linear equations in K unknowns thus obtained, and using the solution as current point in the next iteration. Show that the Newton optimization method is identical to the Newton–Raphson method applied to the system of equations obtained by equating the gradient of a function to the corresponding null vector.

6.12 The observations $w = (w_1 \ldots w_N)^T$ are independent and binomially distributed with expectations $Ew = g(\theta)$ as in Problem 3.5(a). Derive expressions for the gradient and Hessian matrix of the log-likelihood function.

6.13 The observations $w = (w_1 \ldots w_N)^T$ are independent and exponentially distributed with expectations $Ew = g(\theta)$ as in Problem 3.7(b). Derive expressions for the gradient and Hessian matrix of the log-likelihood function.

6.14 The distribution of the observations in Problem 6.12 is a linear exponential family of distributions. Use this to derive the gradient and Hessian matrix of the log-likelihood function.

6.15 The distribution of the observations in Problem 6.13 is a linear exponential family of distributions. Use this to derive the gradient and Hessian matrix of the log-likelihood function.

6.16 Show that the expressions (6.119) and (6.120) are identical.

6.17 Show that the expressions (6.125) and (6.126) are identical.

6.18 The observations $w = (w_1 \ldots w_N)^T$ are independent and binomially distributed as in Problem 3.5(a). Their expectation is

$$g_n(\theta) = \theta_1 + \theta_2 x_n + \theta_3 x_n^2$$

with $\theta = (\theta_1\, \theta_2\, \theta_3)^T$.

(a) Derive an expression for the gradient and the Hessian matrix of the log-likelihood function.

(b) Write a program for maximum likelihood estimation of the parameters θ using the Newton method and test this program by applying it to exact observations and starting from different, random initial parameter values.

(c) Suppose that in an experiment the observations have been made presented in Table 6.1. Plot these observations and compute from them the maximum likelihood estimates

Table 6.1. Problem 6.18

n	x_n	M_n	w_n
1	0.3	22	17
2	0.6	15	10
3	0.9	20	9
4	1.2	10	5
5	1.5	10	7
6	1.8	17	5
7	2.1	13	2
8	2.4	11	3
9	2.7	13	5
10	3.0	10	9
11	3.3	16	7

of θ using the Newton method. Check the negative definiteness of the Hessian matrix in every step.

(d) Test the absence of a cubic term in the expectation model.

6.19 The observations $w = (w_1 \ldots w_N)^T$ are exponentially distributed and have as expectation the Lorentz line

$$g_n(\theta) = \frac{\theta_1}{1 + \left(\dfrac{x_n - \theta_2}{\theta_3}\right)^2},$$

where $\theta = (\theta_1\ \theta_2\ \theta_3)^T$ are the parameters and the x_n are the measurement points.

(a) Derive an expression for the gradient and the Hessian matrix of the log-likelihood function of the elements of θ.

(b) Use the expressions derived under (a) for the computation of the maximum likelihood estimates of θ with the Newton method from the observations presented in Table 6.2.. The measurement points are $x_n = 12 \times (n-1)/51$ with $n = 1, \ldots, 52$. Plot the observations. Choose a starting point and check its suitability by computing the eigenvalues of the Hessian matrix at that point. Start the Newton method and check the eigenvalues of the Hessian matrix from step to step. Repeat the computation from different starting points.

(c) Compute and plot the residuals for the maximum likelihood estimate found. Comment on the residuals found.

Table 6.2. Problem 6.19

n	w_n	n	w_n	n	w_n
1	1.3338	19	3.4730	37	2.5191
2	0.5642	20	3.3709	38	0.7800
3	0.6901	21	0.5898	39	2.7049
4	4.0942	22	2.1181	40	0.4724
5	0.2113	23	9.5023	41	2.9307
6	0.5159	24	7.7400	42	1.4503
7	7.1769	25	10.334	43	0.7401
8	0.5579	26	3.0674	44	0.8622
9	2.5103	27	0.8371	45	1.9368
10	0.6144	28	6.6037	46	0.5427
11	5.3936	29	16.798	47	2.4374
12	3.8128	30	7.9869	48	0.7519
13	1.9977	31	2.9252	49	1.3685
14	0.4076	32	0.4791	50	0.5280
15	1.7730	33	0.5046	51	1.2266
16	3.0194	34	0.0124	52	0.2134
17	0.7744	35	0.8202		
18	1.4538	36	0.3312		

6.20 Suppose that the expectations of the observations w_n are described by

$$g_n(\theta) = \sin(\theta_1 x_n) + \sin(\theta_2 x_n)$$

where $x_n = -1.5 + 0.15 \times (n-1)$ and $n = 1, \ldots, 21$. Furthermore, $\theta = (\theta_1\ \theta_2)^T = (2\pi\ 2.4\pi)^T$.

(a) Plot the expectations.

(b) Plot contours of the reference ordinary least squares criterion for $t_1, t_2 = 0.5\,\theta_2\ (0.01\theta_2)\ 1.5\,\theta_2$. Plot, in the same figure, the points where the determinant of the Hessian matrix vanishes. Comment on the plot. In particular, comment on the nature of the stationary points and on the (in)definiteness of the Hessian matrix in the various regions of the plot.

CHAPTER 7

SOLUTIONS OR PARTIAL SOLUTIONS TO PROBLEMS

3.1 The proof for discrete stochastic variables is obtained by replacing the integrals in the proof of Theorem 3.1 by the corresponding sums.

3.2 The Poisson distribution satisfies

$$\sum_{\omega} \exp(-\lambda)\frac{\lambda^{\omega}}{\omega!} = 1$$

with $\omega = 0, 1, \ldots$. Differentiating both sides of this equation with respect to λ yields

$$-1 + \frac{1}{\lambda}\sum_{\omega} \omega \exp(-\lambda)\frac{\lambda^{\omega}}{\omega!} = 0.$$

Then,

$$Ew = \sum_{\omega} \omega \exp(-\lambda)\frac{\lambda^{\omega}}{\omega!} = \lambda.$$

Differentiating this expression yields

$$-\sum_{\omega} \omega \exp(-\lambda)\,\frac{\lambda^{\omega}}{\omega!} + \frac{1}{\lambda}\sum_{\omega} \omega^2 \exp(-\lambda)\frac{\lambda^{\omega}}{\omega!} = 1.$$

Then, $Ew^2 = \lambda^2 + \lambda$. Therefore, $\operatorname{var} w = Ew^2 - (Ew)^2 = \lambda$.

3.3

(a) $u = \sum_{m=1}^{M} t_m$ where the stochastic variable $t_m = 1$ with probability φ and $t_m = 0$ with probability $1 - \varphi$. Then, $Et_m = \varphi$ and, therefore, $Eu = \sum_{m=1}^{M} Et_m = M\varphi$. Furthermore, $\text{var}\, t_m = Et_m^2 - (Et_m)^2 = \varphi - \varphi^2 = \varphi(1 - \varphi)$. Since the t_m are independent, $\text{var}\, u = \sum_{m=1}^{M} \text{var}\, t_m = M\varphi(1 - \varphi)$. See Appendix A.

(b) Since the w_n are independent, their joint probability function is the product of their marginal probability functions. Furthermore, $Ew_n = M\varphi_n = g_n(\theta)$. Then,

$$p(\omega;\theta) = \prod_n \binom{M}{\omega_n} \varphi_n^{\omega_n} (1 - \varphi_n)^{M-\omega_n}$$

or, since $\varphi_n = g_n(\theta)/M$,

$$p(\omega;\theta) = \frac{1}{M^{MN}} \prod_n \binom{M}{\omega_n} g_n^{\omega_n}(\theta) \left[M - g_n(\theta)\right]^{M-\omega_n}$$

with $n = 1, \ldots, N$. Therefore,

$$\begin{aligned} q(\omega;\theta) &= N(\ln M! - M \ln M) - \sum_n \ln \omega_n! - \sum_n \ln (M - \omega_n)! \\ &\quad + \sum_n \omega_n \ln g_n(\theta) + \sum_n (M - \omega_n) \ln \left[M - g_n(\theta)\right]. \end{aligned}$$

Only both last terms of this expression are functions of θ.

(c) The observations are independent. Therefore, their covariance matrix is diagonal. Its nth diagonal element is

$$\text{var}\, w_n = \frac{1}{M} g_n(\theta)[M - g_n(\theta)].$$

(d) Differentiating $q(\omega;\theta)$ with respect to θ and substituting w for ω yields

$$s_\theta = M \sum_n \frac{w_n - g_n(\theta)}{g_n(\theta)[M - g_n(\theta)]} \frac{\partial g_n(\theta)}{\partial \theta}.$$

3.4

(a) The probability function derived in Solution 3.3(b) is an exponential family with

$$\begin{aligned} \alpha(\theta) &= \prod_n \left[M - g_n(\theta)\right]^M, \\ \beta(\omega) &= \frac{1}{M^{MN}} \prod_n \binom{M}{\omega_n}, \\ \gamma_n(\theta) &= \ln\left(\frac{g_n(\theta)}{M - g_n(\theta)}\right), \end{aligned}$$

and

$$\delta_n(\omega) = \omega_n.$$

Linearity follows from the last equation.

(b) The Fisher score vector of linear exponential family distributed stochastic variables is described by (3.99):

$$s_\theta = \frac{\partial g^T(\theta)}{\partial \theta} C^{-1} d(\theta)$$

with $d(\theta) = w - g(\theta)$. Substituting the expression for C derived in Solution 3.3(c) produces the Fisher score vector found in 3.3(d).

3.5

(a) Since $Ew_n = g_n(\theta) = M_n \varphi_n$, $\varphi_n = g_n(\theta)/M_n$. Also, since the w_n are independent, their joint probability function is the product of their marginal probability functions:

$$p(\omega;\theta) = \prod_n \binom{M_n}{\omega_n} \varphi_n^{\omega_n} (1-\varphi_n)^{M_n - \omega_n}$$

or

$$p(\omega;\theta) = \prod_n \frac{1}{M_n^{M_n}} \binom{M_n}{\omega_n} g_n^{\omega_n}(\theta) \left[M_n - g_n(\theta)\right]^{M_n - \omega_n}$$

with $n = 1, \ldots, N$. Then,

$$\begin{aligned} q(\omega;\theta) &= \sum_n -M_n \ln M_n + \ln M_n! - \ln \omega_n! - \ln(M_n - \omega_n)! \\ &\quad + \sum_n \omega_n \ln g_n(\theta) + (M_n - \omega_n) \ln\left[M_n - g_n(\theta)\right]. \end{aligned}$$

(b) The observations are independent. Therefore, their covariance matrix is diagonal. Its nth diagonal element is

$$M_n \varphi_n (1-\varphi_n) = \frac{1}{M_n} g_n(\theta)[M_n - g_n(\theta)].$$

(c) Differentiating $q(\omega;\theta)$ with respect to θ and substituting w for ω yields

$$s_\theta = \sum_n \frac{M_n[w_n - g_n(\theta)]}{g_n(\theta)[M_n - g_n(\theta)]} \frac{\partial g_n(\theta)}{\partial \theta}.$$

3.6

(a) The probability function derived in Solution 3.5(a) is an exponential family with

$$\begin{aligned} \alpha(\theta) &= \prod_n \left[M_n - g_n(\theta)\right]^{M_n}, \\ \beta(\omega) &= \prod_n \frac{1}{M_n^{M_n}} \binom{M_n}{\omega_n}, \\ \gamma_n(\theta) &= \ln\left(\frac{g_n(\theta)}{M_n - g_n(\theta)}\right), \end{aligned}$$

and

$$\delta_n(\omega) = \omega_n.$$

Linearity follows from the last equation.

(b) The Fisher score vector of linear exponential family distributed observations is described by (3.99):

$$s_\theta = \frac{\partial g^T(\theta)}{\partial \theta} C^{-1} d(\theta),$$

where $d(\theta) = w - g(\theta)$. Substituting the expression for C derived in Solution 3.5(b) produces the Fisher score vector found in Solution 3.5(c).

3.7

(a) $\lambda \int_0^\infty \exp(-\lambda\omega)\, d\omega = 1$. Differentiating both sides of this equation with respect to λ yields

$$\int_0^\infty \exp(-\lambda\omega)\, d\omega - \lambda \int_0^\infty \omega \exp(-\lambda\omega)\, d\omega = 0.$$

Thus, $Ew = \lambda \int_0^\infty \omega \exp(-\lambda\omega)\, d\omega = 1/\lambda$. Differentiating this result yields

$$-1/\lambda^2 = \frac{1}{\lambda}\lambda \int_0^\infty \omega \exp(-\lambda\omega)\, d\omega - \lambda \int_0^\infty \omega^2 \exp(-\lambda\omega)\, d\omega.$$

Therefore, $-1/\lambda^2 = 1/\lambda^2 - Ew^2$. Then, $Ew^2 = 2/\lambda^2$ and hence, var $w = 1/\lambda^2$.

(b)

$$p(\omega;\theta) = \prod_n \lambda_n \exp(-\lambda_n \omega_n)$$

or

$$p(\omega;\theta) = \prod_n \frac{1}{g_n(\theta)} \exp(-\frac{\omega_n}{g_n(\theta)})$$

with $n = 1, \ldots, N$. Then,

$$q(\omega;\theta) = -\sum_n \ln g_n(\theta) + \frac{\omega_n}{g_n(\theta)}.$$

(c) $C = \text{diag}\,[g_1^2(\theta) \ldots g_N^2(\theta)]$.

(d) Differentiating $q(\omega;\theta)$ with respect to θ and substituting w for ω yields

$$s_\theta = \sum_n \frac{1}{g_n^2(\theta)} [w_n - g_n(\theta)] \frac{\partial g_n(\theta)}{\partial \theta}.$$

3.8

(a) The probability density function derived in Solution 3.7(b) is an exponential family with

$$\begin{aligned}
\alpha(\theta) &= \prod_n \frac{1}{g_n(\theta)}, \\
\beta(\omega) &= 1, \\
\gamma_n(\theta) &= -\frac{1}{g_n(\theta)},
\end{aligned}$$

and

$$\delta_n(\omega) = \omega_n.$$

Linearity follows from the last equation.

(b) The Fisher score vector of linear exponential family distributed observations is described by (3.99):

$$s_\theta = \frac{\partial g^T(\theta)}{\partial \theta} C^{-1} d(\theta),$$

where $d(\theta) = w - g(\theta)$. Substituting the expression for C derived in Solution 3.7(c) produces the Fisher score vector found in Solution 3.7(d).

3.9

(a)

$$p(\omega;\theta) = \left(\frac{32}{\pi^2}\right)^N \prod_n \frac{1}{g_n^3(\theta)} \omega_n^2 \exp - \frac{4\omega_n^2}{\pi g_n^2(\theta)}$$

and

$$q(\omega;\theta) = N \ln(\frac{32}{\pi^2}) + 2\sum_n \ln \omega_n - 3\sum_n \ln g_n(\theta) - \frac{4}{\pi}\sum_n \frac{\omega_n^2}{g_n^2(\theta)}.$$

(b) The probability density derived under (a) is an exponential family with

$$\gamma^T(\theta)\delta(\omega) = -\frac{4}{\pi}\sum_n \frac{\omega_n^2}{g_n^2(\theta)}.$$

Therefore,

$$\gamma_n(\theta) = -\frac{4}{\pi g_n^2(\theta)}$$

and $\delta_n(\omega) = \omega_n^2$. This shows that the exponential family is nonlinear.

3.10 The log-probability density function is described by (3.31):

$$q(\omega,\theta) = -\frac{N}{2}\ln 2\pi - \frac{1}{2}\ln \det C(\theta) - \frac{1}{2}[\omega - g(\theta)]^T C^{-1}(\theta)[\omega - g(\theta)].$$

It is an exponential family if it can transformed into

$$\ln \alpha(\theta) + \ln \beta(\omega) + \gamma^T(\theta)\delta(\omega).$$

This would imply that vector functions $\gamma(\theta)$ and $\delta(\omega)$ have to be found such that

$$\begin{aligned}\gamma^T(\theta)\delta(\omega) &= \omega^T C^{-1}(\theta) g(\theta) - \frac{1}{2}\omega^T C^{-1}(\theta)\omega \\ &= \sum_p \left(\sum_q c_{pq}^- g_q(\theta)\right)\omega_p - \frac{1}{2}\sum_p\sum_q c_{pq}^- \omega_p \omega_q\end{aligned}$$

where $p, q = 1, \ldots, N$ and c_{pq}^- is the (p,q)th element of $C^{-1}(\theta)$. This expression shows that these requirements are met by a vector $\gamma(\theta)$ composed of N elements

$$\sum_q c_{pq}^- g_q(\theta),$$

with $p = 1,\ldots,N$, followed by N elements $-\frac{1}{2}c^{-}_{pp}$, $p = 1,\ldots,N$, and $\frac{1}{2}N(N-1)$ elements $-c^{-}_{pq}$, $p < q = 1,\ldots,N$ combined with a $\delta(\omega)$ vector composed of N elements ω_p, $p = 1,\ldots,N$ followed by N elements ω_p^2, $p = 1,\ldots,N$, and $\frac{1}{2}N(N-1)$ elements $\omega_p\omega_q, p < q = 1,\ldots,N$. This shows that $p(\omega;\theta)$ is not a *linear* exponential family.

3.11 The right-hand member of (3.69) is equal to

$$\begin{aligned}
&\frac{1}{g_N}\left(\sum_{n'=1}^{N-1} w_{n'} - \sum_{n'=1}^{N-1} g_{n'} + \frac{g_N}{g_n}(w_n - g_n)\right) \\
&= \frac{1}{g_N}\left(M - w_N - (M - g_N) + \frac{g_N}{g_n}(w_n - g_n)\right) \\
&= \frac{w_n}{g_n} - \frac{w_N}{g_N}.
\end{aligned}$$

This completes the proof.

3.12 The the pth row of the matrix R is

$$\frac{1}{g_N}(1 \ldots 1 \;\; \frac{g_N}{g_p} + 1 \;\; 1 \ldots 1)$$

and the qth column of the matrix C is

$$\frac{g_q}{M}(-g_1 \ldots - g_{q-1} \;\; (M - g_q) \;\; - g_{q+1} \ldots - g_{N-1})^T.$$

For $p \neq q$ their product is

$$\frac{g_q}{M\,g_N}\left(-\sum_{n=1}^{N-1} g_n + M - \frac{g_N}{g_p}g_p\right) = \frac{g_q}{M\,g_N}[-(M - g_N) + M - g_N] = 0$$

and for $p = q$:

$$\begin{aligned}
&\frac{g_p}{M\,g_N}\left(-\sum_{n=1}^{N-1} g_n + M + \frac{g_N}{g_p}(M - g_p)\right) \\
&= \frac{g_p}{M\,g_N}\left(-(M - g_N) + M + M\frac{g_N}{g_p} - g_N\right) = 1.
\end{aligned}$$

This completes the proof.

3.13 $\sum_{n=1}^{N}(w_n - Ew_n) = 0$. Then, $\sum_{n=1}^{N}(w_{n'} - Ew_{n'})(w_n - Ew_n) = 0$ and hence,

$$\sum_{n=1}^{N} \operatorname{cov}(w_{n'}, w_n) = 0$$

with $n' = 1,\ldots,N$. This shows that the columns of the covariance matrix are linearly dependent. Therefore, the covariance matrix is singular.

3.14 The Fisher score vector of multinomially distributed observations is described by (3.70). Then, the kth element of this vector is equal to

$$s_{\theta_k} = \frac{\partial g^T(\theta)}{\partial \theta_k} R\, d(\theta).$$

The matrix R in this expression is defined by equation (3.71) which shows that it may be written as the sum

$$R = \text{diag}\left(\frac{1}{g_1(\theta)} \cdots \frac{1}{g_{N-1}(\theta)}\right) + \frac{1}{g_N(\theta)} U$$

where all elements of the $(N-1) \times (N-1)$ matrix U are equal to one. Substituting this expression for R in (3.70) and rearranging produces the desired result.

3.15 $\sigma_z^2 = \sigma_u^2 + \sigma_v^2$. See (3.132).

3.16 $\text{cov}(z, r) = E[(z - \mu_z)(r - \mu_r)] = E[\{(u - \mu_u) + j(v - \mu_v)\}(r - \mu_r)] = \text{cov}(u, r) + j\,\text{cov}(v, r)$. Therefore, $\text{cov}(z, r) = 0$ if $\text{cov}(u, r) = 0$ and $\text{cov}(v, r) = 0$. Furthermore, if $\text{cov}(z, r) = 0$, its real and imaginary part must be equal to zero. That is, $\text{cov}(u, r) = \text{cov}(v, r) = 0$. This completes the proof.

3.17 $(z - \mu_z)^2 = (u - \mu_u)^2 + 2j\,(u - \mu_u)(v - \mu_v) - (v - \mu_v)^2$. Therefore, $E\,(z - \mu_z)^2 = \text{var}\, u + 2j\,\text{cov}(u, v) - \text{var}\, v$.

3.18 Let $z_\ell = u_\ell + jv_\ell$. Then, the (k, ℓ)-th element of the covariance matrix $\text{cov}(z, z^*)$ is described by $\text{cov}(z_k, z_\ell^*) = E[(z_k - Ez_k)(z_\ell - Ez_\ell)] = \text{cov}(u_k, u_\ell) - \text{cov}(v_k, v_\ell) + j\,[\text{cov}(u_k, v_\ell) + \text{cov}(v_k, u_\ell)]$. Therefore, $\text{cov}(z, z^*)$ is a null matrix if $\text{cov}(u_k, u_\ell) = \text{cov}(v_k, v_\ell)$ and $\text{cov}(u_k, v_\ell) = -\text{cov}(v_k, u_\ell)$ or, equivalently, if the covariance matrix of u is equal to the covariance matrix of v and the covariance matrix of u and v is equal to the additive inverse of its transpose. The former condition implies that $\text{var}\, u_k = \text{var}\, v_k$ and the latter that $\text{cov}(u_k, v_k) = 0$.

3.19

$$\begin{aligned} p(w) &= \frac{1}{(2\pi)^{\frac{K}{2}} (\det R)^{\frac{1}{2}}} \exp\left(-\frac{1}{2}\left[(\rho - \mu_r)^T R^{-1} (\rho - \mu_r)\right]\right) \\ &\times \frac{1}{\pi^L (\det T)^{\frac{1}{2}}} \exp\left(-\frac{1}{2}(\tau - \mu_t)^H T^{-1} (\tau - \mu_t)\right), \end{aligned}$$

where $R = \text{cov}(r, r)$ and $T = \text{cov}(t, t)$ while the elements of ρ correspond to those of r and the elements of τ to those of $t = (z^T\ z^H)^T$. See (3.165).

4.1

$$R = D^T C D$$

with

$$D = \text{diag}\left(\frac{1}{\sigma_1} \cdots \frac{1}{\sigma_N}\right).$$

Since $C \succ O$, R has the form $P^T A P$ with $A \succ O$ and P nonsingular. Then, by Theorem C.4, $R \succ O$.

4.2

(a) The result is the quadratic form: $c_{pp}x_p^2 + 2c_{pq}x_px_q + c_{qq}x_q^2$.

(b) Suppose that $c_{pp} \neq 0$. Then, the quadratic form derived under (a) may alternatively be written

$$c_{pp}\left(x_p + \frac{c_{pq}}{c_{pp}}x_q\right)^2 + \left(c_{qq} - \frac{c_{pq}^2}{c_{pp}}\right)x_q^2.$$

Since $C \succeq O$, this quadratic form is nonnegative. Then, $c_{pp}c_{qq} \geq c_{pq}^2$. Therefore, if $c_{qq} = 0$, then $c_{pq} = 0$. If also $c_{pp} = 0$, the quadratic form vanishes for all (x_p, x_q) and $c_{pp}c_{qq} \geq c_{pq}^2$ is still true. If $C \succ O$, then $c_{pp} > 0$ and $c_{qq} > 0$, and $c_{pp}c_{qq} > c_{pq}^2$ since, if $c_{pp}c_{qq} = c_{pq}^2$ would be included, all nonzero (x_p, x_q) satisfying $c_{pp}x_p + c_{pq}x_q = 0$ would make the quadratic form equal to zero which is contrary to the assumed positive definiteness.

4.3 Suppose that $c_{qq} = 0$. Then, if $C \succeq O$, $c_{pq}^2 \leq 0$. Therefore, $c_{pq} = 0$ and, hence, $c_{qp} = 0$ for any value of p.

4.4 The diagonal elements of a correlation matrix are equal to one. Furthermore, by Solution 4.1, the correlation matrix is positive definite. Then, by Solution 4.2 (b), $c_{pq}^2/(c_{pp}c_{qq}) = r_{pq}^2 < 1$, or $|r_{pq}| < 1$ for $p \neq q$.

4.5 A matrix should be symmetric and positive semidefinite to qualify as a covariance matrix.

(a) Since the matrix is symmetric and the quadratic form associated with it is nonnegative, it could be a covariance matrix.

(b) and **(c)** The matrices are symmetric but the associated quadratic forms are not nonnegative. Therefore, they do not qualify.

(d) Although the quadratic form associated with this matrix is nonnegative, the matrix does not qualify since it is not symmetric.

(e) and **(f)** The matrices qualify since they are symmetric and the associated quadratic forms are nonnegative.

4.6 The (m, n)th element of the matrix $\text{cov}(u, v)$ is equal to $E[(u_m - Eu_m)(v_n - Ev_n)] = E[u_mv_n] - Eu_mEv_n = E[u_mv_n]$. This is the (m, n)th element of the matrix Euv^T.

4.7

$$R = D^H C D$$

with

$$D = \text{diag}\left(\frac{1}{\sigma_1} \cdots \frac{1}{\sigma_N}\right).$$

Since $C \succ O$, R has the form $P^H AP$ with $A \succ O$ and P nonsingular. Then, by Theorem C.11, $R \succ O$.

4.8

(a) The quadratic form is: $c_{pp}z_pz_p^* + c_{pq}z_pz_q^* + c_{qp}z_qz_p^* + c_{qq}z_qz_q^*$.

(b) Suppose that $c_{pp} \neq 0$. Then, the quadratic form derived under (a) may alternatively be written

$$\begin{aligned} & c_{pp}\left(z_p + \frac{c_{qp}}{c_{pp}}z_q\right)\left(z_p^* + \frac{c_{pq}}{c_{pp}}z_q^*\right) + \left(c_{qq} - \frac{c_{pq}c_{qp}}{c_{pp}}\right)z_qz_q^* \\ = \; & c_{pp}\left|z_p + \frac{c_{qp}}{c_{pp}}z_q\right|^2 + \left(c_{qq} - \frac{|c_{pq}|^2}{c_{pp}}\right)|z_q|^2. \end{aligned}$$

Since $C \succeq O$, this quadratic form is nonnegative. Then, $c_{pp}c_{qq} \geq |c_{pq}|^2$. Therefore, if $c_{qq} = 0$, then $c_{pq} = 0$. If also $c_{pp} = 0$, the quadratic form vanishes for all (z_p, z_q) and $c_{pp}c_{qq} \geq |c_{pq}|^2$ is still true. If $C \succ O$, then $c_{pp} > 0$ and $c_{qq} > 0$, and $c_{pp}c_{qq} > |c_{pq}|^2$ since, if $c_{pp}c_{qq} = |c_{pq}|^2$ would be included, all nonzero (z_p, z_q) satisfying $c_{pp}z_p + c_{qp}z_q = 0$ would make the quadratic form equal to zero which is contrary to the assumed positive definiteness.

4.9 Suppose that $c_{qq} = 0$. Then, by Solution 4.8 (b), $|c_{pq}|^2 = 0$. This means that the real and the imaginary part of c_{pq} are equal to zero. Then, $c_{pq} = 0$ and, therefore, $c_{pq}^* = c_{qp} = 0$ for any value of p.

4.10 The diagonal elements of R are equal to one. Furthermore, by Solution 4.7, the correlation matrix is positive definite. Then, by Solution 4.8 (b), $|c_{pq}|^2/(c_{pp}c_{qq}) = |r_{pq}|^2 < 1$, or $|r_{pq}| < 1$ for $p \neq q$.

4.11 A complex matrix should be Hermitian and positive semidefinite to qualify as a covariance matrix.

(a) Since the matrix is Hermitian and the associated quadratic form is nonnegative, it could be a covariance matrix.

(b) The matrix is not Hermitian. Therefore, it cannot be a covariance matrix.

(c) The matrix is Hermitian. The associated quadratic form is

$$\begin{pmatrix} z_1^* & z_2^* \end{pmatrix}\begin{pmatrix} 9 & 2+j \\ 2-j & 5 \end{pmatrix}\begin{pmatrix} z_1 \\ z_2 \end{pmatrix}.$$

This polynomial is positive unless $z_1 = z_2 = 0$. See Solution 4.8(b). It vanishes for

$$\begin{pmatrix} z_1 & z_2 & z_3 \end{pmatrix} = \begin{pmatrix} 0 & 0 & z_3 \end{pmatrix}$$

where z_3 is any complex number. Then, the matrix is positive semidefinite and may be a covariance matrix.

(d) The matrix is Hermitian but cannot be positive semidefinite since $c_{11}c_{22} < |c_{12}|^2$, see Solution 4.8(b).

4.12

$$\begin{aligned} F_\theta &= -E\frac{\partial^2 q(w;\theta)}{\partial\theta^T\partial\theta} = -E\frac{\partial}{\partial\theta^T}\left(\frac{\partial q(w;\theta)}{\partial\theta}\right) \\ &= -E\frac{\partial s_\theta}{\partial\theta^T} = -E\frac{\partial s_\theta^T}{\partial\theta}. \end{aligned}$$

4.13 Proof of Lemma C.2: see the proof of Lemma C.1. Show that here the condition for nonpositivity of the discriminant is $\text{Re}(b^H Va) = 0$ for all b or, equivalently, $Va = o$. Proof of Theorem C.9: follow proof of Theorem C.2 and use Lemma C.2.

4.14 Prove that

$$Es_\theta = E\frac{\partial q(w;\theta)}{\partial \theta^*} = \int \frac{\partial q(\omega;\theta)}{\partial \theta^*} p(\omega;\theta)\, d\omega = o.$$

Subsequently, use this equation to show that

$$-E\frac{\partial^2 q(w;\theta)}{\partial \theta^* \partial \theta^T} = E\left[\frac{\partial q(w;\theta)}{\partial \theta^*}\frac{\partial q(w;\theta)}{\partial \theta^T}\right].$$

4.15

(a)

$$F_\theta = \sum_n \frac{x_n^2}{\sigma_n^2},$$

(b)

$$F_\theta = \frac{1}{\theta}\sum_n x_n,$$

(c)

$$F_\theta = \frac{M}{\theta}\sum_n \frac{x_n}{M - \theta x_n},$$

(d)

$$F_\theta = \frac{N}{\theta^2},$$

and

(e)

$$F_\theta = \frac{M\sum_n x_n}{\theta(M - \theta\sum_n x_n)},$$

where $n = 1, \ldots, N$ in (a)–(c) and $n = 1, \ldots, N-1$ in (e).

4.16

(a) The Fisher information matrix is described by

$$\begin{pmatrix} \sum_n \frac{x_n^2}{\theta_1 x_n + \theta_2} & \sum_n \frac{x_n}{\theta_1 x_n + \theta_2} \\ \sum_n \frac{x_n}{\theta_1 x_n + \theta_2} & \sum_n \frac{1}{\theta_1 x_n + \theta_2} \end{pmatrix}.$$

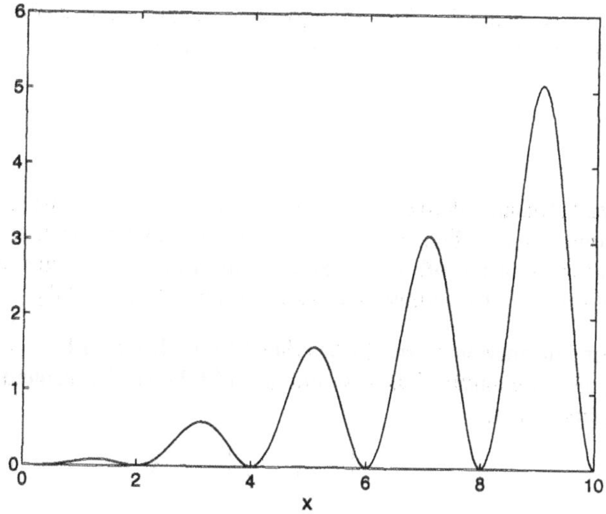

Figure 7.1. Contribution to the Fisher information for the angular frequency per measurement point. (Problem 4.17)

(b) To maximize the Fisher information for θ_1, that is, the (1,1)th element of the Fisher information matrix, the x_n should be chosen as large as allowed. On the other hand, to maximize the Fisher information for θ_2, the (2,2)th element, the x_n should be equal to zero. The conclusion is that the optimal design for θ_1 and that for θ_2 are conflicting.

4.17

(a)

$$\frac{1}{\sigma^2}\begin{pmatrix} \sum \cos^2(\theta_2 x_n) & -\theta_1 \sum x_n \cos(\theta_2 x_n)\sin(\theta_2 x_n) \\ -\theta_1 \sum x_n \cos(\theta_2 x_n)\sin(\theta_2 x_n) & \theta_1^2 \sum x_n^2 \sin^2(\theta_2 x_n) \end{pmatrix}$$

where all summations are over n.

(b) The points where $\cos(\theta_2 x_n)$ is maximum or minimum contribute most and to the same extent to the Fisher information f_{11} for the amplitude θ_1. Figure 7.1. shows the summand in the expression for f_{22} as a function of x_n in units of $\pi/2$. For increasing x_n, the locations of the maxima of the summand approach the points where x_n is an odd multiple of $\pi/2$. These points are the zero-crossings of $\cos(\theta_2 x_n)$. Thus, they contribute most to the Fisher information f_{22} for the angular frequency θ_2, those for large x_n in particular. The conclusion is that the optimal design for θ_1 and that for θ_2 are conflicting.

4.18

(a) First show that the diagonal elements f_{kk} of the Fisher information matrix F_θ are described by

$$\sum_n \frac{1}{g_n(\theta)}\left(\frac{\partial g_n(\theta)}{\partial \theta_k}\right)^2,$$

where $g_n(\theta) = g(x_n;\theta)$. See Example 4.2. Then, the contribution of the nth measurement point x_n to the Fisher information for θ_k is

$$\frac{1}{g_n(\theta)}\left(\frac{\partial g_n(\theta)}{\partial \theta_k}\right)^2 .$$

(b) For this expectation model, the contribution of the nth measurement point to the Fisher information for θ_1 is $(1/\theta_1)\exp(-\theta_2 x_n)$. This contribution is maximum for $x_n = 0$. The contribution of the nth measurement point to the Fisher information for θ_2 is $\theta_1 x_n^2 \exp(-\theta_2 x_n)$. This contribution is maximum for $x_n = 2/\theta_2$.

4.19 The Fisher information matrix is described by (4.46) with $C^{-1} = R$ defined by (3.71). Decomposing the matrix R as in Solution 3.14 followed by elementary calculations produces the desired result.

4.20

(a) Since the w_n are independent, they are also uncorrelated and their covariance matrix $C(\theta)$ is diagonal. Then, (4.68) shows that

$$f_{kk} = \sum_n \frac{1}{c_{nn}(\theta)}\left(\frac{\partial g_n(\theta)}{\partial \theta_k}\right)^2 ,$$

where f_{kk} is the kth diagonal element of F_θ and $c_{nn}(\theta)$ is the nth diagonal element of $C(\theta)$. Therefore, the contribution of an observation w_{N+1} to the Fisher information for θ_k is

$$\frac{1}{c_{N+1,N+1}(\theta)}\left(\frac{\partial g_{N+1}(\theta)}{\partial \theta_k}\right)^2 .$$

(b) Equating c and d in (4.84) and (4.85) to the corresponding null vector and to $c_{N+1,N+1}(\theta)$, respectively, shows the compatibility of the results.

4.21

(a) The necessary and sufficient conditions met by $t(w)$ are described by (4.149):

$$F_\theta^{-1} s_\theta - t(w) + \theta = o ,$$

where o is the $K \times 1$ null vector. Consider the linear transformation $\rho(\theta) = R\theta$, where R is an $L \times K$ matrix independent of θ. Then,

$$\begin{aligned}\frac{\partial \rho(\theta)}{\partial \theta^T} F_\theta^{-1} s_\theta - r(w) + \rho(\theta) &= RF_\theta^{-1} s_\theta - Rt(w) + R\theta \\ &= R\left[F_\theta^{-1} s_\theta - t(w) + \theta\right] = o .\end{aligned}$$

Therefore, by (4.139), $Rt(w)$ is efficient unbiased for $\rho(\theta) = R\theta$.

(b) Since $t(w)$ satisfies the conditions for efficiency and unbiasedness, its covariance matrix is equal to F_θ^{-1}. Then, the covariance matrix of $Rt(w)$ is equal to

$$RF_\theta^{-1} R^T .$$

Furthermore, since $r(w)$ estimates a function of θ instead of θ itself, the Cramér–Rao lower bound for $r(w)$ equals

$$\frac{\partial \rho}{\partial \theta^T} F_\theta^{-1} \frac{\partial \rho^T}{\partial \theta} = R F_\theta^{-1} R^T.$$

Therefore, $Rt(w)$ attains the Cramér–Rao lower bound. Furthermore, $Et(w) = \theta$. Hence, $E[Rt(w)] = R\,Et(w) = R\theta$.

4.22 Generally, the necessary and sufficient conditions for efficiency and unbiasedness are described by (4.139). If the observations are normally distributed, the Fisher information matrix F_θ and the Fisher score vector s_θ in this expression are given by (4.40) and (3.35), respectively. Substituting these expressions in (4.139) yields

$$\frac{\partial \rho(\theta)}{\partial \theta^T} \left(\frac{\partial g^T(\theta)}{\partial \theta} C^{-1} \frac{\partial g(\theta)}{\partial \theta^T} \right)^{-1} \frac{\partial g^T(\theta)}{\partial \theta} C^{-1} \left[w - g(\theta) \right] = r(w) - \rho(\theta)\,.$$

(a) In this special case, $\partial \rho(\theta)/\partial \theta^T = R$, $\partial g(\theta)/\partial \theta^T = X$ since $\rho(\theta) = R\theta$ and $g(\theta) = X\theta$ with R and X known.

(b) $r(w) = R(X^T C^{-1} X)^{-1} X^T C^{-1} w$.

(c) Then, $R = I$ and, therefore, $r(w) = (X^T C^{-1} X)^{-1} X^T C^{-1} w$.

4.23 The necessary and sufficient conditions for efficiency and unbiasedness are: $s_\theta = F_\theta[t(w) - \theta]$, where F_θ is the Fisher information matrix and s_θ the Fisher score vector. Generally, the Fisher information matrix and the Fisher score vector of linear exponential family distributed observations are described by (4.68):

$$F_\theta = \frac{\partial g^T(\theta)}{\partial \theta} C^{-1} \frac{\partial g(\theta)}{\partial \theta^T}$$

and (3.99):

$$s_\theta = \frac{\partial g^T(\theta)}{\partial \theta} C^{-1} \left[w - g(\theta) \right],$$

respectively. Here, $g_n(\theta) = \theta$, $g(\theta) = \theta u$ with $u = (1 \ldots 1)^T$, $\partial g(\theta)/\partial \theta^T = u$ and $C = \sigma^2(\theta) I$, where I is the identity matrix of order N. Therefore, $F_\theta = N/\sigma^2(\theta)$ and $s_\theta = \left(\sum_n w_n - N\theta\right)/\sigma^2(\theta)$. Then, $\overline{w}_n - \theta = t(w) - \theta$ and, therefore, $t(w) = \overline{w}_n$, where $\overline{w}_n$, is the average of the w_n.

4.24

(a) The Fisher information matrices follow from (4.42) and (4.43).

Case 1: If the knowledge of λ is *not* used, the vector of parameters to be estimated is $\theta = (\alpha_1\ \alpha_2\ \beta_1\ \beta_2)^T$ and the Jacobian matrix $\partial g(\theta)/\partial \theta^T$ is 51×4 with $\partial g_n(\theta)/\partial \alpha_k = \kappa \exp(-\beta_k x_n)$ and $\partial g_n(\theta)/\partial \beta_k = -\kappa \alpha_k x_n \exp(-\beta_k x_n)$, $k = 1, 2$. The resulting Fisher information matrix is 4×4.

Case 2: If the knowledge of λ *is* used, the vector of parameters to be estimated is $(\alpha_1\ \beta_1\ \beta_2)$ and the Jacobian matrix is 51×3 with

$$\frac{\partial g_n(\theta)}{\partial \alpha_1} = \kappa \left[\exp(-\beta_1 x_n) + \lambda \exp(-\beta_2 x_n) \right]$$

and

$$\frac{\partial g_n(\theta)}{\partial \beta_k} = -\kappa \alpha_k x_n \exp(-\beta_k x_n),$$

$k = 1, 2$, with $\alpha_2 = \lambda \alpha_1$. The resulting Fisher information matrix is 3×3. Multiplying the Cramér–Rao standard deviation for α_1 by λ produces the corresponding standard deviation for α_2.

The results are presented in Table 7.1.

Table 7.1. Parameter values and Cramér–Rao standard deviations (Solution 4.24)

Parameter	Value	Cramér–Rao Standard Deviations Case 1	Cramér–Rao Standard Deviations Case 2
α_1	1.6	0.7	0.009
α_2	0.9	0.7	0.005
β_1	1	0.14	0.016
β_2	0.6	0.11	0.007

(b) In Case 1, the Cramér–Rao standard deviations for the parameters α_1 and α_2 are so large compared with the magnitude of these parameters that even a hypothetical estimator that has these standard deviations is not precise enough to allow meaningful conclusions about the true magnitude of the parameters. On the other hand, in Case 2, the reduction of the number of unknown parameters causes the Cramér–Rao standard deviations to decrease to such an extent that relative standard deviations are obtained of less than two percent.

4.25

(a)

$$\frac{\partial \rho}{\partial \theta^T} F_\theta^{-1} \frac{\partial \rho^T}{\partial \theta} = \frac{1}{2} \begin{pmatrix} 1 & 1 \\ 1 & -1 \end{pmatrix} \begin{pmatrix} \operatorname{var} \check{t}_1 & \operatorname{cov}(\check{t}_1, \check{t}_2) \\ \operatorname{cov}(\check{t}_1, \check{t}_2) & \operatorname{var} \check{t}_2 \end{pmatrix} \begin{pmatrix} 1 & 1 \\ 1 & -1 \end{pmatrix}.$$

Then,

$$\operatorname{var} \check{r}_1 = \frac{1}{2} \left[\operatorname{var} \check{t}_1 + 2 \operatorname{cov}(\check{t}_1, \check{t}_2) + \operatorname{var} \check{t}_2\right],$$
$$\operatorname{var} \check{r}_2 = \frac{1}{2} \left[\operatorname{var} \check{t}_1 - 2 \operatorname{cov}(\check{t}_1, \check{t}_2) + \operatorname{var} \check{t}_2\right],$$

and

$$\operatorname{cov}(\check{r}_1, \check{r}_2) = \frac{1}{2} \left[\operatorname{var} \check{t}_1 - \operatorname{var} \check{t}_2\right].$$

Therefore, $\operatorname{var} \check{t}_1 + \operatorname{var} \check{t}_2 = \operatorname{var} \check{r}_1 + \operatorname{var} \check{r}_2$.

(b) If the correlation coefficient is close to one and $\operatorname{var} \check{t}_1 \approx \operatorname{var} \check{t}_2$, $\operatorname{cov}(\check{t}_1, \check{t}_2) \approx (\operatorname{var} \check{t}_1)^{\frac{1}{2}} (\operatorname{var} \check{t}_2)^{\frac{1}{2}} \approx \operatorname{var} \check{t}_1$ and, therefore,

$$\operatorname{var} \check{r}_1 = \frac{1}{2} \left[\operatorname{var} \check{t}_1 + 2 \operatorname{cov}(\check{t}_1, \check{t}_2) + \operatorname{var} \check{t}_2\right] \approx 2 \operatorname{var} \check{t}_1.$$

Furthermore, under these conditions, $\operatorname{var} \check{r}_2 \approx 0$ and $\operatorname{cov}(\check{r}_1, \check{r}_2) \approx 0$. The proof for a correlation coefficient close to minus one is analogous.

4.26 The complex parameter vector $\theta = (\varphi_1 \;\; \varphi_2 + j\varphi_3 \;\; \varphi_2 - j\varphi_3)^T$. Then,

$$\frac{\partial \theta}{\partial \varphi^T} = \begin{pmatrix} 1 & 0 & 0 \\ 0 & 1 & j \\ 0 & 1 & -j \end{pmatrix}.$$

Premultiplying Ψ by this matrix followed by postmultiplying by its complex conjugate transpose yields the Cramér–Rao lower bound matrix for θ with ψ_{11} as its first diagonal element and $\psi_{22} + \psi_{33}$ as both its second and third diagonal element.

4.27 The complex parameter vector corresponding to φ is described by

$$\begin{aligned} \theta = A\varphi &= (\alpha_y + j\beta_y \;\; \alpha_u + j\beta_u \;\; \alpha_y - j\beta_y \;\; \alpha_u - j\beta_u)^T \\ &= (\gamma_y \; \gamma_u \; \gamma_y^* \; \gamma_u^*)^T, \end{aligned}$$

where A is a 4×4 matrix defined as (4.197). Then, the Cramér–Rao lower bound matrix for unbiased estimation of θ is equal to $A\Psi A^H$. The scalar function to be estimated is $\rho = \rho(\theta) = \gamma_y/\gamma_u$. Then, $\partial\rho/\partial\theta = (1/\gamma_u \;\; -\gamma_y/\gamma_u^2 \;\; 0 \;\; 0)^T$. Therefore, the scalar Cramér–Rao lower bound for unbiased estimation of ρ is $\partial\rho/\partial\theta^T A\Psi A^H (\partial\rho/\partial\theta^T)^H$. Elaborating yields

$$\frac{1}{|\gamma_u|^2}\left[\psi_{11} + \psi_{33} + (\psi_{22} + \psi_{44})\,|H|^2 - 2(\psi_{12} + \psi_{34})\operatorname{Re} H + 2(\psi_{14} - \psi_{23})\operatorname{Im} H\right],$$

where $H = H(j\omega)$.

4.28

(a) The original model

$$g_n(\theta) = \alpha \cos[\omega(x_n - \tau)] + \beta \sin[\omega(x_n - \tau)]$$

with $\theta = (\alpha \, \beta \, \omega \, \tau)^T$, may be reparameterized into

$$h_n(\psi) = \gamma \cos(\omega x_n + \varphi)$$

with $\gamma = (\alpha^2 + \beta^2)^{\frac{1}{2}} > 0$, $\varphi = -\omega\tau - \arcsin(\beta/\gamma)$, and $\psi = (\gamma \; \omega \; \varphi)^T$.

(b) For the original model, the three columns of the Jacobian matrix corresponding to α, β, and τ are linearly dependent. For the reparameterized model, the elements of the columns are described by: $\cos(\omega x_n + \varphi)$, $-\gamma x_n \sin(\omega x_n + \varphi)$, and $-\gamma \sin(\omega x_n + \varphi)$. Therefore, these columns are linearly independent.

4.29 The form of the Fisher score vector is the same for the three distributions in Example 4.18 and is described by

$$s_\theta = \frac{\partial g^T(\theta)}{\partial \theta} C^{-1} d(\theta).$$

If $g_n(\theta)$ may be reparameterized by a number of parameters smaller than the dimension K of θ, then the columns of $\partial g(\theta)/\partial \theta^T$ are linearly dependent and can be nontrivially linearly combined into an $N \times 1$ null vector:

$$\sum_k \mu_k \frac{\partial g(\theta)}{\partial \theta_k} = o,$$

where the μ_k are scalars. Since s_{θ_k} is scalar and C^{-1} is symmetric,

$$\begin{aligned}\sum_k \mu_k s_{\theta_k} &= \sum_k \mu_k \frac{\partial g^T(\theta)}{\partial \theta_k} C^{-1} d(\theta) = \sum_k \mu_k d^T(\theta) C^{-1} \frac{\partial g(\theta)}{\partial \theta_k} \\ &= d^T(\theta) C^{-1} \sum_k \mu_k \frac{\partial g(\theta)}{\partial \theta_k} = 0.\end{aligned}$$

Because this is a constant, nontrivial linear combination of the elements of s_θ, this completes the proof.

4.30 The equations (4.244), (4.246), (4.250), and (4.251) show that for the computation of the gradient and the Hessian matrix of Q with respect to $x = (x_1 \ldots x_N)^T$, expressions for the elements of F_θ, $\partial F_\theta / \partial x_n$, and $\partial^2 F_\theta / \partial x_m \partial x_n$ are needed.

For Poisson distributed observations, the elements of F_θ are described by (4.44):

$$f_{k\ell} = \sum_n \frac{1}{g_n(\theta)} \frac{\partial g_n(\theta)}{\partial \theta_k} \frac{\partial g_n(\theta)}{\partial \theta_\ell}$$

with

$$\begin{aligned}\frac{\partial g_n(\theta)}{\partial \alpha_k} &= \exp(-\beta_k x_n), \\ \frac{\partial g_n(\theta)}{\partial \beta_k} &= -\alpha_k x_n \exp(-\beta_k x_n),\end{aligned}$$

for $k = 1, 2$, and

$$g_n(\theta) = \alpha_1 \exp(-\beta_1 x_n) + \alpha_2 \exp(-\beta_2 x_n).$$

Then, the (k, ℓ)th element of $\partial F_\theta / \partial x_n$ is equal to

$$\frac{\partial f_{k\ell}}{\partial x_n} = \frac{\partial}{\partial x_n} \left(\frac{1}{g_n(\theta)} \frac{\partial g_n(\theta)}{\partial \theta_k} \frac{\partial g_n(\theta)}{\partial \theta_\ell} \right).$$

This expression shows that the following derivatives are needed:

$$\begin{aligned}\frac{\partial g_n(\theta)}{\partial x_n} &= -\alpha_1 \beta_1 \exp(-\beta_1 x_n) - \alpha_2 \beta_2 \exp(-\beta_2 x_n), \\ \frac{\partial^2 g_n(\theta)}{\partial \alpha_k \partial x_n} &= -\beta_k \exp(-\beta_k x_n),\end{aligned}$$

and

$$\frac{\partial^2 g_n(\theta)}{\partial \beta_k \partial x_n} = -\alpha_k (1 - \beta_k x_n) \exp(-\beta_k x_n)$$

for $k = 1, 2$.

The expression for $\partial f_{k\ell} / \partial x_n$ shows that $\partial^2 f_{k\ell} / \partial x_n \partial x_m$ is equal to zero if $n \neq m$. Then, $\partial^2 F_\theta / \partial x_n \partial x_m$ is equal to the null matrix. For $n = m$, the computation of $\partial^2 f_{k\ell} / \partial x_n^2$ requires, in addition to the derivatives already computed, the derivatives

$$\begin{aligned}\frac{\partial^2 g_n(\theta)}{\partial x_n^2} &= \alpha_1 \beta_1^2 \exp(-\beta_1 x_n) + \alpha_2 \beta_2^2 \exp(-\beta_2 x_n), \\ \frac{\partial^3 g_n(\theta)}{\partial \alpha_k \partial x_n^2} &= \beta_k^2 \exp(-\beta_k x_n),\end{aligned}$$

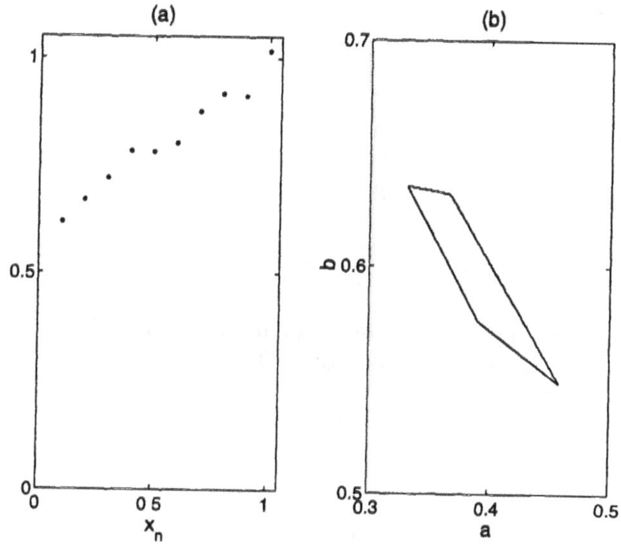

Figure 7.2. (a) Uniformly distributed straight-line observations. (b) Corresponding possible maximum likelihood estimates represented by all points inside contour. (Problem 5.3)

and

$$\frac{\partial^3 g_n(\theta)}{\partial \beta_k \partial x_n^2} = \alpha_k \beta_k (2 - \beta_k x_n) \exp(-\beta_k x_n)$$

for $k = 1, 2$. All of these computations are simple operations on the arrays $\exp(-\beta_1 x_n)$ and $\exp(-\beta_2 x_n)$, $n = 1, \ldots, N$.

5.1 An expression for the binomial log-probability function has been derived in Solution 3.3. This expression shows that the log-likelihood function is described by

$$\begin{aligned} q(w;t) &= N(\ln M! - M \ln M) - \sum_n \ln w_n! - \sum_n \ln (M - w_n)! \\ &\quad + \sum_n w_n \ln g_n(t) + \sum_n (M - w_n) \ln [M - g_n(t)]. \end{aligned}$$

Only both last terms are dependent on t.

5.2 An expression for the exponential log-probability density function has been derived in Solution 3.7. This expression shows that the log-likelihood function is described by

$$q(w;t) = -\sum_n \ln g_n(t) - \sum_n \frac{w_n}{g_n(t)}.$$

5.3 Figure 7.2.(a) shows the observations. These indicate that $\alpha \approx 0.4$ and $\beta \approx 0.6$. On a sufficiently dense grid around these values in the (a, b)-plane, the boundary is computed of the collection of all points such that $|w_n - a x_n + b| \leq 0.05$ for $n = 1, \ldots, 10$. All interior points of this boundary qualify as maximum likelihood estimates of (α, β). Figure 7.2.(b) shows the boundary.

5.4

(a) The likelihood equations are obtained by substituting the variables $t = (a\ b)^T$ for θ in the Fisher score vector and equating the result to the null vector. For Poisson distributed observations, this yields

$$\frac{\partial g^T(t)}{\partial t} C^{-1} d(t) = o$$

with $d(t) = w - g(t)$, $\partial g_n(t)/\partial a = (1/a) g_n(t)$, $\partial g_n(t)/\partial b = -x_n g_n(t)$, $C = \operatorname{diag} g(t)$, where $g_n(t) = \gamma a \exp(-b x_n)$. Or,

$$\sum_n d_n(t) = 0$$

and

$$\sum_n x_n\, d_n(t) = 0.$$

(b) The first likelihood equation shows that

$$a = \frac{\sum_n w_n}{\gamma \sum_n \exp(-b x_n)}.$$

Substituting this expression for a in the second equation yields

$$\sum_n \exp(-b x_n) \sum_n w_n x_n - \sum_n x_n \exp(-b x_n) \sum_n w_n = 0\,.$$

This scalar equation may be solved for b by a numerical root finding method. The solution is the maximum likelihood estimate $\tilde{b}$ of β. The maximum likelihood estimate $\tilde{a}$ is obtained by substituting $\tilde{b}$ for b in the expression for a.

(c) Figure 7.3. shows the observations. These indicate that $\beta \approx 1$. Therefore, as a guess, the solution is supposed to lie on the interval [0.8, 1.2]. Specification of such an interval is required by most root finding methods. The solution found was $\tilde{b} = 1.03$ resulting in $\tilde{a} = 1.02$.

5.5 First, show that the likelihood equations may be written as

$$\sum_n \frac{w_n}{a x_n + b} = N$$

and

$$\sum_n \frac{w_n x_n}{a x_n + b} = \sum_n x_n\,.$$

Next, reparameterize these equations as follows:

$$\frac{1}{a} \sum_n \frac{w_n}{x_n + c} = N$$

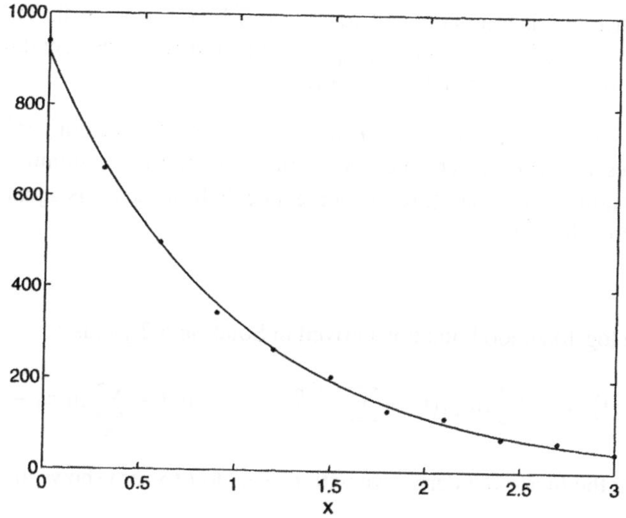

Figure 7.3. Poisson distributed monoexponential decay observations (dots) and the corresponding maximum likelihood estimate of the expectation model (solid line). (Problem 5.4)

and

$$\frac{1}{a}\sum_n \frac{w_n x_n}{x_n + c} = \sum_n x_n$$

with $c = b/a$. From the first of these equations:

$$a = \frac{1}{N}\sum_n \frac{w_n}{x_n + c}.$$

Substituting this expression in the second equation yields an equation in the parameter c only that can be solved numerically by root finding. The result $\tilde{c}$ is substituted for c in the expression for a to yield the maximum likelihood estimate $\tilde{a}$ of the parameter α. Finally, the maximum likelihood estimate $\tilde{b}$ of β is $\tilde{a}\tilde{c}$.

5.6

(a) The covariance matrix is the covariance matrix $(X^T C^{-1} X)^{-1}$ of the maximum likelihood estimator of the parameters of linear models from normally distributed observations (5.93) with

$$X = \begin{pmatrix} x_1 & 1 \\ x_2 & 1 \\ \dots & \dots \\ x_N & 1 \end{pmatrix}$$

and $C = \sigma^2 I$. Then, substituting X and C in the expression for the covariance matrix shows that the variances of $\tilde{a}$ and $\tilde{b}$ are equal to $N\sigma^2/D$ and $\sigma^2 \sum_n x_n^2/D$, respectively, with the scalar $D = N\sum_n x_n^2 - (\sum_n x_n)^2$.

(b) The variance of $\tilde{b}$ is minimum and equal to σ^2/N if $\sum_n x_n = 0$. The variance of $\tilde{a}$ is minimum if $\sum_n x_n = 0$ and $\sum_n x_n^2$ is maximum. These conditions are met if one–half of the x_n is equal to A and the rest to $-A$.

(c) The x_n lie on a sphere with radius R if $\sum_n x_n^2 = R^2$. Substituting this expression in the expressions for the variances shows that the latter are minimum if $\sum_n x_n = 0$. These conditions are met if, for example, one–half of the x_n is equal to $R/\sqrt{N}$ and the rest to $-R/\sqrt{N}$.

5.7

(a) Using the log-likelihood function derived in Solution 5.2 yields

$$q(w;t) = -\sum_n \ln g_n(t) - \sum_n \frac{w_n}{g_n(t)} = -N\ln t - \sum_n \ln x_n - \frac{1}{t}\sum_n \frac{w_n}{x_n}.$$

Equating the first-order derivative with respect to t to zero shows that

$$\tilde{t} = \frac{1}{N}\sum_n \frac{w_n}{x_n}.$$

This is a maximum of $q(w;t)$ since the second-order derivative at $t = \tilde{t}$ is $-N/\tilde{t}^2 < 0$.

(b)

$$E\tilde{t} = E\frac{1}{N}\sum_n \frac{w_n}{x_n} = \frac{1}{N}\sum_n \frac{Ew_n}{x_n} = \frac{1}{N}\sum_n \frac{\theta x_n}{x_n} = \theta.$$

(c) The necessary and sufficient condition for unbiasedness and efficiency is

$$\frac{d\rho(\theta)}{d\theta} F_\theta^{-1} s_\theta = r(w) - \rho(\theta).$$

Next, use $d\rho(\theta)/d\theta = 1$, $F_\theta = -E\,\partial^2 q(w;\theta)/\partial\theta^2 = N/\theta^2$ and

$$s_\theta = -\frac{N}{\theta} + \frac{1}{\theta^2}\sum_n \frac{w_n}{x_n}$$

to show that $\tilde{t}$ meets the condition.

5.8 Show that the log-likelihood function of the parameters $t = (a\,b\,c)^T$ is described by

$$q(w;t) = -\frac{N}{2}\ln 2\pi - \sum_n \ln x_n^2 - N\ln c - \frac{1}{2}\sum_n \left(\frac{w_n - ax_n - b}{cx_n^2}\right)^2.$$

The values $\tilde{a}$ and $\tilde{b}$ of a and b that maximize this log-likelihood function are independent of the value of c and are equal to the least squares solution $(X^TRX)^{-1}X^TRw$ with $R = \mathrm{diag}\,(1/x_1^4 \dots 1/x_N^4)$ and X defined as in Solution 5.6(a). Subsequently equating the derivative of $q(w;t)$ with respect to c to zero yields

$$c^2 = \frac{1}{N}\sum_n \left(\frac{w_n - ax_n - b}{x_n^2}\right)^2.$$

Substituting $\tilde{a}$ and $\tilde{b}$ for a and b in this expression yields the maximum likelihood estimate $\tilde{c}$ of γ.

5.9

(a) Show that the estimator $\hat{t}$ maximizes the log-likelihood function for uncorrelated, normally distributed observations with equal variance. Subsequently, show that the estimator $\hat{t}$ is the best linear unbiased estimator under the conditions mentioned.

(b) For the straight line, this is the $N \times 2$ matrix X defined in Solution 5.6 (a). For the constant, X is the $N \times 1$ vector $(1\,1\ldots 1)^T$.

(c) See Section 5.4.3.2. For the straight-line model, the number of parameters K is equal to two. Then, the unbiased estimator of the variance is

$$\frac{1}{N-2}(w - X\hat{t}\,)^T(w - X\hat{t}\,) = \frac{1}{N-2}\sum_n (w_n - \hat{a}x_n - \hat{b})^2,$$

where $\hat{t} = (\hat{a}\;\hat{b}\,)^T$. For the constant, the number of parameters K is equal to one and, therefore,

$$\frac{1}{N-1}(w - X\hat{t})^T(w - X\hat{t}) = \frac{1}{N-1}\sum_n (w_n - \hat{b})^2$$

is unbiased.

(d) Show that the maximum likelihood estimators are, respectively,

$$\tilde{s^2} = \frac{1}{N}\sum_n (w_n - \hat{a}x_n - \hat{b})^2$$

and

$$\tilde{s^2} = \frac{1}{N}\sum_n (w_n - \hat{b})^2 .$$

Both are different from the described unbiased estimators.

5.10

(a) The Cramér–Rao lower bound matrix is the inverse of the expression (4.42) for the Fisher information matrix for Poisson distributed observations. This expression produces the Cramér–Rao variances 0.00234 and 0.00214 for α and β, respectively.

(b) Repeating the experiment 10 000 times, we found for the maximum likelihood method $m_a = 1.0008$, $s_a^2 = 0.00238$ and $s_a = 0.05$ where m_a, s_a^2 and s_a are the sample mean, variance, and standard deviation of the estimates of α, respectively. Then, the estimated standard deviation of m_a is $s_a/100 = 0.0005$. With this standard deviation, the estimated bias of 0.0008 is not significant if tested at the 0.95 level. For the maximum likelihood estimates of β, $m_b = 1.0012$, $s_b^2 = 0.00217$ and $s_b = 0.05$. Therefore, the standard deviation of m_b is also 0.0005. The estimated bias of 0.0012 is significant if tested at the 0.95 level. However, in both cases, the bias is approximately two percent of the standard deviation and may, therefore, be neglected in the computation of the mean squared error. Then, the estimated efficiencies are equal to $0.00234/0.00238 = 0.98$ and $0.00214/0.00217 = 0.99$, respectively. The corresponding quantities for the straight-line estimates from the same observations are as follows. For the estimates of α, $m_a = 1.0070$, $s_a^2 = 0.00513$, and $s_a = 0.0716$, respectively. For the estimates of β, $m_b = 1.0136$, $s_b^2 = 0.00443$, and $s_b = 0.0666$, respectively. In both cases, the bias is significant when tested at the 0.95 level and is approximately ten times larger than the bias of the corresponding maximum likelihood estimates. As a result, it may not be neglected in the computation of the mean squared errors and the efficiencies derived from them. These efficiencies are equal to $0.00234/(0.00513 + 0.00005) = 0.45$ and $0.00214/(0.00443 + 0.00019) = 0.46$, respectively. The conclusion is that the straight-line estimator is much more biased and much less efficient than the maximum likelihood estimator.

5.11 First, show that $\tilde{m} = w$ is the maximum likelihood estimator of $Ew = \mu$. Then, the log-likelihood ratio for the model is described by

$$\begin{aligned} \ell &= q\left(w; g(\tilde{t})\right) - q(w; w) \\ &= \sum_n w_n \ln g(\tilde{t}) + (M - w_n) \ln[M - g(\tilde{t})] \\ &\quad - \sum_n w_n \ln w_n + (M - w_n) \ln(M - w_n) \\ &= \sum_n w_n \ln \left(\frac{g_n(\tilde{t})}{w_n}\right) + (M - w_n) \ln \left(\frac{M - g(\tilde{t})}{M - w_n}\right), \end{aligned}$$

where the binomial log-likelihood function has been used derived in Solution 5.1.

5.12 First, show that $\tilde{m} = w$ is the maximum likelihood estimator of $Ew = \mu$. Then, the log-likelihood ratio for the model is described by

$$\begin{aligned} \ell &= q\left(w; g(\tilde{t})\right) - q(w; w) \\ &= -\sum_n \ln g_n(\tilde{t}) + \frac{w_n}{g_n(\tilde{t})} + \sum_n \ln w_n + \frac{w_n}{w_n} \\ &= N - \sum_n \ln \left(\frac{g_n(\tilde{t})}{w_n}\right) + \frac{w_n}{g_n(\tilde{t})} \end{aligned}$$

where the exponential log-likelihood function has been used derived in Solution 5.2.

5.13 The general expression for the log-likelihood ratio for distributions that are linear exponential families is (5.162):

$$\ell = \ln \frac{\alpha\left(g(\tilde{t})\right)}{\alpha(w)} + \left[\gamma\left(g(\tilde{t})\right) - \gamma(w)\right]^T w.$$

Use Solution 3.4 (a) and rearrange to prove identity with Solution 5.11.

5.14 Like in Solution 5.13, use (5.162) combined with Solution 3.8(a) and rearrange to prove identity with Solution 5.12.

5.15 The normal equations with respect to the vector a are described by (5.181):

$$\frac{\partial g^T(t)}{\partial a} R[w - g(t)] = 0\,,$$

where $g(t) = [g_1(t) \ldots g_N(t)]^T = Ha$. Then, $H^T R(w - Ha) = o$ or

$$\hat{a} = Pw$$

with $P = (H^T RH)^{-1} H^T R$. Substituting this expression for a in the least squares criterion $[w - g(t)]^T R[w - g(t) = (w - Ha)^T R(w - Ha)$ yields

$$w^T(I - P^T H^T)R(I - HP)w\,,$$

where I is the identity matrix of order N. Substituting $P = (H^T RH)^{-1} H^T R$, rearranging, and using the symmetry of both R and $(H^T RH)^{-1}$, yields the desired expression for the criterion. Then, the value $\hat{b}$ of b minimizing this criterion may be substituted in the expression for $\hat{a}$ to produce the least squares solution for α.

5.16 The (k, ℓ)th element of the Hessian matrix follows directly from the gradient (5.180) and is described by

$$\frac{\partial^2 J(t)}{\partial t_k \partial t_\ell} = 2 \sum_{p,q} r_{pq} \frac{\partial g_p(t)}{\partial t_\ell} \frac{\partial g_q(t)}{\partial t_k} - 2 \sum_{p,q} r_{pq} d_p(t) \frac{\partial^2 g_q(t)}{\partial t_k \partial t_\ell}\,.$$

Therefore,

$$\frac{\partial^2 J(t)}{\partial t \partial t^T} = 2 \frac{\partial g^T(t)}{\partial t} R \frac{\partial g(t)}{\partial t^T} - 2 \sum_{p,q} r_{pq} d_p(t) \frac{\partial^2 g_q(t)}{\partial t\,, \partial t^T}\,,$$

where $d_p(t) = w_p - g_p(t)$ and $p, q = 1, \ldots, N$.

5.17 The ordinary least squares estimator of the vector of Fourier coefficients $(\alpha_1\, \beta_1 \ldots \alpha_K\, \beta_K)^T$ is $(X^T X)^{-1} X^T w$ where w is the $JM \times 1$ vector of the observations w_n and X is the $JM \times 2K$ matrix with odd and even numbered columns described by, respectively,

$$(\,1\; \cos(2\pi k/M)\; \cos(2\pi k2/M) \ldots \cos[2\pi k(JM - 1)/M]\,)^T$$

and

$$(\,0\; \sin(2\pi k/M)\; \sin(2\pi k2/M) \ldots \sin[2\pi k(JM - 1)/M]\,)^T\,.$$

Then, as a consequence of the described orthogonalities, the diagonal elements of $(X^T X)$ are different from zero only. The even and odd numbered diagonal elements are described by, respectively,

$$\sum_{n=0}^{JM-1} \cos^2(2\pi kn/M) \quad \text{and} \quad \sum_{n=0}^{JM-1} \sin^2(2\pi kn/M)$$

that are both equal to $JM/2$ for all k. Then, the least squares estimator is equal to

$$\frac{2}{JM} X^T w.$$

In view of the definition of the columns of the matrix X, this completes the proof.

5.18

(a) The estimator $\check{t} = (FX)^{-1}Fw$ is linear in w. Furthermore, $E[(FX)^{-1}Fw] = (FX)^{-1}F\,Ew = (FX)^{-1}FX\theta = \theta$. Therefore, $\check{t}$ is linear unbiased.

(b) $\text{cov}(\check{t}, \check{t}) = (FX)^{-1}FCF^T(FX)^{-T}$.

(c) If $F = X^TC^{-1}$, the covariance matrix of $\check{t}$ is equal to $(X^TC^{-1}X)^{-1}$, which is the covariance matrix of the best linear unbiased estimator.

5.19 For $N' + N - 1 > n \geq N'$, the least squares estimator $\hat{a}_n$ for α is described by

$$\begin{aligned} \hat{a}_n &= \frac{\sum_{m=n-N+1}^{n} w_m x_m}{\sum_{m=n-N+1}^{n} x_m^2} = \frac{\sum_{m=n-N+1}^{N'-1} w_m x_m + \sum_{m=N'}^{n} w_m x_m}{\sum_{m=n-N+1}^{N'-1} x_m^2 + \sum_{m=N'}^{n} x_m^2} \\ &= \hat{a}'\frac{p}{1+p} + \hat{a}''\frac{1}{1+p} \end{aligned}$$

with

$$p = \frac{\sum_{m=n-N+1}^{N'-1} x_m^2}{\sum_{m=N'}^{n} x_m^2},$$

$$\hat{a}' = \frac{\sum_{m=n-N+1}^{N'-1} w_m x_m}{\sum_{m=n-N+1}^{N'-1} x_m^2},$$

and

$$\hat{a}'' = \frac{\sum_{m=N'}^{n} w_m x_m}{\sum_{m=N'}^{n} x_m^2}.$$

The quantity p decreases monotonically with n since its numerator decreases and its denominator increases because $x_m \neq 0$. The expectation of $\hat{a}$ is equal to

$$E\hat{a} = \alpha'\frac{p}{1+p} + \alpha''\frac{1}{1+p}.$$

Then, for $N' + N - 1 > n \geq N'$, the derivative of $E\hat{a}$ with respect to p is equal to $(\alpha' - \alpha'')/(1+p)^2$ and, therefore, does not change sign. Then, since p decreases monotonically with n, the transition of $E\hat{a}$ on $N' - 1 \leq n \leq N' + N - 1$ from α' to α'' is also monotonic.

6.1 The gradient and the Hessian matrix of the Poisson log-likelihood function are described by (6.118) and (6.120), respectively. Substitute $g(\theta)$ for w in these expressions and show that at $t = \theta$ the gradient vanishes and the Hessian matrix is negative definite if $\partial g(t)/\partial t^T$ is nonsingular. Then, $t = \theta$ maximizes the log-likelihood function.

6.2

(a) An expression for the log-likelihood function concerned has been derived in Solution 5.1. It shows that the reference log-likelihood function is

$$\begin{aligned} q(g(\theta);t) &= N(\ln M! - M\ln M) \\ &\quad - \sum_n \ln\Gamma\left(g_n(\theta)+1\right) - \sum_n \ln\Gamma\left(M - g_n(\theta)+1\right) \\ &\quad + \sum_n g_n(\theta)\ln g_n(t) + \sum_n \left[M - g_n(\theta)\right]\ln\left[M - g_n(t)\right]. \end{aligned}$$

(b) Then, the gradient of $q(g(\theta);t)$ described by

$$s_t = \frac{\partial q(g(\theta);t)}{\partial t} = M\sum_n \frac{g_n(\theta) - g_n(t)}{g_n(t)\left[M - g_n(t)\right]}\frac{\partial g_n(t)}{\partial t}$$

vanishes at $t = \theta$. Therefore, $t = \theta$ is a stationary point. Subsequently, show by differentiating s_t with respect to t^T that the Hessian matrix at $t = \theta$ is equal to

$$-\frac{\partial g^T(t)}{\partial t}C^{-1}(t)\frac{\partial g(t)}{\partial t^T}\bigg|_{t=\theta},$$

where $C(\theta)$ is the covariance matrix of w. See Solution 3.3(c). The Hessian matrix is negative definite if $\partial g(t)/\partial t^T$ is nonsingular at $t = \theta$. Then, the stationary point $t = \theta$ is a maximum.

(c) Binomial observations are integers while their expectations are typically not. Therefore, exact binomial observations do not exist.

6.3

(a) An expression for the log-likelihood function concerned has been derived in Solution 5.2. This expression shows that the reference log-likelihood function is

$$q(g(\theta);t) = -\sum_n \ln g_n(t) - \sum_n \frac{g_n(\theta)}{g_n(t)}.$$

(b) Then, the gradient of $q(g(\theta);t)$ described by

$$s_t = \frac{\partial q(g(\theta);t)}{\partial t} = \sum_n \frac{1}{g_n^2(t)}\left[g_n(\theta) - g_n(t)\right]\frac{\partial g_n(t)}{\partial t}$$

vanishes at $t = \theta$. Therefore, $t = \theta$ is a stationary point. Next, show by differentiating s_t with respect to t^T that the Hessian matrix of $q(g(\theta);t)$ at $t = \theta$ is

$$-\frac{\partial g^T(t)}{\partial t}C^{-1}(t)\frac{\partial g(t)}{\partial t^T}\bigg|_{t=\theta},$$

where $C(\theta)$ is the covariance matrix of w. See Solution 3.7(c). The Hessian matrix is negative definite if it is assumed that $\partial g(t)/\partial t^T$ is nonsingular at $t = \theta$. Then, the stationary point $t = \theta$ is a maximum.

6.4 Expression (5.129) shows that the gradient of the reference log-likelihood function for a linear exponential family of distributions is described by

$$\frac{\partial g^T(t)}{\partial t} C^{-1}(t) \left[g(\theta) - g(t) \right] ,$$

where $C(t)$ is the covariance matrix of the observations w at $t = \theta$. Then, $t = \theta$ is a stationary point. Subsequently, show that the Hessian matrix of the reference log-likelihood function at $t = \theta$ is equal to

$$- \frac{\partial g^T(t)}{\partial t} C^{-1}(t) \frac{\partial g(t)}{\partial t^T} \bigg|_{t=\theta} .$$

Since $C^{-1}(\theta)$ is positive definite, the Hessian matrix is negative definite if $\partial g(t)/\partial t^T$ is nonsingular at $t = \theta$. Then, the stationary point $t = \theta$ is a maximum.

6.5

(a) For the computation of the gradient of the criterion Q described in Section 4.10.2, expressions for the Fisher information matrix F_θ and its first-order derivatives with respect to the x_n are needed. For uncorrelated, normally distributed observations with equal variance,

$$F_\theta = \frac{1}{\sigma^2} \frac{\partial g^T(\theta)}{\partial \theta} \frac{\partial g(\theta)}{\partial \theta^T} ,$$

where $g(\theta) = [g_1(\theta) \ldots g_N(\theta)]^T$ with

$$g_n(\theta) = g(x_n; \theta) = \alpha_1 \exp(-\beta_k x_n) + \alpha_2 \exp(-\beta_2 x_n).$$

Then,

$$\frac{\partial F_\theta}{\partial x_n} = \frac{1}{\sigma^2} \left(\frac{\partial (\frac{\partial g^T}{\partial x_n})}{\partial \theta} \frac{\partial g}{\partial \theta^T} + \frac{\partial g^T}{\partial \theta} \frac{\partial (\frac{\partial g}{\partial x_n})}{\partial \theta^T} \right) ,$$

where $g = g(\theta)$. Continued differentiation with respect to x_m produces the derivatives $\partial^2 F_\theta / \partial x_n \partial x_m$. The elements of $\partial F_\theta / \partial x_n$ are functions of x_n only. Therefore, $\partial^2 F_\theta / \partial x_n \partial x_m$ is a null matrix if $n \neq m$. Using (4.244), (4.246), (4.250), and (4.251), computing $\partial Q / \partial x$ and $\partial^2 Q / \partial x \partial x^T$, and from these $\partial Q / \partial r$ and $\partial^2 Q / \partial r \partial r^T$, is relatively straightforward. Together, these are the ingredients for steepest descent minimizing Q and checking the nature of the solution using the eigenvalues of $\partial^2 Q / \partial r \partial r^T$. Our numerical solution is: $x_1 = 0,\ x_2 = x_3 = 0.43, x_4 = x_5 = 1.91,\ x_6 = x_7 = x_8 = x_9 = x_{10} = 5.38$ while for $\sigma = 1$ the criterion value $Q = 1.31 \times 10^6$. It is the best solution we obtained starting from a number of random $x = (x_1 \ldots x_{10})^T$ with elements uniformly distributed on $[0,\ 6]$. The Cramér–Rao variances are, respectively: $1.55 \times 10^6,\ 1.55 \times 10^6,\ 0.04 \times 10^6$, and 0.13×10^6. The eigenvalues of $\partial^2 Q / \partial r \partial r^T$ are all positive.

(b) For the uniform design concerned, the criterion value is $Q = 7.35 \times 10^6$. The Cramér–Rao variances are, respectively: $8.67 \times 10^6,\ 8.67 \times 10^6,\ 0.22 \times 10^6$, and 0.78×10^6. These results illustrate the improvement by the optimal design.

(c) The Cramér–Rao standard deviations for the optimal design are: $1.25 \times 10^3\sigma$, $1.25 \times 10^3\sigma$, $0.2 \times 10^3\sigma$, and $0.36 \times 10^3\sigma$. Simple calculations then show that, for a maximum 10% relative Cramér–Rao standard deviation, σ should be as small as 2.4×10^{-5}. For the uniform design, the maximum allowable value of σ is 1.0×10^{-5}.

6.6

(a) The gradient of $f = f(x)$ is defined by $\partial f/\partial x_1 = 2x_1x_2^2 - 8x_1x_2 - 2x_2^2 + 10x_1 + 8x_2 - 10$ and $\partial f/\partial x_2 = 2x_1^2x_2 - 4x_1^2 - 4x_1x_2 + 6x_2 + 8x_1 - 12$. The Hessian matrix is defined by $\partial^2 f/\partial x_1^2 = 2x_2^2 - 8x_2 + 10$, $\partial^2 f/\partial x_1 \partial x_2 = 4x_1x_2 - 8x_1 - 4x_2 + 8$ and $\partial^2 f/\partial x_2^2 = 2x_1^2 - 4x_1 + 6$. Together, these partial derivatives define the Newton step.

The eigenvalues of the Hessian matrix at $x = (2 \;\; 3)^T$, are 0.88 and 9.12. The procedure arrives after the first step at $x = (2 \;\; 2)^T$, where the eigenvalues are equal to 2 and 6. The next point is $x = (1 \;\; 2)^T$, where the gradient vanishes and the procedure stops. Since the eigenvalues at $x = (1 \;\; 2)^T$ are equal to 2 and 4, this stationary point is a minimum.

The eigenvalues at $x = (2 \;\; 4)^T$ are 16.25 and -0.25, respectively. This point is, therefore, not suitable as a starting point of a Newton minimization procedure. This is demonstrated by the subsequent points arrived at. These are $x = (-7 \;\; 14)^T$ with eigenvalues 603.0 and -181.0, and $x = (-4.2 \;\; 10.0)^T$ with eigenvalues 267.5 and -77.2. The gradient at $x = (-4.2 \;\; 10.0)^T$ is equal to $(-6.9 \;\; 4.7)^T \times 10^2$. This point is, therefore, not a stationary point.

At $x = (2 \;\; 2 + \sqrt{3})^T$, the Hessian matrix is singular since one of its eigenvalues vanishes. Therefore, the Newton step is not defined.

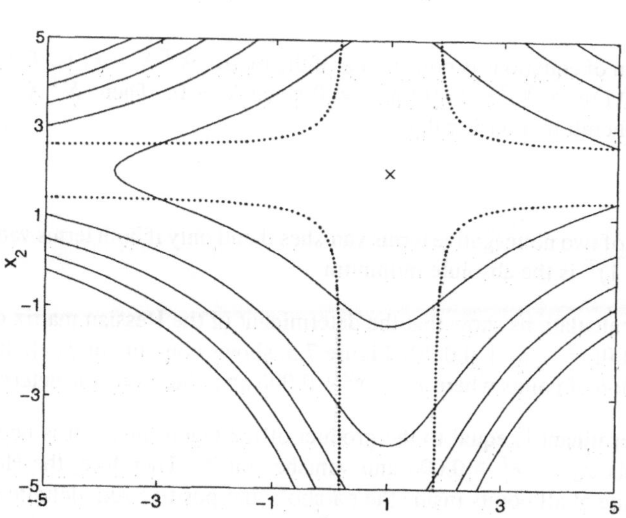

Figure 7.4. Function contours (solid lines) and the collection of points where the Hessian matrix is singular (dotted lines). The Hessian matrix is positive definite in the region bounded by the dotted lines that includes the minimum (cross). It is indefinite outside this region. (Problem 6.6)

(b) Figure 7.4. shows the plots. Points in the region enclosed by dotted lines are suitable starting points for Newton minimization.

6.7

(a) The nth element of the $N \times 1$ vector $\gamma(\theta)$ for binomially distributed observations is equal to

$$\gamma_n(\theta) = \ln\left(\frac{g_n(\theta)}{M - g_n(\theta)}\right).$$

See Solution 3.4(a). Equating this to $\theta^T x_n$ yields

$$g_n(\theta) = \frac{M\exp(\theta^T x_n)}{1 + \exp(\theta^T x_n)},$$

where $\theta = (\theta_1 \ldots \theta_K)^T$ and $x_n = (x_{n1} \ldots x_{nK})^T$ is the nth vector measurement point.

(b) The nth element of the $N \times 1$ vector $\gamma(\theta)$ for exponentially distributed observations is equal to

$$\gamma_n(\theta) = -\frac{1}{g_n(\theta)}.$$

See Solution 3.8(a). Equating this to $\theta^T x_n$ shows that

$$g_n(\theta) = -\frac{1}{\theta^T x_n},$$

where $\theta = (\theta_1 \ldots \theta_K)^T$ and $x_n = (x_{n1} \ldots x_{nK})^T$ is the nth vector measurement point.

6.8 The system of equations is nonsingular if the matrix $X^TX + \lambda I$ is. The quadratic form $y^T(X^TX + \lambda I)y \geq \lambda y^Ty > 0$ for $y \neq 0$ since $\lambda > 0$. Then, $X^TX + \lambda I$ is positive definite and, therefore, nonsingular.

6.9

(a) The sum of two nonnegative terms vanishes if and only if both terms vanish. Therefore, $x = (1\ 1)^T$ is the absolute minimum.

(b) Simple calculations show that the determinant of the Hessian matrix of $f(x)$ is equal to $80\,000(x_1^2 - x_2 + 0.005)$. Figure 7.5. shows contours of $f(x)$. It also shows the collection of points where $x_2 = x_1^2 + 0.005$ and, therefore, the determinant vanishes.

(c) The determinant is equal to the product of the eigenvalues. It is negative inside the parabola $x_2 = x_1^2 + 0.005$ and vanishes on it. Therefore, the Hessian matrix is indefinite at all points inside the parabola and positive semidefinite on it. Then, the points outside the parabola, where the Hessian matrix is positive definite, are suitable starting points.

6.10 For the generalized linear expectation model, the Newton step is identical to the generalized Gauss–Newton step. Then, its direction is, typically, an ascent direction.

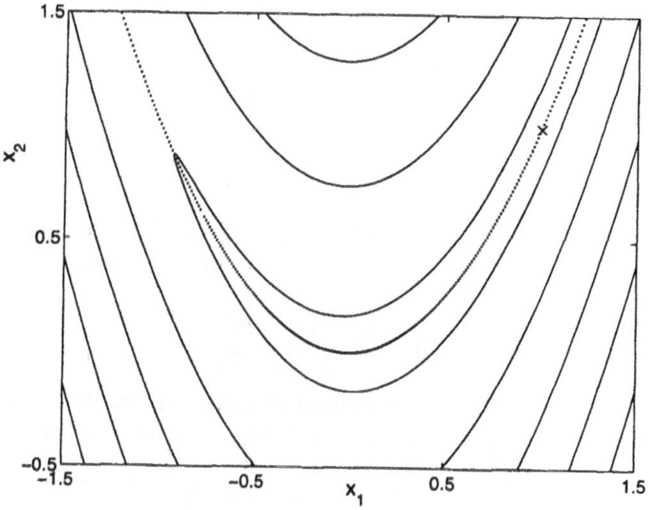

Figure 7.5. Contours of Rosenbrock's function (solid lines) and the collection of points where the Hessian matrix is singular (dotted line). The minimum is also shown (cross). (Problem 6.9)

6.11 Let $f(x)$ be a function of the elements of $x = (x_1 \ldots x_K)^T$. Then, the linear Taylor polynomial of the kth element of the gradient $\partial f(x)/\partial x$ is described by

$$\overline{\frac{\partial f(x)}{\partial x_k}} = \frac{\partial f(x)}{\partial x_k} + \frac{\partial^2 f(x)}{\partial x_1 \partial x_k}\Delta x_1 + \cdots + \frac{\partial^2 f(x)}{\partial x_K \partial x_k}\Delta x_K$$

for $k = 1, \ldots, K$, where all derivatives are taken at $x = x_c$ and $\Delta x_k = x_k - x_{ck}$. Or,

$$\overline{\frac{\partial f(x)}{\partial x}} = \frac{\partial f(x)}{\partial x} + \frac{\partial^2 f(x)}{\partial x \partial x^T}\Delta x$$

where $\Delta x = (\Delta x_1 \ldots \Delta x_K)^T$. Equating this gradient to the corresponding null vector shows that

$$\Delta x = -\left(\frac{\partial^2 f(x)}{\partial x \partial x^T}\right)^{-1} \frac{\partial f(x)}{\partial x}.$$

This is the Newton step.

6.12 An expression for the Fisher score vector of binomial observations has been derived in Solution 3.5(c). It shows that the kth element of the gradient vector is described by

$$s_{t_k} = \sum_n \frac{M_n \, d_n(t)}{g_n(t) \, m_n(t)} \frac{\partial g_n(t)}{\partial t_k}$$

with $d_n(t) = w_n - g_n(t)$ and $m_n(t) = M_n - g_n(t)$. The (k, ℓ)th element of the Hessian matrix of the log-likelihood function follows from differentiating s_{t_k} with respect to t_ℓ. It is described by

$$\frac{\partial^2 q(w;t)}{\partial t_k \partial_{t_\ell}} = \sum_n M_n \left(\frac{-M_n w_n + 2w_n g_n(t) - g_n^2(t)}{g_n^2(t)\, m_n^2(t)} \right) \frac{\partial g_n(t)}{\partial t_k} \frac{\partial g_n(t)}{\partial t_\ell}$$
$$+ \sum_n \frac{M_n\, d_n(t)}{g_n(t)\, m_n(t)} \frac{\partial^2 g_n(t)}{\partial t_k \partial t_\ell}.$$

6.13 An expression for the Fisher score vector of exponentially distributed observations has been derived in Solution 3.7(d). It shows that the kth element of the gradient vector is described by

$$s_{t_k} = \sum_n \frac{d_n(t)}{g_n^2(t)} \frac{\partial g_n(t)}{\partial t_k}$$

with $d_n(t) = w_n - g_n(t)$. The (k, ℓ)th element of the Hessian matrix of the log-likelihood function follows from differentiating s_{t_k} with respect to t_ℓ. It is described by

$$\frac{\partial^2 q(w;t)}{\partial t_k \partial_{t_\ell}} = - \sum_n \left(\frac{2w_n - g_n(t)}{g_n^3(t)} \right) \frac{\partial g_n(t)}{\partial t_k} \frac{\partial g_n(t)}{\partial t_\ell}$$
$$+ \sum_n \frac{d_n(t)}{g_n^2(t)} \frac{\partial^2 g_n(t)}{\partial t_k \partial t_\ell}.$$

6.14 Since the distribution of the observations is a linear exponential family, the gradient and the Hessian matrix of the log-likelihood function are fully defined by the first-order and the second-order partial derivatives of the elements of the vector function $\gamma(t)$ with respect to the parameters t and by the covariance matrix C of the observations. See Section 6.10. For the binomial distribution, the nth element of the vector $\gamma(t)$ is

$$\gamma_n(t) = \ln \frac{g_n(t)}{m_n(t)}$$

with $m_n(t) = M_n - g_n(t)$. See Solution 3.6 (a). Then,

$$\frac{\partial \gamma_n(t)}{\partial t_k} = \frac{M_n}{g_n(t)\, m_n(t)} \frac{\partial g(t)}{\partial t_k}$$

and

$$\frac{\partial^2 \gamma_n(t)}{\partial t_k \partial t_\ell} = M_n \frac{g_n(t) - m_n(t)}{g_n^2(t)\, m_n^2(t)} \frac{\partial g_n(t)}{\partial t_k} \frac{\partial g_n(t)}{\partial t_\ell} + \frac{M_n}{g_n(t)\, m_n(t)} \frac{\partial^2 g_n(t)}{\partial t_k \partial t_\ell}.$$

The covariance matrix of the observations has been derived in Solution 3.5(b) and is the diagonal matrix of order N with as nth diagonal element:

$$\frac{1}{M_n} g_n(t)[M_n - g_n(t)].$$

6.15 Since the distribution of the observations is a linear exponential family, the gradient and the Hessian matrix of the log-likelihood function are fully defined by the first-order and the second-order partial derivatives of the elements of the vector function $\gamma(t)$ with respect to the parameters t and by the covariance matrix C of the observations. See Section 6.10. For the exponential distribution, the nth element of the vector $\gamma(t)$ is

$$\gamma_n(\theta) = -\frac{1}{g_n(\theta)}.$$

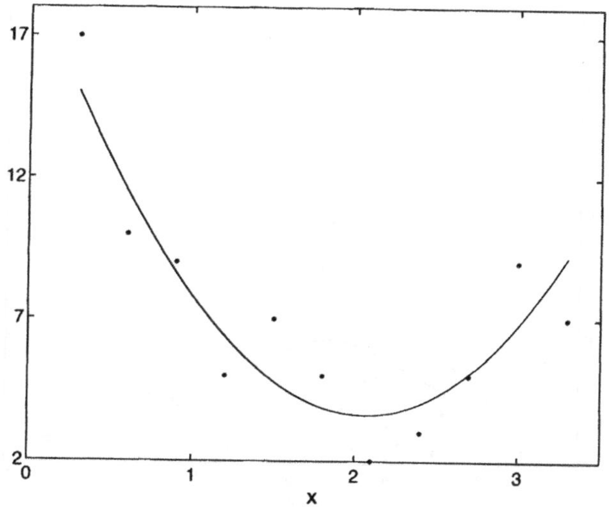

Figure 7.6. Binomially distributed observations (dots) and the corresponding maximum likelihood estimate of the parabolic expectation model (solid line). (Problem 6.18)

See Solution 3.8(a). Then,

$$\frac{\partial \gamma_n(t)}{\partial t_k} = \frac{1}{g_n^2(t)} \frac{\partial g_n(t)}{\partial t_k}$$

and

$$\frac{\partial^2 \gamma_n(t)}{\partial t_k \partial t_\ell} = -\frac{2}{g_n^3(t)} \frac{\partial g_n(t)}{\partial t_k} \frac{\partial g_n(t)}{\partial t_\ell} + \frac{1}{g_n^2(t)} \frac{\partial^2 g_n(t)}{\partial t_k \partial t_\ell}.$$

The covariance matrix of the observations has been derived in Solution 3.7(c) and is the diagonal matrix of order N with nth diagonal element $g_n^2(t)$.

6.16 Use the expression (5.115) for the covariance matrix C of independent Poisson distributed observations to compute C^{-1} and $\partial C^{-1}/\partial t_\ell$. Substitute the result in (6.120) and elaborate.

6.17 Use the closed-form expression (5.123) for the inverse C^{-1} of the covariance matrix C of the multinomial observations to compute $\partial C^{-1}/\partial t_\ell$. Substitute the result and the expression for C^{-1} in (6.126), and elaborate.

6.18

(a) The gradient and the Hessian matrix of the log-likelihood function concerned have been derived in Solution 6.12. All derivatives $\partial g_n(t)/\partial t_k$ are constants and all derivatives $\partial^2 g_n(t)/\partial t_k \partial t_\ell$ vanish as a consequence of the linearity of $g(t)$ in the elements of t.

(b) —

(c) Figure 7.6. shows the observations. Using the linearity of the expectation model, we first compute the ordinary least squares estimates of the parameters of the quadratic

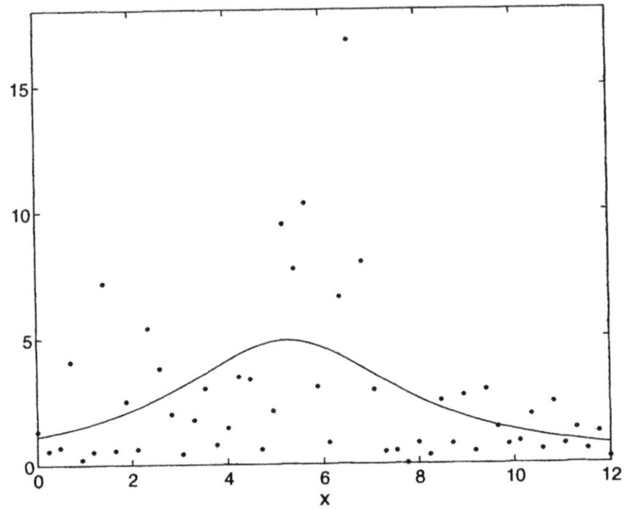

Figure 7.7. Exponentially distributed observations (dots) and the corresponding maximum likelihood estimate of the Lorentz line expectation model (solid line). (Problem 6.19)

model by means of the closed-form estimator (5.275). The result is $\hat{t}_{\text{quad}} = (19.8 \ -15.4 \ \ 3.6)^T$, where the subscript "quad" refers to the quadratic polynomial model. This is not a maximum likelihood estimate since the observations are not *iind* around their expectations. Next, $\hat{t}_{\text{quad}}$ is used as starting point for Newton maximization of the log-likelihood function. Our maximum likelihood estimate of the parameters θ is $\tilde{t}_{\text{quad}} = (19.3 \ \ -15.2 \ \ 3.7)^T$. The eigenvalues of the Hessian matrix are $-(114.2 \ \ 2.2 \ \ 0.09)^T$. The log-likelihood function follows directly from the log-probability function derived in Problem 3.5. The parameter dependent part of this log-likelihood function is described by

$$\sum_n w_n \ln g_n(t) + (M_n - w_n) \ln[M_n - g_n(t)]$$

and is, at $t = \tilde{t}_{\text{quad}}$, equal to 324.7168.

(d) Our maximum likelihood estimates of the parameters of the cubic polynomial from the same observations are : $\tilde{t}_{\text{cub}} = (20.9 \ \ -19.3 \ 6.4 \ \ -0.5)^T$. The corresponding value of the parameter dependent part of the log-likelihood function is: $q(w; \tilde{t}_{\text{cub}}) =$ 324.8720. Then,

$$\begin{aligned} -2\ell &= -2[q(w; \tilde{t}_{\text{quad}}) - q(w; \tilde{t}_{\text{cub}})] \\ &= -2(324.7168 - 324.8720) = 0.3104 \end{aligned}$$

is a sample value of a stochastic variable with a chi-square distribution with one degree of freedom. Since the 0.05 quantile of this distribution is 3.8143, there is no reason to reject the quadratic model.

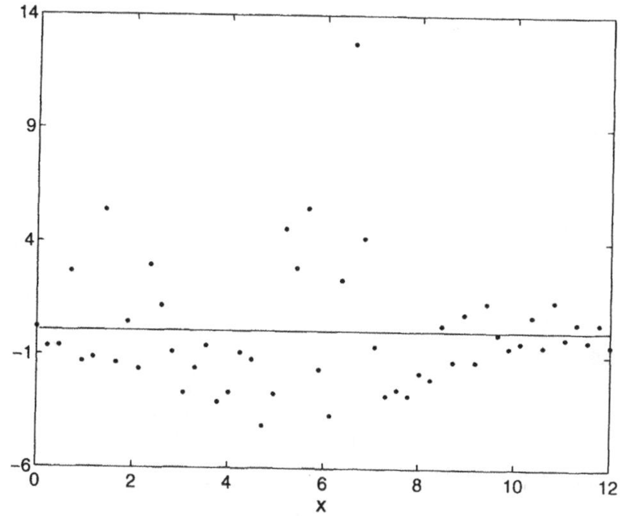

Figure 7.8. Residuals. (Problem 6.19)

6.19

(a) Expressions for the gradient and the Hessian matrix of the log-likelihood function for independent exponentially distributed observations have been derived in Solution 6.13. Computing the partial derivatives of the expectation model with respect to the parameters appearing in these expressions is straightforward and left out.

(b) The observations are plotted in Fig. 7.7. Inspecting these observations, we take $t = (6\ 5\ 2)^T$ as starting point. At this point, the Hessian matrix of $q(w;t)$ is negative definite and remains so until the Newton method using the full step converges to our maximum likelihood estimate $\hat{t} = (4.91\ 5.26\ 2.89)^T$. Starting from $t = (5\ \ 5.5\ \ 3)^T$ produces the same result. The plot of the maximum likelihood estimate of the expectation model has been added to the plot of the observations.

(c) The residuals are shown in Fig. 7.8. The qualitative similarity of the residuals to the observations shown in Fig. 7.7. is striking. This is because the expectation and the standard deviation of an exponentially distributed observation are equal and the exponential distribution is long-tailed. As a result, the observations have a much larger range than the expectations.

6.20

(a) Figure 7.9. shows the expectation model and its values at the measurement points.

(b) The plot in Fig. 7.10. shows two equivalent absolute minima resulting from the symmetry of the criterion in the line $t_1 = t_2$. They are located at $t = (2\pi\ \ 2.4\pi)^T$ and $t = (2.4\pi\ \ 2\pi)^T$, respectively. The plot also shows the relatively small regions around these absolute minima where the Hessian matrix of the criterion is positive definite. These regions are surrounded by a region of complicated shape where the Hessian matrix is indefinite. The latter region also includes a saddle point with

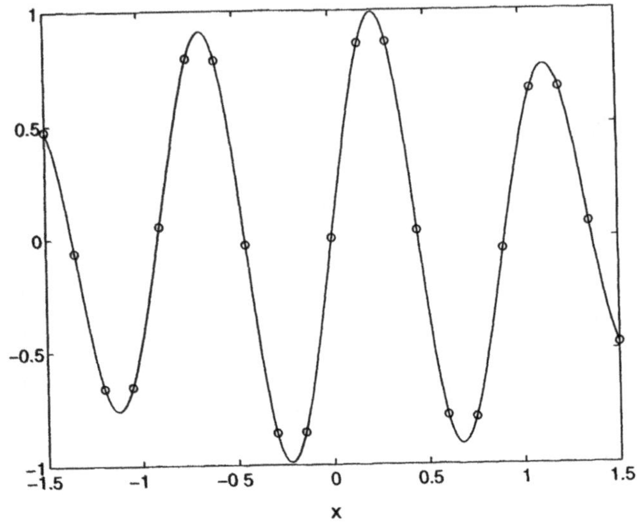

Figure 7.9. Bisinusoidal expectation model (solid line) and its values at the measurement points (circles). (Problem 6.20)

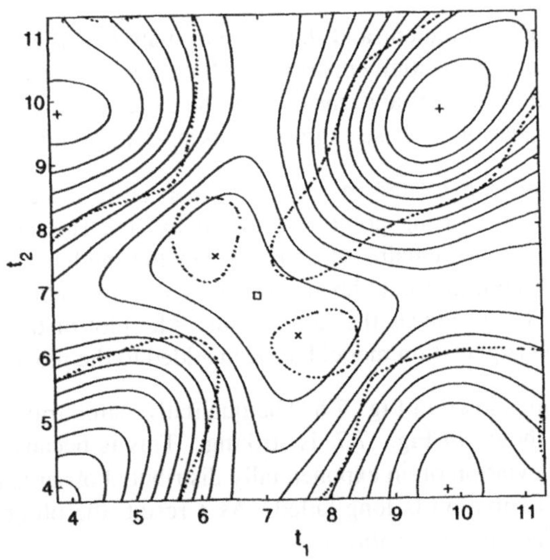

Figure 7.10. Contours of the reference ordinary least squares criterion and the collection of points where the Hessian matrix is singular (dotted line). Two equivalent absolute minima (crosses), three maxima (pluses), and one saddle point (square) are also shown. (Problem 6.20)

coordinates $(2.2\pi,\ 2.2\pi)$. This region is surrounded by four regions each containing a maximum, where the Hessian matrix is negative definite. The conclusion is that in most points of the plot it would be advisable to start with a steepest descent method until the Hessian matrix becomes positive definite. Subsequently, the Newton method or the Gauss–Newton method may be used till convergence.

APPENDIX A

STATISTICAL RESULTS

A.1 STATISTICAL PROPERTIES OF LINEAR COMBINATIONS OF STOCHASTIC VARIABLES

Suppose that

$$w = (w_1 \dots w_N)^T \tag{A.1}$$

is a vector of stochastic variables with vector expectation μ defined as

$$\mu = (\mu_1 \dots \mu_N)^T, \tag{A.2}$$

where $\mu_n = Ew_n$ and covariances

$$c_{mn} = \text{cov}(w_m, w_n). \tag{A.3}$$

Then, the covariance matrix C of the vector w is described by

$$C = \text{cov}(w, w) = [\text{cov}(w_m, w_n)]. \tag{A.4}$$

Definition A.1 *(Linear combination) Let $a = (a_1 \dots a_N)^T$ be a vector of real constants. Then, the scalar stochastic variable*

$$y = a^T w = \sum_n a_n w_n\,, \tag{A.5}$$

with $n = 1, \ldots, N$, is said to be a linear combination of the elements of w. The elements of a are the coefficients of the linear combination.

With respect to linear combinations, we have the following definition.

Definition A.2 *(Linear dependence and linear independence of stochastic variables) The elements w_n of the vector of stochastic variables $w = (w_1 \ldots w_N)^T$ are said to be linearly dependent if a vector of coefficients $a = (a_1 \ldots a_N)^T \neq o$ exists such that the linear combination $y = a^T w$ is a constant. Otherwise, the elements of w are linearly independent.*

Let $y = a^T w$ be a constant. Then, Ey must be equal to this constant. Therefore, $\operatorname{var} y = 0$. Also, if $\operatorname{var} y = 0$, then y is constant and equal to Ey with probability one. See [20]. If the elements of w are linearly independent, there exists no $a \neq o$ such that y is a constant. Then, $\operatorname{var} y > 0$. Conversely, if $\operatorname{var} y > 0$ for all $a \neq o$, y cannot be a constant. Then, the elements of w are linearly independent.

Let

$$a = (a_1 \ldots a_N)^T \tag{A.6}$$

be an arbitrary vector of real constants and consider the linear combination

$$y = a^T w = \sum_n a_n w_n \tag{A.7}$$

with $n = 1, \ldots, N$. Then,

$$Ey = E \sum_n a_n w_n = \sum_n a_n \mu_n \tag{A.8}$$

and

$$y - Ey = \sum_n a_n (w_n - \mu_n) . \tag{A.9}$$

The variance of y is

$$\operatorname{var} y = E(y - Ey)^2. \tag{A.10}$$

Then, from (A.9) and (A.10),

$$\begin{aligned} \operatorname{var} y &= E \sum_m \sum_n a_m a_n (w_m - \mu_m)(w_n - \mu_n) \\ &= \sum_m \sum_n a_m a_n E[(w_m - \mu_m)(w_n - \mu_n)] \\ &= \sum_m \sum_n a_m a_n c_{mn} . \end{aligned} \tag{A.11}$$

This expression shows that $\operatorname{var} y$ may also be written

$$\operatorname{var} y = a^T C a. \tag{A.12}$$

We next consider a special case. Suppose that the w_n are uncorrelated, that is, their covariances are equal to zero. Then, $c_{mn} = 0$ if $m \neq n$ and (A.11) simplifies to

$$\operatorname{var} y = \sum_n a_n^2 \sigma_n^2 . \tag{A.13}$$

If y is the *sum* of all w_n, all a_n are equal to one. Then,

$$\operatorname{var} y = \sum_n \sigma_n^2 . \tag{A.14}$$

Therefore, if, in addition, all variances are equal to σ^2 :

$$\operatorname{var} y = N\sigma^2 . \tag{A.15}$$

The results (A.13), (A.14), and (A.15) may be summarized as follows. The variance of a weighted sum of uncorrelated stochastic variables is equal to the quadratically weighted sum of the individual variances. Therefore, the variance of a sum of uncorrelated stochastic variables is equal to the sum of the individual variances. If, in addition, the individual variances are equal, the variance of the sum is equal to the product of the number of stochastic variables and the common variance. Since being independent implies being uncorrelated, these results also apply if the stochastic variables are independent.

A.2 THE CAUCHY–SCHWARZ INEQUALITY FOR EXPECTATIONS

Theorem A.1 *(Cauchy–Schwarz inequality for expectations) Suppose that x and y are scalar stochastic variables. Then,*

$$(E[xy])^2 \leq Ex^2\, Ey^2 . \tag{A.16}$$

Equality holds if and only if x and y are proportional with probability one.

Proof. Consider the stochastic variable $cx - y$ where c is a scalar constant. Then,

$$\begin{aligned} E(cx-y)^2 &= E[c^2x^2 - 2cxy + y^2] \\ &= c^2 Ex^2 - 2c\,E[xy] + Ey^2 \end{aligned} \tag{A.17}$$

is the expectation of a quadratic scalar and is, therefore, nonnegative. Consequently, its discriminant meets the condition

$$4(E[xy])^2 - 4Ex^2\, Ey^2 \leq 0. \tag{A.18}$$

Equality is seen to hold if and only if $cx - y = 0$ with probability one. This completes the proof. ■

Corollary A.1 *Suppose that x and y are scalar stochastic variables. Then,*

$$\operatorname{cov}^2(x,y) \leq \sigma_x^2 \sigma_y^2 . \tag{A.19}$$

Equality holds if and only if $x - \mu_x$ and $y - \mu_y$ are proportional with probability one.

Proof. Replacing in Theorem A.1 x by $x - \mu_x$ and y by $y - \mu_y$ yields the proof. ■

APPENDIX B

VECTORS AND MATRICES

B.1 VECTORS

Small case letters are used to denote vectors.

Definition B.1 *(Column vector) An $M \times 1$ column vector a is defined as*

$$a = \begin{pmatrix} a_1 \\ a_2 \\ \cdots \\ \cdots \\ a_M \end{pmatrix}, \tag{B.1}$$

where the scalar a_m is the mth element of the vector a.

Definition B.2 *(Transpose) The transpose of the $M \times 1$ column vector a is the $1 \times M$ row vector a^T defined as*

$$a^T = \begin{pmatrix} a_1 & a_2 & \ldots & \ldots & a_M \end{pmatrix}. \tag{B.2}$$

Definition B.3 *(Complex conjugate transpose) The complex conjugate transpose of a vector is the transpose of the complex conjugate vector. Complex conjugate transposition is indicated by the superscript H. Thus, the complex conjugate a^H of the column vector a is defined as*

$$a^H = \begin{pmatrix} a_1^* & a_2^* & \ldots & \ldots & a_M^* \end{pmatrix}, \tag{B.3}$$

where a_m^* is the complex conjugate of a_m.

Vectors are always column vectors unless specified otherwise. A vector is called *null vector* if all its elements are equal to zero. The notation o is used for column null vectors of all dimensions.

Definition B.4 *(Inner product) The inner product of two real $M \times 1$ vectors a and b is defined as*

$$a^T b = a_1 b_1 + \ldots + a_M b_M \,, \tag{B.4}$$

that is, the sum of the products of their corresponding elements. The nonnegative scalar

$$\|a\| = (a^T a)^{\frac{1}{2}} = \left(\sum_m a_m^2 \right)^{\frac{1}{2}} \tag{B.5}$$

is called the length or the (Euclidean) norm of the vector a. The length of the normalized vector

$$a' = \frac{a}{\|a\|}, \tag{B.6}$$

with $\|a\| \neq o$, is equal to one.

Theorem B.1 *(Sign of inner product) Let a and b be real and nonzero vectors of the same dimension. Suppose that b is decomposed into a vector proportional to a and a vector orthogonal to it. Then, a and the component of b proportional to it, point in the same direction if and only if $a^T b > 0$ and in opposite directions if and only if $a^T b < 0$. If $a^T b = 0$, a and b are, by definition, orthogonal. The component of b orthogonal to a is $b - pa$ with $p = a^T b / a^T a$.*

Proof. Let $b = pa + c$ where p is a scalar proportionality constant and the vector c is orthogonal to a. Then, $a^T b = pa^T a + a^T c = pa^T a$. Therefore, $p = a^T b / a^T a$. Then, $p > 0$ if and only if $a^T b > 0$ and $p < 0$ if and only if $a^T b < 0$. Also, by definition, $c = b - pa$ with $p = a^T b / a^T a$. This completes the proof. ∎

B.2 MATRICES

Capitals are used to denote matrices.

Definition B.5 *(Matrix) An $M \times N$ matrix A is defined as*

$$A = [a_{mn}] = \begin{pmatrix} a_{11} & a_{12} & \cdots & \cdots & a_{1N} \\ a_{21} & a_{22} & \cdots & \cdots & a_{2N} \\ \cdots & & & & \cdots \\ \cdots & & & & \cdots \\ \cdots & & & & \cdots \\ a_{M1} & a_{M2} & \cdots & \cdots & a_{MN} \end{pmatrix} \tag{B.7}$$

where the scalar a_{mn} is the (m,n)th element of the matrix A. The vector

$$\begin{pmatrix} a_{m1} & a_{m2} & \cdots & \cdots & a_{mN} \end{pmatrix} \tag{B.8}$$

is called the mth *row of the matrix A. The vector*

$$\begin{pmatrix} a_{1n} \\ a_{2n} \\ \cdots \\ \cdots \\ a_{Mn} \end{pmatrix} \tag{B.9}$$

is the nth *column of the matrix.*

Definition B.6 *(Transpose) The transpose of a matrix is obtained by interchanging its rows and columns. Transposition is indicated by the superscript* T. *Therefore, the transpose* A^T *of the matrix A is defined as*

$$A^T = [a_{nm}] = \begin{pmatrix} a_{11} & a_{21} & \cdots & \cdots & a_{M1} \\ a_{12} & a_{22} & & & a_{M2} \\ \cdots & & & & \cdots \\ \cdots & & & & \cdots \\ \cdots & & & & \cdots \\ a_{1N} & a_{2N} & \cdots & \cdots & a_{MN} \end{pmatrix}. \tag{B.10}$$

Definition B.7 *(Complex conjugate transpose) The complex conjugate transpose of a matrix is the transpose of the complex conjugate matrix. Complex conjugate transposition is indicated by the superscript* H. *Thus, the complex conjugate transpose* A^H *of the matrix A is defined as*

$$A^H = [a^*_{nm}] = \begin{pmatrix} a^*_{11} & a^*_{21} & \cdots & \cdots & a^*_{M1} \\ a^*_{12} & a^*_{22} & & & a^*_{M2} \\ \cdots & & & & \cdots \\ \cdots & & & & \cdots \\ \cdots & & & & \cdots \\ a^*_{1N} & a^*_{2N} & \cdots & \cdots & a^*_{MN} \end{pmatrix}. \tag{B.11}$$

Definition B.8 *(Null matrix) A matrix is called null matrix if all its elements are equal to zero. The notation O is used for null matrices of all dimensions.*

Definition B.9 *(Multiplication of matrices) A matrix is defined as the product AB of the matrices A and B if its* (m, n)th *element is equal to the inner product of the* mth *row of A and the* nth *column of B. This definition implies that the number of columns of the matrix A is supposed to be equal to the number of rows of the matrix B.*

Corollary B.1 *A direct consequence of the definition of the matrix product is that*

$$(AB)^T = B^T A^T. \tag{B.12}$$

Definition B.10 *(Addition of matrices) A matrix is defined as the sum A+B of the matrices A and B if its* (m, n)th *element is equal to* $a_{mn} + b_{mn}$. *The sum is defined only if the dimensions, that is, the number of rows and number of columns of A and B, agree.*

Definition B.11 *(Square matrix) The matrix A is square if the number of its rows is equal to the number of its columns. The elements* a_{mn} *of a square matrix with* $m = n$ *are called diagonal elements. The remaining elements are called off-diagonal.*

Definition B.12 *(Symmetric matrix) A square matrix A is symmetric if it is equal to its transpose A^T, or, equivalently, if $a_{mn} = a_{nm}$.*

Definition B.13 *(Hermitian matrix) A square matrix A is Hermitian if it is equal to its complex conjugate transpose A^H, or, equivalently, if $a_{mn} = a^*_{nm}$.*

Definition B.14 *(Diagonal matrix) If all off-diagonal elements of a square matrix are equal to zero, the matrix is called diagonal. The notation for an $M \times M$ diagonal matrix is*

$$\text{diag}\ (d_1\, d_2\, \ldots\, d_M) \tag{B.13}$$

or

$$\text{diag}\, d\,, \tag{B.14}$$

where d is defined as the vector of diagonal elements $(d_1 \ldots d_M)^T$.

Definition B.15 *(Block diagonal matrix) A block diagonal matrix A with K blocks has the form*

$$A = \begin{pmatrix} D_{11} & & & O \\ & D_{22} & & \\ & & \ddots & \\ O & & & D_{KK} \end{pmatrix}. \tag{B.15}$$

The blocks $D_{11}, \ldots, D_{KK}$ are square matrices of not necessarily equal dimensions. All elements of A not belonging to the blocks are equal to zero. The notation is

$$A = \text{diag}\ (D_{11}\, D_{22} \ldots D_{KK})\,. \tag{B.16}$$

Definition B.16 *(Identity matrix) The $N \times N$ diagonal matrix*

$$\begin{aligned} I &= \text{diag}\ (\ 1\ \ \ldots\ \ \ldots\ \ \ldots\ \ 1\ ) \\ &= \begin{pmatrix} 1 & 0 & \ldots & \ldots & 0 \\ 0 & 1 & \ddots & & \vdots \\ \vdots & \ddots & \ddots & \ddots & \vdots \\ \vdots & & \ddots & 1 & 0 \\ 0 & \ldots & \ldots & 0 & 1 \end{pmatrix} \end{aligned} \tag{B.17}$$

is called the identity matrix of order N.

Definition B.17 *(Singularity) Let A be an $M \times N$ matrix and x an $N \times 1$ vector. Then, A is called nonsingular if*

$$Ax = o \tag{B.18}$$

only if $x = o$. Otherwise, A is singular.

Remark B.1 *If $M \geq N$, nonsingularity of A is equivalent to linear independence of the columns of A. If $M < N$, the matrix A is singular.*

Definition B.18 *(Inverse matrix) If A is square and nonsingular, there exists a square matrix A^{-1}, called the inverse of A, such that*

$$AA^{-1} = A^{-1}A = I. \tag{B.19}$$

The inverse of a singular matrix does not exist.

Definition B.19 *(Determinant of a matrix) The determinant* $\det A$ *of a square $M \times M$ matrix A is a scalar defined by the recursion*

$$\det A = \sum_{n=1}^{M} (-1)^{m+n} a_{mn} \det A_{mn} \tag{B.20}$$

for any chosen row number m or, equivalently, by

$$\det A = \sum_{m=1}^{M} (-1)^{m+n} a_{mn} \det A_{mn} \tag{B.21}$$

for any chosen column number n, where A_{mn} is the $(M-1) \times (M-1)$ matrix obtained by deleting mth row and the nth column of the $M \times M$ matrix A and $1 \leq m, n \leq M$. The determinant of a scalar is equal to that scalar:

$$\det a_{mn} = a_{mn}. \tag{B.22}$$

The descriptions (B.20) and (B.21) are called *Laplace expansion* along the mth row and along the nth column, respectively.

Remark B.2 *By (B.20)*

$$\det\begin{pmatrix} a_{11} & a_{12} \\ a_{21} & a_{22} \end{pmatrix} = a_{11}a_{22} - a_{12}a_{21}, \tag{B.23}$$

etc.

Definition B.20 *(Trace of a matrix) The trace* $\operatorname{tr} A$ *of a square $M \times M$ matrix A is defined as the sum of the diagonal elements:*

$$\operatorname{tr} A = \sum_{m=1}^{M} a_{mm}. \tag{B.24}$$

Theorem B.2 *(Trace of sum) The trace of the sum of two matrices is equal to the sum of the traces:*

$$\operatorname{tr}(A + B) = \operatorname{tr} A + \operatorname{tr} B. \tag{B.25}$$

Proof.

$$\sum_{m=1}^{M} a_{mm} + b_{mm} = \sum_{m=1}^{M} a_{mm} + \sum_{m=1}^{M} b_{mm}. \tag{B.26}$$

■

Theorem B.3 *Let AB be the product of the matrices A and B. Then,* $\operatorname{tr}(AB) = \operatorname{tr}(BA)$.

Proof. Let A and B be an $M \times N$ and an $N \times M$ matrix, respectively. Then,

$$\mathrm{tr}\,\mathrm{AB} = \sum_{m=1}^{M}\sum_{n=1}^{N} a_{mn}b_{nm} = \sum_{n=1}^{N}\sum_{m=1}^{M} b_{nm}a_{mn} = \mathrm{tr}\,BA. \tag{B.27}$$

■

Definition B.21 *(Eigenvalues) Let A be a square matrix. Then, the eigenvalues of A are the solutions for the scalar λ of the system of equations*

$$Ax = \lambda x \tag{B.28}$$

where the solutions for x are different from the null vector.

Theorem B.4 *Let A be a square matrix. Then, A is singular if and only if one or more of its eigenvalues are equal to zero.*

Proof. If $\lambda = 0$, $Ax = o$ for $x \neq o$. Then, A is singular. Next assume that A is singular. By definition, a square matrix A is singular if and only if a vector x different from the null vector exists such that $Ax = o$. Then, $\lambda = 0$ is an eigenvalue. ■

The matrix $A - \lambda I$ is singular since there exists a nonzero vector x such that $(A - \lambda I)x = o$. Then, $\det(A - \lambda I) = 0$ because x linearly combines the columns of $A - \lambda I$ into a zero column. An elementary property of determinants is that their value does not change if a linear combination of other rows or columns is added to a particular row or column. Also, the expansion along column (B.21) shows that a zero column implies that this value is zero. Fully expanding the determinant shows that

$$\det(A - \lambda I) \tag{B.29}$$

is an Mth degree polynomial in λ. The roots of this polynomial are the eigenvalues.

Definition B.22 *(Characteristic polynomial) Let A be an $M \times M$ matrix. Then, the M-th degree polynomial* $\det(A - \lambda I)$ *is called the characteristic polynomial of A.*

Fully expanding the determinant shows that the coefficient of λ^{M-1} in the characteristic polynomial is equal to $\mathrm{tr}A$. This proves the following theorem:

Theorem B.5 *The trace of a matrix is equal to the sum of its eigenvalues.*

Similarly, the constant term of the characteristic polynomial is equal to $\det A$. This proves the theorem:

Theorem B.6 *The determinant of a matrix is equal to the product of all eigenvalues.*

A further useful theorem is:

Theorem B.7 *(Determinant of a matrix product) Let A and B be square matrices. Then,*

$$\det(AB) = \det A \det B\,. \tag{B.30}$$

Proof. See [1]. ■

Definition B.23 *(Matrix partitioning) Matrix partitioning is the decomposition of a matrix into rectangular blocks such that horizontally and vertically the number of rows and the number of columns of the partitions agree.*

For example, an $M \times M$ matrix A may be partitioned into

$$A = \begin{pmatrix} P & Q \\ R & S \end{pmatrix}, \tag{B.31}$$

where P is $M_1 \times M_1$, S is $M_2 \times M_2$, Q is $M_1 \times M_2$, and R is $M_2 \times M_1$ with $M_1 + M_2 = M$.

Corollary B.2 *(Transpose of partitioned matrix) A direct consequence of Definition B.23 is*

$$A^T = \begin{pmatrix} P^T & R^T \\ Q^T & S^T \end{pmatrix}. \tag{B.32}$$

Corollary B.3 *(Multiplication of partitioned matrices) Multiplication of partitioned matrices follows the same rules as usual matrix multiplication. For example,*

$$\begin{pmatrix} P & Q \\ R & S \end{pmatrix} \begin{pmatrix} C & D \\ E & F \end{pmatrix} = \begin{pmatrix} PC + QE & PD + QF \\ RC + SE & RD + SF \end{pmatrix} \tag{B.33}$$

if the dimensions of the partitions are such that all matrix products are defined.

Theorem B.8 *(Determinant of partitioned matrix) Suppose that in (B.31) the matrix A and the matrix P are square and that P is nonsingular. Then,*

$$\det A = \det P \det(S - RP^{-1}Q). \tag{B.34}$$

Proof. See [37]. ■

Lemma B.1 *(Inverse of partitioned matrix: Frobenius formula) Suppose that in (B.31) the matrix A and the matrix P are square. Then, the inverse of A is described by the Frobenius formula:*

$$A^{-1} = \begin{pmatrix} P^{-1} + P^{-1}QU^{-1}RP^{-1} & -P^{-1}QU^{-1} \\ & \\ -U^{-1}RP^{-1} & U^{-1} \end{pmatrix}, \tag{B.35}$$

where U is equal to $S - RP^{-1}Q$ and all inverses are assumed to exist.

Proof. Multiplication of the right-hand member of (B.35) by (B.31) yields the identity matrix of order M. ■

Corollary B.4 *An important special case of (B.35) is that both Q and R^T are equal to the $(M-1) \times 1$ vector q and, consequently, S is a scalar that will be called s. Then,*

$$A = \begin{pmatrix} P & q \\ q^T & s \end{pmatrix} \tag{B.36}$$

and

$$A^{-1} = \begin{pmatrix} P^{-1} + \dfrac{1}{u} P^{-1} q q^T P^{-1} & -\dfrac{1}{u} P^{-1} q \\ & \\ -\dfrac{1}{u} q^T P^{-1} & \dfrac{1}{u} \end{pmatrix} \tag{B.37}$$

with $u = s - q^T P^{-1} q$.

Lemma B.2 *(Matrix Inversion Lemma) Let P be an $M \times M$ matrix and let Q and R be $M \times N$ and $N \times M$ matrices, respectively. Then,*

$$(P + QR)^{-1} = P^{-1} - P^{-1}Q(I + RP^{-1}Q)^{-1}RP^{-1}, \tag{B.38}$$

where I is the identity matrix of order N and all inverses are assumed to exist.

Proof. Multiplication of the right-hand member of (B.38) by the matrix $P + QR$ yields the identity matrix of order M. ■

Corollary B.5 *If in (B.38) Q is an $M \times 1$ vector q and R is an $1 \times M$ vector r^T, then*

$$(P + qr^T)^{-1} = P^{-1} - P^{-1}\frac{qr^T}{1 + r^T P^{-1} q}P^{-1}. \tag{B.39}$$

Proof. This result follows directly from (B.38) if q is substituted for Q and r^T for R, respectively, and the fact is used that I and $RP^{-1}Q$ are now scalars equal to 1 and to $r^T P^{-1} q$, respectively. ■

APPENDIX C

POSITIVE SEMIDEFINITE AND POSITIVE DEFINITE MATRICES

C.1 REAL POSITIVE SEMIDEFINITE AND POSITIVE DEFINITE MATRICES

Definition C.1 *The real symmetric $N \times N$ matrix V is said to be positive semidefinite if*

$$a^T V a \geq 0 \tag{C.1}$$

for any real $N \times 1$ vector a.

Definition C.2 *The real symmetric $N \times N$ matrix V is said to be positive definite if*

$$a^T V a > 0 \tag{C.2}$$

for any real $N \times 1$ vector a different from the null vector.

The notation

$$V \succcurlyeq O \tag{C.3}$$

and

$$V \succ O, \tag{C.4}$$

with O the $N \times N$ null matrix, is used if V is positive semidefinite or positive definite, respectively. A positive definite matrix is also positive semidefinite but a positive semidefinite matrix is not necessarily positive definite. Of course, the expressions (C.3) and (C.4) do not imply that all elements of V are nonnegative or positive .

Definition C.3 *The real symmetric matrix V is said to be negative semidefinite if $-V$ is positive semidefinite. It is said to be negative definite if $-V$ is positive definite.*

Thus, for any property of positive semidefinite or positive definite matrices there exists a negative semidefinite or negative definite counterpart.

Positive (semi)definite and negative (semi)definite matrices together are called *definite* matrices. A symmetric matrix that is not definite is said to be *indefinite*.

With respect to the diagonal elements of real symmetric and positive (semi)definite matrices we have the following theorem.

Theorem C.1 *If V is positive semidefinite, the diagonal elements v_{nn} are nonnegative and if V is positive definite they are positive.*

Proof. Take the vector a as

$$(0\ldots 0\,1\,0\ldots 0)^T \tag{C.5}$$

where the element equal to one is in the nth position. Then,

$$a^T V a = v_{nn}\,. \tag{C.6}$$

The conclusion from (C.6), (C.1) and (C.2) is that $v_{nn} \geq 0$ if V is positive semidefinite and $v_{nn} > 0$ if V is positive definite. ■

Lemma C.1 *Let the real symmetric matrix V be positive semidefinite. Then, $a^T V a = 0$ if and only if $V a = o$.*

Proof. If $Va = o$, then $a^T V a = 0$. To show the converse, define the quadratic polynomial $p(\lambda)$ as

$$\begin{aligned} p(\lambda) &= (a + \lambda b)^T\; V\; (a + \lambda b) \\ &= a^T V a + 2\lambda b^T V a + \lambda^2 b^T V b\,, \end{aligned} \tag{C.7}$$

where a and b are vectors of appropriate dimensions and λ is scalar. Then, for all λ, a, and b,

$$p(\lambda) \geq 0\,. \tag{C.8}$$

The discriminant of $p(\lambda)$ described by

$$4\left[\left(b^T V a\right)^2 - \left(b^T V b\right)\left(a^T V a\right)\right] \tag{C.9}$$

must be nonpositive. This expression shows that, if $a^T V a = 0$, the discriminant is nonpositive only if

$$b^T\; V\; a = 0 \tag{C.10}$$

for all b or, equivalently, if $V a = o$. This completes the proof. ■

Theorem C.2 *A real symmetric and positive semidefinite matrix V is nonsingular if and only if it is positive definite.*

Proof. If V is positive semidefinite,

$$a^T V a \geq 0\,. \tag{C.11}$$

By Lemma C.1, $a^TVa = 0$ if and only if $Va = o$. Suppose that V is positive definite. Then, $a^TVa = 0$ only if $a = o$. Therefore, Va is equal to o only if $a = o$. Then, V is nonsingular. Conversely, if V is nonsingular, $Va = o$ only if $a = o$. Then, $a^TVa = 0$ only if $a = o$. Therefore, V is positive definite. ■

Corollary C.1 *If a real symmetric matrix V is positive semidefinite but not positive definite, it is singular.*

Proof. By Theorem C.2, the matrix V is nonsingular only if it is positive definite. Therefore, if it is not positive definite, it is singular. ■

Theorem C.3 *Suppose that the real symmetric matrix V is positive definite. Then, V^{-1} is real symmetric and positive definite.*

Proof. Since $VV^{-1} = I$, $V^{-T}V^T = V^{-T}V = I$ with $V^{-T} = (V^{-1})^T$. Therefore, $V^{-1} = V^{-T}$. Furthermore, by definition,

$$a^TVa > 0 \tag{C.12}$$

for any $a \neq 0$. Then,

$$a^TVV^{-1}Va = (Va)^T V^{-1}Va > 0. \tag{C.13}$$

Since V is nonsingular, $b = Va$ may represent any vector different from the null vector if $a \neq o$ and is the null vector for $a = o$ only. This completes the proof. ■

Theorem C.4 *Let the real symmetric $M \times M$ matrix V be positive definite and let P be a real $M \times N$ matrix. Then, the $N \times N$ matrix P^TVP is real symmetric and positive semidefinite. The matrix P^TVP is positive definite if and only if P is nonsingular.*

Proof. Transposition of P^TVP shows that this matrix is symmetric. Furthermore, if a is any $N \times 1$ vector, the scalar

$$a^TP^TVPa = b^TVb, \tag{C.14}$$

with $b = Pa$, is positive if $Pa \neq o$ and is equal to zero if $Pa = o$ since V is positive definite. Since P may be singular, Pa may be equal to o for $a \neq o$. Therefore, P^TVP is positive semidefinite. If P is nonsingular, $Pa = o$ only if $a = o$. Therefore, a^TP^TVPa is positive if $a \neq o$ and equal to zero if $a = o$ since V is positive definite. Therefore, P^TVP is positive definite if P is nonsingular. It cannot be positive definite if P is singular since then a may be chosen such that $Pa = o$ and, hence, $a^TP^TVPa = 0$ for $a \neq o$. This completes the proof. ■

Corollary C.2 *Let P be a real $M \times N$ matrix. Then, the $N \times N$ matrix P^TP is real symmetric and positive semidefinite. The matrix P^TP is real symmetric and positive definite if and only if P is nonsingular.*

Proof. The proof follows from Theorem C.4 by taking the positive definite $M \times M$ matrix V as the identity matrix of order M. ■

Theorem C.5 *Let the real symmetric $M \times M$ matrix V be positive semidefinite and let P be a real $M \times N$ matrix. Then, the $N \times N$ matrix P^TVP is real symmetric and positive semidefinite.*

Proof. Transposition of $P^T V P$ shows that this matrix is symmetric. Furthermore, if a is any $N \times 1$ vector, the scalar

$$a^T P^T V P a = b^T V b, \tag{C.15}$$

with $b = Pa$, is larger than or equal to zero since V is positive semidefinite. This completes the proof. ■

Theorem C.6 *The real symmetric matrix V is positive definite if and only if its eigenvalues are positive. It is positive semidefinite if and only if its eigenvalues are nonnegative.*

Proof. See [15]. ■

Theorem C.7 *Suppose that A and B are real symmetric and positive definite matrices. Furthermore, suppose that $A - B$ is positive semidefinite. Then, $B^{-1} - A^{-1}$ is real symmetric and positive semidefinite.*

Proof. See [7]. ■

Theorem C.8 *Suppose that the $(N+1) \times (N+1)$ matrix A and the $N \times N$ matrix P are real symmetric and positive definite, and related by*

$$A = \begin{pmatrix} P & q \\ q^T & r \end{pmatrix}, \tag{C.16}$$

where q is an $N \times 1$ vector and r is scalar. Then, if $q \neq o$, the first N diagonal elements of the inverse matrix A^{-1} are larger than or equal to the corresponding diagonal elements of P^{-1}. If $q = o$, these elements are equal. Also the $(N+1)$th diagonal element of A^{-1} is larger than $1/r$ if $q \neq o$. If $q = o$ they are equal.

Proof. For the computation of A^{-1}, we use the special form of the Frobenius formula (B.37):

$$A^{-1} = \begin{pmatrix} P^{-1} + \frac{1}{u} P^{-1} q q^T P^{-1} & -\frac{1}{u} P^{-1} q \\ -\frac{1}{u} q^T P^{-1} & \frac{1}{u} \end{pmatrix} \tag{C.17}$$

with

$$u = r - q^T P^{-1} q \, . \tag{C.18}$$

By Theorem C.3, the matrix A^{-1} is positive definite since A is. Therefore, $1/u$ is positive. Also, the matrix P^{-1} is positive definite since P is. Then, the vector $P^{-1}q$ is equal to the null vector if q is only. The matrix

$$\frac{1}{u} P^{-1} q q^T P^{-1} \tag{C.19}$$

is positive semidefinite by Theorem C.5. Then, its diagonal elements are nonnegative. Therefore, the first N diagonal elements of A^{-1} are larger than or equal to the corresponding elements of P^{-1}. If $q = o$, these elements are equal.

Furthermore, since P^{-1} is positive definite, the scalar quantity

$$q^T P^{-1} q \tag{C.20}$$

is positive unless $q = o$. Then, u is smaller than r and, therefore, $1/u$ is larger than $1/r$ if $q \neq o$. If $q = o$, $1/u$ and $1/r$ are equal. This completes the proof. ■

C.2 COMPLEX POSITIVE SEMIDEFINITE AND POSITIVE DEFINITE MATRICES

Definition C.4 *The $N \times N$ Hermitian matrix V is said to be positive semidefinite if*

$$a^H V a \geq 0 \tag{C.21}$$

for any complex $N \times 1$ vector a where the superscript H denotes complex conjugate transposition.

Definition C.5 *The $N \times N$ Hermitian matrix V is said to be positive definite if*

$$a^H V a > 0 \tag{C.22}$$

for any complex $N \times 1$ vector a different from the null vector.

Without proof, we now mention the complex equivalents to some of the results presented in Section C.1.

Lemma C.2 *Let the Hermitian matrix V be positive semidefinite. Then, $a^H V a = 0$ if and only if $V a = o$.*

Theorem C.9 *An Hermitian and positive semidefinite matrix V is nonsingular if and only if it is positive definite.*

Corollary C.3 *If an Hermitian matrix V is positive semidefinite but not positive definite, it is singular.*

Theorem C.10 *Suppose that the matrix V is Hermitian and positive definite. Then, V^{-1} is Hermitian and positive definite.*

Theorem C.11 *Let the Hermitian $M \times M$ matrix V be positive definite and let P be a complex $M \times N$ matrix. Then, the $N \times N$ matrix $P^H V P$ is Hermitian and positive semidefinite. The matrix $P^H V P$ is positive definite if and only if P is nonsingular.*

Corollary C.4 *Let P be a complex $M \times N$ matrix. Then, the $N \times N$ matrix $P^H P$ is positive semidefinite. The matrix $P^H P$ is positive definite if and only if P is nonsingular.*

Theorem C.12 *Let the Hermitian $M \times M$ matrix V be positive semidefinite and let P be a complex $M \times N$ matrix. Then, the $N \times N$ matrix $P^H V P$ is Hermitian and positive semidefinite.*

Theorem C.13 *The Hermitian matrix V is positive definite if and only if its eigenvalues are positive. It is positive semidefinite if and only if its eigenvalues are nonnegative.*

Proof. See [15]. ■

APPENDIX D
VECTOR AND MATRIX DIFFERENTIATION

Definition D.1 *(Gradient) Let $f(x)$ be a scalar function of the elements of the vector $x = (x_1 \dots x_N)^T$. Then, the gradient (vector) of $f(x)$ with respect to x is defined as*

$$\frac{\partial f(x)}{\partial x} = \begin{pmatrix} \dfrac{\partial f(x)}{\partial x_1} \\ \dfrac{\partial f(x)}{\partial x_2} \\ \vdots \\ \dfrac{\partial f(x)}{\partial x_N} \end{pmatrix}. \tag{D.1}$$

The transpose of the gradient is the column vector

$$\left(\frac{\partial f(x)}{\partial x}\right)^T = \frac{\partial f(x)}{\partial x^T} = \begin{pmatrix} \dfrac{\partial f(x)}{\partial x_1} & \dfrac{\partial f(x)}{\partial x_2} & \cdots & \dfrac{\partial f(x)}{\partial x_N} \end{pmatrix}. \tag{D.2}$$

Definition D.2 *(Hessian matrix) Let $f(x)$ be a twice continuously differentiable scalar function of the elements of the vector $x = (x_1 \dots x_N)^T$. Then, the Hessian matrix of $f(x)$*

with respect to x is defined as

$$\frac{\partial^2 f(x)}{\partial x \partial x^T} = \begin{pmatrix} \frac{\partial^2 f(x)}{\partial x_1^2} & \frac{\partial^2 f(x)}{\partial x_1 \partial x_2} & \cdots & \frac{\partial^2 f(x)}{\partial x_1 \partial x_N} \\ \frac{\partial^2 f(x)}{\partial x_2 \partial x_1} & \frac{\partial^2 f(x)}{\partial x_2^2} & & \vdots \\ \vdots & & \ddots & \vdots \\ \frac{\partial^2 f(x)}{\partial x_N \partial x_1} & \cdots & \cdots & \frac{\partial^2 f(x)}{\partial x_N^2} \end{pmatrix}. \tag{D.3}$$

Since, under the assumptions made, $\partial^2 f(x)/\partial x_p \partial x_q = \partial^2 f(x)/\partial x_q \partial x_p$, the Hessian matrix is symmetric.

Definition D.3 *(Jacobian matrix) Let $f(x)$ be a $K \times 1$ vector function*

$$f(x) = \begin{pmatrix} f_1(x) \\ f_2(x) \\ \vdots \\ f_K(x) \end{pmatrix} \tag{D.4}$$

of the elements of the $L \times 1$ vector x. Then, the $K \times L$ Jacobian matrix of $f(x)$ with respect to x is defined as

$$\frac{\partial f(x)}{\partial x^T} = \begin{pmatrix} \frac{\partial f_1(x)}{\partial x_1} & \frac{\partial f_1(x)}{\partial x_2} & \cdots & \frac{\partial f_1(x)}{\partial x_L} \\ \frac{\partial f_2(x)}{\partial x_1} & \frac{\partial f_2(x)}{\partial x_2} & \cdots & \frac{\partial f_2(x)}{\partial x_L} \\ \vdots & \vdots & & \vdots \\ \frac{\partial f_K(x)}{\partial x_1} & \frac{\partial f_K(x)}{\partial x_2} & \cdots & \frac{\partial f_K(x)}{\partial x_L} \end{pmatrix}. \tag{D.5}$$

The transpose of the Jacobian matrix is

$$\left(\frac{\partial f(x)}{\partial x^T}\right)^T = \frac{\partial f^T(x)}{\partial x}. \tag{D.6}$$

Definition D.4 *Let the elements of the $M \times N$ matrix A be functions of the elements x_q of a vector x. Then, the $M \times N$ matrix $\partial A/\partial x_q$ is defined as*

$$\frac{\partial A}{\partial x_q} = \begin{pmatrix} \frac{\partial a_{11}}{\partial x_q} & \cdots & \frac{\partial a_{1N}}{\partial x_q} \\ \vdots & \ddots & \vdots \\ \frac{\partial a_{M1}}{\partial x_q} & \cdots & \frac{\partial a_{MN}}{\partial x_q} \end{pmatrix} \tag{D.7}$$

and the matrix of second-order derivatives as

$$\frac{\partial^2 A}{\partial x_p \partial x_q} = \begin{pmatrix} \frac{\partial^2 a_{11}}{\partial x_p \partial x_q} & \cdots & \frac{\partial^2 a_{1N}}{\partial x_p \partial x_q} \\ \vdots & \ddots & \vdots \\ \frac{\partial^2 a_{M1}}{\partial x_p \partial x_q} & \cdots & \frac{\partial^2 a_{MN}}{\partial x_p \partial x_q} \end{pmatrix}. \tag{D.8}$$

Thus, the derivative of a matrix is the matrix of the derivatives.

Theorem D.1 *(Product differentiation rule for matrices) Let A and B be an $K \times M$ and an $M \times L$ matrix, respectively, and let C be the product matrix AB. Furthermore, suppose that the elements of A and B are functions of the elements x_p of a vector x. Then,*

$$\frac{\partial C}{\partial x_p} = \frac{\partial A}{\partial x_p} B + A \frac{\partial B}{\partial x_p}. \tag{D.9}$$

Proof. By definition, the (k, ℓ)-th element of the matrix C is described by

$$c_{k\ell} = \sum_{m=1}^{M} a_{km} b_{m\ell}. \tag{D.10}$$

Then, the product rule for differentiation yields

$$\frac{\partial c_{k\ell}}{\partial x_p} = \sum_{m=1}^{M} \frac{\partial a_{km}}{\partial x_p} b_{m\ell} + a_{km} \frac{\partial b_{m\ell}}{\partial x_p} \tag{D.11}$$

and hence, by (D.7),

$$\frac{\partial C}{\partial x_p} = \frac{\partial A}{\partial x_p} B + A \frac{\partial B}{\partial x_p}. \tag{D.12}$$

This completes the proof. ■

Theorem D.2 *Let the $N \times N$ matrix A be nonsingular and let the elements of A be functions of the elements x_q of a vector x. Then, the first-order and the second-order derivatives of the inverse A^{-1} with respect to the elements of x are equal to, respectively,*

$$\frac{\partial A^{-1}}{\partial x_q} = -A^{-1} \frac{\partial A}{\partial x_q} A^{-1} \tag{D.13}$$

and

$$\frac{\partial^2 A^{-1}}{\partial x_p \partial x_q} = A^{-1} \left(\frac{\partial A}{\partial x_p} A^{-1} \frac{\partial A}{\partial x_q} - \frac{\partial^2 A}{\partial x_p \partial x_q} + \frac{\partial A}{\partial x_q} A^{-1} \frac{\partial A}{\partial x_p} \right) A^{-1}. \tag{D.14}$$

Proof. Differentiating $AA^{-1} = I$, where I is the identity matrix of order N, yields

$$A \frac{\partial A^{-1}}{\partial x_q} + \frac{\partial A}{\partial x_q} A^{-1} = \frac{\partial I}{\partial x_q} = O, \tag{D.15}$$

where O is the $N \times N$ null matrix. Then,

$$\frac{\partial A^{-1}}{\partial x_q} = -A^{-1} \frac{\partial A}{\partial x_q} A^{-1}. \tag{D.16}$$

This expression shows that

$$\frac{\partial^2 A^{-1}}{\partial x_p \partial x_q} = -\frac{\partial}{\partial x_p}\left(A^{-1}\frac{\partial A}{\partial x_q}A^{-1}\right). \tag{D.17}$$

Applying Theorem D.1 to this expression yields

$$\frac{\partial^2 A^{-1}}{\partial x_p \partial x_q} = -\frac{\partial A^{-1}}{\partial x_p}\frac{\partial A}{\partial x_q}A^{-1} - A^{-1}\frac{\partial^2 A}{\partial x_p \partial x_q}A^{-1} - A^{-1}\frac{\partial A}{\partial x_q}\frac{\partial A^{-1}}{\partial x_p}. \tag{D.18}$$

Subsequently substituting the first-order derivatives (D.16) of A^{-1}in this expression shows that

$$\frac{\partial^2 A^{-1}}{\partial x_p \partial x_q} = A^{-1}\left(\frac{\partial A}{\partial x_p}A^{-1}\frac{\partial A}{\partial x_q} - \frac{\partial^2 A}{\partial x_p \partial x_q} + \frac{\partial A}{\partial x_q}A^{-1}\frac{\partial A}{\partial x_p}\right)A^{-1}. \tag{D.19}$$

This completes the proof. ■

REFERENCES

1. A.C. Aitken. *Determinants and Matrices.* New York: Interscience, 1954.
2. A.C. Atkinson and A.N. Donev. *Optimum Experimental Designs.* Oxford: Clarendon, 1992.
3. O. E. Barndorff-Nielsen. *Information and Exponential Families in Statistical Theory.* Chichester: Wiley, 1978.
4. D.M. Bates and D.G. Watts. *Nonlinear Regression Analysis and its Applications.* New York: Wiley, 1988.
5. H. Cramér. *Mathematical Methods of Statistics.* Princeton: Princeton University Press, 1999.
6. A.J. den Dekker, S. Van Aert, A. van den Bos, and D. Van Dyck. Maximum likelihood estimation of structure parameters from high resolution electron microscopy images. Part 1: A theoretical framework. *Ultramicroscopy*, 104(2):83–106, 2005.
7. P.J. Dhrymes. *Econometrics —Statistical Foundations and Applications.* New York: Harper and Row, 1970.
8. S.L. Fagin. Recursive linear regression theory, optimal filter theory, and error analyses of optimal systems. *IEEE International Convention Record*, 12(Part 1):216–240, 1964.
9. V.V. Fedorov. *Theory of Optimal Experiments.* New York: Academic Press, 1972.
10. R.A. Fisher. Theory of statistical estimation. *Proceedings Cambridge Philosophical Society*, 12:700–725, 1925.
11. R.A. Fisher. *Statistical Methods for Research Workers.* Edinburgh: Oliver and Boyd, 1970.
12. P.E. Gill, W. Murray, and M.H. Wright. *Practical Optimization.* London: Academic Press, 1981.
13. A.S. Goldberger. *Econometric Theory.* New York: Wiley, 1964.
14. G.C. Goodwin and R.L. Payne. *Dynamic System Identification—Experiment Design and Data Analysis.* New York: Academic Press, 1977.

15. R.A. Horn and C.R. Johnson. *Matrix Analysis*. Cambridge: Cambridge University Press, 1992.

16. R.I. Jennrich. Asymptotic properties of non-linear least squares estimators. *The Annals of Mathematical Statistics*, 40(2):633–643, 1969.

17. R.I. Jennrich. *An Introduction to Computational Statistics: Regression Analysis*. Englewood Cliffs, NJ: Prentice-Hall, 1995.

18. R.I. Jennrich and R.H. Moore. Maximum likelihood estimation by means of nonlinear least squares. In *Proceedings of the Statistical Computing Section of the American Statistical Association*, pages 57–65, 1975.

19. N.L. Johnson, S. Kotz, and N. Balakrishnan. *Discrete Multivariate Distributions*. New York: Wiley, 1996.

20. K. Knight. *Mathematical Statistics*. Boca Raton: Chapman and Hall/CRC, 2000.

21. S. Kotz, N. Balakrishnan, and N.L. Johnson. *Continuous Multivariate Distributions, Vol. 1: Models and Applications*. New York: Wiley, 2000.

22. E.L. Lehmann and G. Casella. *Theory of Point Estimation*. New York: Springer, 2001.

23. D.W. Marquardt. An algorithm for least-squares estimation of nonlinear parameters. *Journal of the Society for Industrial and Applied Mathematics*, 11(2):431–441, 1963.

24. A.M. Mood, F.A. Graybill, and D.C. Boes. *Introduction to the Theory of Statistics*. Auckland: McGraw-Hill, 3rd edition, 1987.

25. A. Papoulis and S.U. Pillai. *Probability, Random Variables, and Stochastic Processes*. Boston: McGraw-Hill, 2002.

26. C.R. Rao. *Linear Statistical Inference and its Applications*. New York: Wiley, 2002.

27. L.L. Scharf. *Statistical Signal Processing: Detection, Estimation, and Time Series Analysis*. Reading: Addison-Wesley, 1991.

28. G.A.F. Seber and C.J. Wild. *Nonlinear Regression*. New York: Wiley, 1989.

29. A. Stuart and J. K. Ord. *Kendall's Advanced Theory of Statistics, Vol. 1: Distribution Theory*. London: Arnold, 1994.

30. A. Stuart, J.K. Ord, and S. Arnold. *Kendall's Advanced Theory of Statistics, Vol. 2A: Classical Inference and the Linear Model*. London: Arnold, 1999.

31. A. van den Bos. A Cramér-Rao lower bound for complex parameters. *IEEE Transactions on Signal Processing*, 42(10):2859, 1994.

32. A. van den Bos. The real-complex normal distribution. *IEEE Transactions on Information Theory*, 44(4):1670–1672, 1998.

33. A. Wald. The fitting of straight lines if both varables are subject to error. *The Annals of Mathematical Statistics*, 11:284–300, 1940.

34. S. Zacks. *The Theory of Statistical Inference*. New York: Wiley, 1971.

35. S. Zacks. *Parametric Statistical Inference: Basic Theory and Modern Approaches*. Oxford: Pergamon Press, 1981.

36. P.W. Zehna. Invariance of maximum likelihood estimators. *The Annals of Mathematical Statistics*, 37:744, 1966.

37. F. Zhang. *Matrix Theory: Basic Results and Techniques*. New York: Springer, 1999.

INDEX

Printed and bound by CPI Group (UK) Ltd, Croydon, CR0 4YY

16/09/2026

14963459-0002